PRAISE FOR

THE WINTER FORTRESS

A Saint Louis Post-Dispatch Best Book of 2016
An Amazon Best History Book of 2016
A Goodreads Best History and Biography Book of 2016

"Neal Bascomb's *The Winter Fortress* is a riveting, high-action World War II thriller with nothing less than the fate of planet earth on the line. Just imagine the horror if Hitler had gotten the atomic bomb. Written with great verve and historical acumen, Bascomb hits the mark of excellence. Highly recommended!" — Douglas Brinkley,
New York Times best-selling author of *The Great Deluge* and *Cronkite*

"Bascomb brings this overlooked tale of wartime nuclear sabotage to life while taking care to explain the science behind the story." — *Scientific American*

"This book is a must-read! A small band of spies commit themselves, their lives, and their families' lives to literally saving the world from the Nazis. An exciting and accurate story detailing a very dark time in world history, when the world pendulum could have tipped either way. If you liked *Bridge of Spies,* you are going to love this!" — Scott McEwen,
coauthor of the #1 *New York Times* bestseller *American Sniper*

"What would have happened if Hitler had managed to develop nuclear weapons? In *The Winter Fortress,* Neal Bascomb brilliantly tells the extraordinary true story of arguably the most important and daring commando raid of World War II: how an amazing band of men on skis made sure Hitler never got to drop the ultimate bomb." — Alex Kershaw,
New York Times best-selling author of *The Longest Winter*

The
Winter Fortress

THE EPIC MISSION TO SABOTAGE
HITLER'S ATOMIC BOMB

Neal Bascomb

Mariner Books
Houghton Mifflin Harcourt
BOSTON   NEW YORK

First Mariner Books edition 2017
Copyright © 2016 by Neal Bascomb

For information about permission to reproduce selections from this book,
write to trade.permissions@hmhco.com or to Permissions, Houghton Mifflin Harcourt
Publishing Company, 3 Park Avenue, 19th Floor, New York, New York 10016.

www.hmhco.com

Library of Congress Cataloging-in-Publication Data
Names: Bascomb, Neal.
Title: The winter fortress : the epic mission to sabotage Hitler's atomic bomb
/ Neal Bascomb
Description: Boston : Houghton Mifflin Harcourt, 2016. | Includes
bibliographical references and index. | Description based on print version
record and CIP data provided by the publisher; resource not viewed.
Identifiers: LCCN 2015048287 (print) | LCN 2015042716 (ebook)
| ISBN 9780544368064 (ebook) | ISBN 9780544368057 (hardcover)
ISBN 9780544947290 (pbk.)
Subjects: LCSH: World War, 1939–1945 — Commando operations — Norway. | World
War, 1939–1945 — Underground movements — Norway. |
Sabotage — Norway — History — 20th century. | Atomic bomb — Germany — History.
| World War, 1939–1945 — Germany — Technology.
Classification: LCC D794.5 (print) | LCC D794.5.B373 2016 (ebook) | DDC
940.54/86481094828 — dc23
LC record available at http://lccn.loc.gov/2015048287

Printed in the United States of America
DOC 10 9 8 7 6 5 4 3
4500745634

All maps © Svein Vetle Trae/Fossøy; interior maps rendered by Jim McMahon/Scholastic.
Map source notes: **Attack on Vemork** Jens-Anton Poulsson, Knut Werner Hagen; **Grouse's
Arrival in Norway** Jens-Anton Poulsson, Knut Werner Hagen; **Operation Freshman** Per
Johnsen; **Grouse Hideouts** Jens-Anton Poulsson, Knut Werner Hagen; **Gunnerside's Retreat
to Sweden** Joachim Rønneberg; **Bombing of Vemork** Norsk Hydro Archive; **Sinking of the
D/F *Hydro*** Knut Haukelid, Knut Lier-Hansen.

To those who brave the struggle

Contents

HARDANGERVIDDA

● Grouse landing zone
Oct. 18, 1942

Lake Møs

Måna
*Lake Møs
Dam*
Vemork Rjukan

*Skoland
Marshes*
● Sand Lake cabin
Nov. 5, 1942

**Grouse's
Arrival in
Norway**

0   5 MI
0   5 KM

Map © Svein Vetle Trae/Fossøy

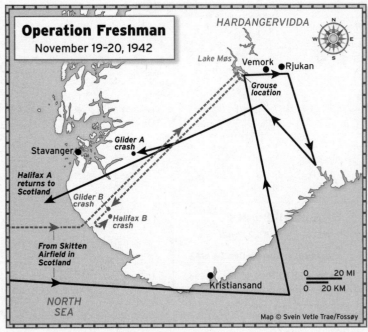

Operation Freshman
November 19-20, 1942

HARDANGERVIDDA

Lake Møs Vemork ● Rjukan

*Grouse
location*

*Glider A
crash*

Stavanger ●

*Halifax A
returns to
Scotland*

*Glider B
crash*
● *Halifax B
crash*

*From Skitten
Airfield
in Scotland*

0   20 MI
0   20 KM

● Kristiansand

NORTH
SEA

Map © Svein Vetle Trae/Fossøy

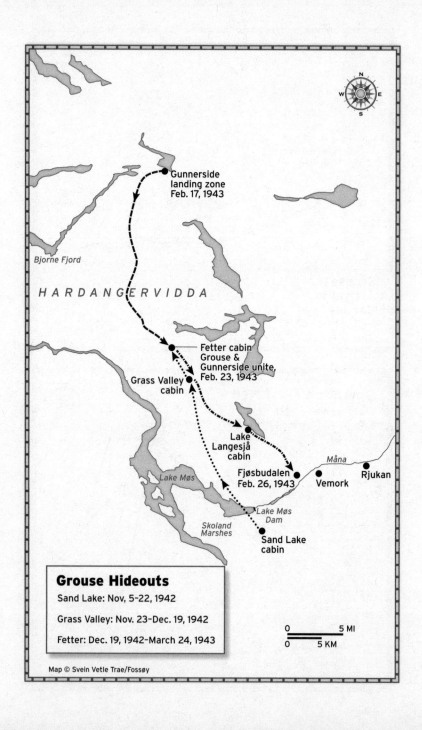

Grouse Hideouts

Sand Lake: Nov. 5–22, 1942

Grass Valley: Nov. 23–Dec. 19, 1942

Fetter: Dec. 19, 1942–March 24, 1943

Gunnerside landing zone Feb. 17, 1943

Bjorne Fjord

HARDANGERVIDDA

Fetter cabin Grouse & Gunnerside unite, Feb. 23, 1943

Grass Valley cabin

Lake Langesjå cabin

Lake Møs

Fjøsbudalen Feb. 26, 1943

Måna

Vemork

Rjukan

Lake Møs Dam

Skoland Marshes

Sand Lake cabin

0   5 MI

0   5 KM

Map © Svein Vetle Trae/Fossøy

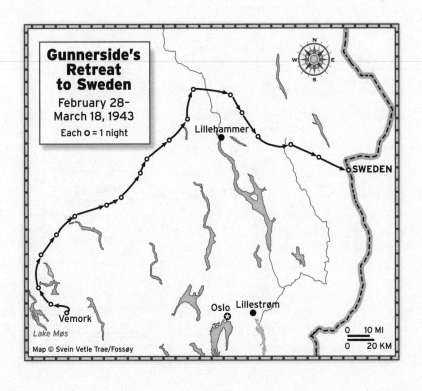

Gunnerside's Retreat to Sweden

February 28–
March 18, 1943

Each ○ = 1 night

Lillehammer

SWEDEN

N
W　E
S

Vemork

Lake Møs

Oslo
Lillestrøm

0　10 MI
0　20 KM

Map © Svein Vetle Trae/Fossøy

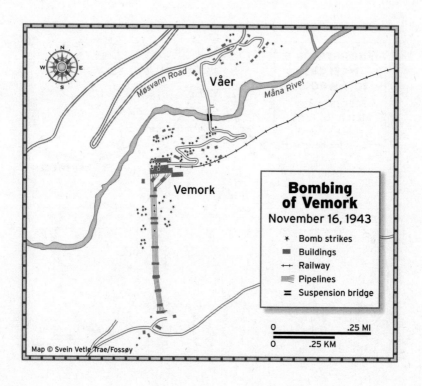

Bombing of Vemork
November 16, 1943

* Bomb strikes
■ Buildings
+―+ Railway
≋ Pipelines
= Suspension bridge

Våer

Møsvann Road

Måna River

Vemork

Map © Svein Vetle Trae/Fossøy

0   .25 MI
0   .25 KM

HARDANGERVIDDA

Lake Tinnsjø

Mæl ❸

D/F Hydro ❹

Rjukan ❷

Måna River

Vemork ❶

Gaustatoppen

N
W · E
S

**Sinking of the
D/F Hydro**
February 19-20, 1944

+—+ Railway
‐‐ Ferry route
❶ Loading of heavy water
❷ Railroad cars under night watch
❸ D/F Hydro embarkation point
❹ D/F Hydro sinks

Map © Svein Vetle Trae/Fossøy

0   3 MI
0   5 KM

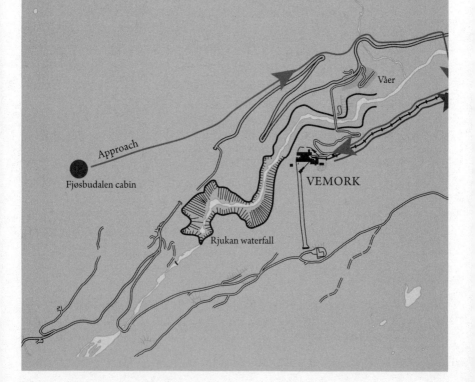

ATTACK ON VEMORK
February 27–28, 1943

—— Approach route to Vemork
—— Escape route from Vemork
—— Mosvann Road
+++ Railway

Våer

Approach

Fjøsbudalen cabin

VEMORK

Rjukan waterfall

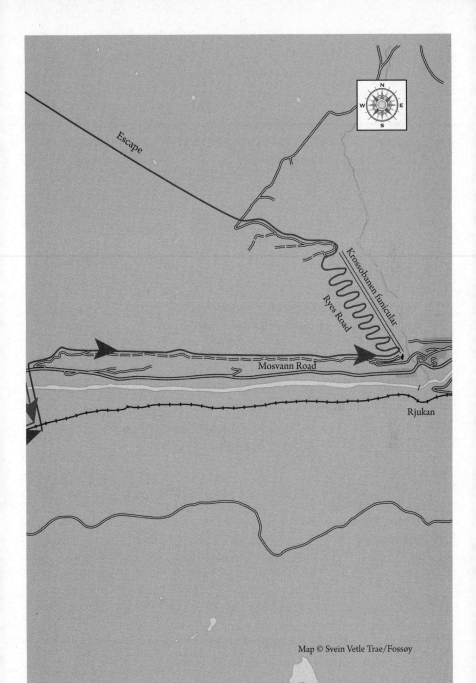

Escape

Krossobanen funicular

Ryes Road

Mosvann Road

Rjukan

Map © Svein Vetle Trae/Fossøy

List of Participants

Operation Grouse/Swallow
Jens-Anton Poulsson, leader of Grouse
Knut Haugland, radio operator
Claus Helberg
Arne Kjelstrup
Einar Skinnarland

Operation Gunnerside
Joachim Rønneberg, leader of Gunnerside
Knut Haukelid, second in command
Birger Strømsheim
Fredrik Kayser
Kasper Idland
Hans Storhaug

D/F Hydro *Sinking*
Alf Larsen, engineer at Vemork
Knut Lier-Hansen, Milorg resistance fighter
Gunnar Syverstad, laboratory assistant at Vemork
Rolf Sørlie, construction engineer at Vemork
Kjell Nielsen, transport manager at Vemork
Ditlev Diseth, Norsk Hydro pensioner

Norwegians

Leif Tronstad, scientist and Kompani Linge leader

Jomar Brun, chief engineer at Vemork

Torstein Skinnarland, brother of Einar

Olav Skogen, leader of local Rjukan Milorg

Lillian Syverstad, courier for Einar Skinnarland

Hamaren, Hovden, and Skindalen families, farmers who aided
  Skinnarland

Allies

Winston Churchill, prime minister of Great Britain

Franklin D. Roosevelt, president of the United States

Eric Welsh, head of the Norwegian branch of the British Secret
  Intelligence Service (SIS)

John Wilson, Norwegian section chief of British Special Operations
  Executive (SOE)

Wallace Akers, head of the Directorate of Tube Alloys

Mark Henniker, commanding officer of Operation Freshman

Owen Roane, American Air Force pilot

Nazis and Collaborators in Norway

Josef Terboven, Reichskommissar in Norway

General Nikolaus von Falkenhorst, head of German military forces in
  Norway

Lieutenant Colonel Heinrich Fehlis, head of the Gestapo and security
  forces in Norway

Captain Siegfried Fehmer, Gestapo bloodhound in Oslo

Second Lieutenant Muggenthaler, Fehlis's SS official in Rjukan

Vidkun Quisling, leader of the Nasjonal Samling, Norwegian fascist party

German Scientists

Kurt Diebner

Werner Heisenberg

Paul Harteck

Abraham Esau

Walther Gerlach

You have to fight for your freedom and for peace. You have to fight for it every day, to keep it. It's like a glass boat; it's easy to break. It's easy to lose.

— JOACHIM RØNNEBERG, Gunnerside leader

Prologue

I N   A   S T A G G E R E D   line, the nine saboteurs cut across the mountain slope. Instinct, more than the dim light of the moon, guided the young men. They threaded through the stands of pine and traversed down the sharp, uneven terrain, much of it pocked with empty hollows and thick drifts of snow. Dressed in white camouflage suits over their British Army uniforms, the men looked like phantoms haunting the woods. They moved as quietly as ghosts, the silence broken only by the swoosh of their skis and the occasional slap of a pole against an unseen branch. The warm, steady wind that blew through the Vestfjord Valley dampered even these sounds. It was the same wind that would eventually, they hoped, blow their tracks away.

A mile into the trek from their base hut, the woods became too dense and steep for them to continue by any means other than on foot. The young Norwegians unfastened their skis and hoisted them to their shoulders. It was still tough going. Carrying rucksacks filled with thirty-five pounds of gear, and armed with submachine guns, grenades, pistols, explosives, and knives, they waded, slid, and clambered their way down through the heavy, wet snow. Under the weight of their equipment they occasionally sank to their waists in the drifts. The darkness, thickening when the low clouds hid the moon, didn't help matters.

Finally the forest cleared. The men came onto the road that ran across the northern side of Vestfjord Valley toward Lake Møs to the west and the town of Rjukan a few miles to the east. Directly south, an eagle's swoop over the precipitous Måna River gorge, stood Vemork, their target.

Despite the distance across the gorge and the wind singing in their ears, the commandos could hear the low hum of the hydroelectric plant. The power station and eight-story hydrogen plant in front of it were perched on a ledge overhanging the gorge. From there it was a six-hundred-foot drop to the Måna River, which snaked through the valley below. It was a valley so deep, the sun rarely reached its base.

Had Hitler not invaded Norway, had the Germans not seized control of the plant, Vemork would have been lit up like a beacon. But now, its windows were blacked out to deter nighttime raids by Allied bombers. Three sets of cables stretched across the valley to discourage low-flying air attacks during the day as well.

In dark silhouette, the plant looked an imposing fortress on an icy crag of rock. A single-lane suspension bridge provided the only point of entry for workers and vehicles, and it was closely guarded. Mines were scattered about the surrounding hillsides. Patrols frequently swept the grounds. Searchlights, sirens, machine-gun nests, and a troop barracks were also at the ready.

And now the commandos were going to break into it.

Standing at the edge of the road, they were mesmerized by their first sight of Vemork. They did not need the bright of day to know its legion of defenses. They had studied scores of reconnaissance photographs, read reams of intelligence, memorized blueprints, and practiced setting their explosive charges dozens of times on a dummy model of the target. Each man could navigate every path, corridor, and stairwell of the plant in his mind's eye.

They were not the first to try to blow up Vemork. Many had already died in the attempt. While war raged across Europe, Russia, North Africa, and in the Pacific, while battalions of tanks, squadrons of bombers, fleets of submarines and destroyers, and millions of soldiers faced off against each other in a global conflict, it was this plant, hidden away deep in the rugged Norwegian wilds, that Allied leaders believed lay on the thin line separating victory and defeat.

For all their intricate knowledge of Vemork, the nine were still not exactly sure how this target could possibly be of such value. They had been told that the plant produced something called heavy water, and that with this mysterious substance the Nazis might be able "to blow up a good part of London." The saboteurs assumed this was an exaggeration to ensure their full commitment to the job.

And they were committed, no matter the price, which would likely include their own lives. From the start, they had known that the odds of their survival were long. They might get inside the plant and complete their mission, but getting out and away would be another story. If necessary, they would try to fight their way out, but escape was unlikely. Resolved not to be captured alive, each of them carried a cyanide pill encased in rubber, stashed in a lapel or waistband.

There were nerves about the operation, for sure, but a sense of fatalism prevailed. For many months now they had been away from their homes, training, planning, and preparing. Now at least they were about to act. If they died, if they "went west," as many in their special company already had in other operations, so be it. At least they would have had their chance to fight. In a war such as this one, most expected to die, sooner or later.

Back in England, the mastermind of the operation, Leif Tronstad, was awaiting news of the operation. Before the commandos left for their mission, he had promised them that their feats would be remembered for a hundred years. But none of the men were there for history. If you went to the heart of the question, none of them were there for heavy water, or for London. They had seen their country invaded by the Germans, their friends killed and humiliated, their families starved, their rights curtailed. They were there for Norway, for the freedom of its lands and people from Nazi rule.

Their moment now at hand, the saboteurs refastened their skis and started down the road through the darkness.

Part I

1

The Water

ON FEBRUARY 14, 1940, Jacques Allier, a middle-aged, nattily dressed banker, hurried through the doors of the Hotel Majestic, on rue la Pérouse. Situated near the Arc de Triomphe, the landmark hotel had welcomed everyone from diplomats attending the Versailles peace talks in 1919 to the influx of artists who made the City of Light famous in the decade that followed. Now, with all of France braced for a German invasion, likely to begin with a thrust through Belgium, and Paris largely evacuated, a shell of its former self, conversation at the hotel was once again all about war. Allier crossed the lobby. He was not there on bank business but rather as an agent of the Deuxième Bureau, the French internal spy agency. Raoul Dautry, the minister of armaments, and physicist Frédéric Joliot-Curie were waiting for him, and their discussion involved the waging of a very different kind of war.

Joliot-Curie, who with his wife, Irène, had won the Nobel Prize for the discovery that stable elements could be made radioactive by artificial or induced methods, explained to Allier that he was now in the middle of constructing a machine to exploit the energy held within atoms. Most likely it would serve to power submarines, but it had the potential for developing an unsurpassed explosive. He needed Allier's help. It was the same pitch Joliot-Curie had given Dautry months before, one made all the more forceful by the suggestion that the energy held within an ordinary kitchen table, if unlocked, could turn the world into a ball of fire. Allier offered to do whatever he could to help the scientist.

Joliot-Curie explained that he needed a special ingredient for his experiments — heavy water — and that there was only one company in the world that produced it to any quantity: Norsk Hydro, in Norway. As an official at the Banque de Paris et des Pays-Bas, which owned a majority stake in the Norwegian concern, Allier was ideally positioned to obtain whatever supplies Norsk Hydro had at its Vemork plant as quickly and discreetly as possible. The French prime minister himself, Édouard Daladier, had already signed off on the mission.

There was one problem, Allier said. Only a month before, the chief lawyer for Norsk Hydro, Bjarne Eriksen, had visited him in his Paris office. According to Eriksen, the Germans were also interested in the production at Vemork. They had placed a number of orders and had suggested that they might need as much as two tons of heavy water in the near future. Startled by the demand for such vast quantities — and denied any information about how the material might be used — Norsk Hydro had yet to fulfill more than twenty-five kilograms of these orders.

This report troubled Dautry and Joliot-Curie deeply. The Germans must be on the same track with their research. Allier needed to move — and fast — to secure the stock before the Germans did. If there was trouble in bringing it out of Norway, he was to see that the heavy water was contaminated, thus rendered useless for experiments.

Two weeks later, Allier headed across the vast hall of Paris's Gare du Nord and boarded a train to Amsterdam. He was traveling under his mother's maiden name, Freiss. Concealed in his briefcase were two documents. One was a letter of credit for up to 1.5 million kroner for the heavy water. The other gave him the authority to recruit any French agents required in smuggling out the supply. Short of the false beard, Allier felt like he had all the accouterments of a hero in the spy novels he loved.

From Amsterdam he flew to Malmö, Sweden, then took a train to Stockholm. There he sat down with three French intelligence agents, tasking them to meet him in Oslo a few days later. Early on March 4, Allier traveled by train to the Norwegian capital, arriving into Eastern Central Station. At the French Legation, he learned that his cover was already blown. An intercepted message from Berlin's spy agency, the Abwehr, had been deciphered. "At any price," it read, "stop a suspect Frenchman traveling under the name of Freiss."

Allier was undeterred. He left the legation and rang Norsk Hydro from a public phone booth. Within the hour, he entered the company headquarters at Solligata 7, a short distance from the royal residence of King Haakon VII. In a meeting with Dr. Axel Aubert, Allier made his offer to buy the company's stocks of heavy water. He said nothing about their intended use, unsure that he could trust Aubert. The tough, long-standing director general, who looked like he chewed stones for breakfast, was clear: his sympathies were with France; he had refused the Germans any great quantities of heavy water, and he would provide Allier with whatever he needed.

The next day, Allier traveled by car to Vemork, one hundred miles from the Norwegian capital. Aubert followed. Their arrival was unannounced.

For thousands of years, water had run plentifully throughout the high wilderness plateau of the Hardangervidda in Telemark, a region west of Oslo. Much of this water, a vast flow, descended from the Vidda into its natural reservoir at Lake Møs. Then the river Måna carried the water for eighteen miles through the steep Vestfjord Valley to Lake Tinnsjø.

The river's flow changed when Norsk Hydro, a burgeoning industrial giant, built a dam at the lake's outlet in 1906. The company redirected the water through tunnels blasted out of the rock, which ran for three miles underground before they reached the Vemork power station. From there, the water fell 920 vertical feet through eleven steel pipelines into turbine generators that produced 145,000 kilowatts of electricity. It was the world's largest hydroelectric power station.

A fraction of the water, roughly sixteen tons an hour, was then directed into a hydrogen plant, also the world's largest, thirty feet away on the edge of the cliff. There it flowed into tens of thousands of steel electrolysis cells, which consumed almost all of the power generated at the station. Currents of electricity running through the cells split the water's two hydrogen atoms from its lone oxygen one. These separated gases were then pumped down to chemical plants in Rjukan, at the base of the valley. A company town, Rjukan had seven thousand residents most of whom worked for Norsk Hydro. The hydrogen was primarily used to make fertilizer — a huge market.

A fraction of *this* water, which had by now coursed from the Vidda to Lake Møs through tunnels, then pipelines, then electrolysis cells, was sent through a cascade of specialized electrolysis cells that terminated in a basement cellar at Vemork. The water was then reduced and further reduced until it amounted to a steady drip similar in output to a leaky faucet. This water was now something unique and precious. It was heavy water.

The American chemist Harold Urey won the Nobel Prize for his 1931 discovery of heavy water. While most hydrogen atoms consist of a single electron orbiting a single proton in the atom's nucleus, Urey showed that there was a variant, or isotope, of hydrogen that carried a neutron in its nucleus as well. He called this isotope deuterium, or heavy hydrogen, because its atomic weight (the sum of an atom's protons and neutrons) was 2 instead of 1. The isotope was extremely rare in nature (.015 percent of all hydrogen), and there was just one molecule of heavy water (D_2O) for every 41 million molecules of ordinary water (H_2O).

Building on Urey's work, several scientists found that the best method for producing heavy water was electrolysis. The substance didn't break down as easily as ordinary water when an electric current ran through it, so any water remaining in a cell after the hydrogen gas was removed was more highly concentrated with heavy water. But generating the substance in any quantity demanded tremendous resources. A scientist noted that in order to produce a single kilogram (2.2 pounds) of heavy water, "50 tons of ordinary water had to be treated for one year, consuming 320,000 kilowatt hours [of electricity], and, then, the output had a purity no better than about ten percent." That was a lot of electricity for a low level of purity in a very small quantity of deuterium.

In 1933 Leif Tronstad, a celebrated young Norwegian professor, and his former college classmate Jomar Brun, who ran the hydrogen plant at Vemork, proposed to Norsk Hydro the idea of a heavy water industrial facility. They weren't exactly sure what the substance might be used for in the end, but as Tronstad frequently said to his students, "Technology first, then industry and applications!" They did know that Vemork, with its inexhaustible supply of cheap power and water, provided the perfect setup for such a facility.

They matched the plant's natural advantages with an ingenious new design for the equipment. An early working plant, designed by Tronstad and Brun, had six stages. Think of a group of cans stacked in a pyramid. Now picture that pyramid upside down, with the single can at the bottom. In the Tronstad/Brun design, water flowed into the top row of cans — really 1,824 electrolysis cells, which treated the water (mixed with potash lye as a conductor) with a current. Some of the water was decomposed into bubbles of hydrogen and oxygen gas that were vacated from the cells, and the remainder, now containing a higher percentage of heavy water, cascaded down to the next row of cans in the pyramid (570 cells). Then it repeated the process through the third (228 cells), fourth (20 cells), and fifth (3 cells) rows of electrolysis cells. However, by the end of the fifth stage, with a huge amount of time and power exhausted, the cells still contained only 10 percent deuterium-rich water.

Then the water cascaded into the bottom can of the pyramid. This sixth and final phase was called the high-concentration stage. Set in the cavernous, brightly lit basement of the hydrogen plant, it actually consisted of seven unique steel electrolysis cells lined up in a row. These specialized cells followed a similar cascade model to concentrate the heavy water in each cell. But they could also recycle the gaseous form of deuterium back into the production process, while it was essentially wasted in the other stages. As a result, the heavy water concentration rose quickly from one cell to the next. By the seventh, final cell in this high-concentration stage, the slow, steady drip had been purified to 99.5 percent heavy water.

When the Vemork plant started production in earnest using this method, scientists around the world heralded it as a breakthrough, even though heavy water's application remained uncertain. Because it froze at four degrees Celsius instead of zero, some joked it was only good for creating better skating rinks. Tronstad, who served as a consultant to Norsk Hydro and left the running of the plant to Brun, believed in the potential of heavy water. He spoke passionately of its use in the burgeoning field of atomic physics, and of its promise for chemical and biomedical research. Researchers found the life processes of mice slowed down when they were given minute amounts of heavy water. Seeds germinated more gradually in a diluted solution — and not at all in a pure one. Some believed that heavy water could lead to a cure for cancer.

Vemork shipped its first containers of heavy water in January 1935 in batches of ten to one hundred grams, but business did not boom. Laboratories in France, Norway, Britain, Germany, the United States, Scandinavia, and Japan ordered no more than a few hundred grams at a time. In 1936 Vemork produced only forty kilograms for sale. Two years later, the amount had increased to eighty kilograms, a trifling amount valued at roughly $40,000. The company placed advertisements in industry magazines to little avail: there simply wasn't sufficient demand.

In June 1939 a Norsk Hydro audit of this small sideline business showed it to be a loser. Nobody wanted heavy water, at least not enough to make it worth the investment, and the company abandoned the venture.

But only months after Brun shut off the lights in the basement and dust started to gather on the seven specialized cells in the high-concentration room, everything changed — and quickly — just as it had in the field of atomic physics.

For decades, scientists had been plumbing the mysteries of "atoms and void," which was how the ancient Greeks described the makeup of the universe. In dark rooms, experimenters bombarded elements with sub-atomic particles. Theoreticians made brilliant deductions on the black-board. Pierre and Marie Curie, Max Planck, Albert Einstein, Enrico Fermi, Niels Bohr, and other scientists discovered an atomic world full of energy and possibilities.

The English physicist Ernest Rutherford observed that heavy, unstable elements such as uranium would break down naturally into lighter ones such as argon. When he calculated the huge amount of energy emitted during this process, he realized what was at stake. "Could a proper deto-nator be found," he suggested to a member of his lab, "a wave of atomic disintegration might be started through matter, which would indeed make this old world vanish in smoke . . . Some fool in a laboratory might blow up the universe unawares."

Then, in 1932, another English scientist, James Chadwick, discovered that proper detonator: the neutron. The neutron had mass, but unlike protons and electrons, which held positive and negative charges respec-tively, it carried no charge to hinder its movement. That made it the per-fect particle to send into the nucleus of the atom. Sometimes the neu-tron was absorbed; sometimes it knocked a proton out, transforming the chemical element. Physicists had discovered a way to manipulate the ba-sic fabric of the world, and with this ability, they could further investigate its many separate strands — and even create some of their own.

Using radon or beryllium as neutron sources, physicists began flinging neutrons at all kinds of elements to produce changes in their nature. Led by the Italian Fermi, they found this process particularly effective when the neutrons had to pass through a "moderator" of some kind, which slowed their progress. Paraffin wax and plain water proved to be the best early moderators. Both contained lots of hydrogen, and when these hydrogen atoms collided with the neutrons (which had the same mass), they stole some of their speed, much like when two billiard balls collide. Bombarding uranium with neutrons in this manner brought the most mysterious results, including the unexpected presence of much lighter elements.

In December 1938 two German chemists, the pioneering Otto Hahn and his young assistant Fritz Strassmann, proved that a neutron colliding with a uranium atom could do more than chip away at its nucleus or become absorbed within it. The neutron could split the atom in two — a process called fission. By early January 1939, word of the discovery had spread, bringing great excitement to the field of atomic research: Why, how, and to what effect had the uranium atom split?

Springboarding off an observation by the Danish theorist Niels Bohr, physicists realized that the uranium atom's nucleus had acted like an overfilled water balloon. Its "skin" was stretched thin by the large number of protons and neutrons inside, and when a neutron was shot into it, it formed a dumbbell: two spheres connected by a thin waist. When the tension on the skin finally became too much, it snapped, and the two spheres — two lighter atoms — were flung apart with tremendous force, an amount equal to the energy that had once held the nucleus together (its binding energy). Researchers were quick to come to a figure, too: 200 million electron volts — enough to bounce a single grain of sand. A tiny amount, perhaps, but given that a single gram of uranium contained roughly 2.5 sextillion atoms (2.5×10^{21}), the numbers alone obscured the potential energy release. One physicist calculated that a cubic meter of uranium ore could provide enough energy to raise a cubic kilometer of water twenty-seven kilometers into the air.

The atom's potential power became even clearer when scientists discovered that splitting the uranium nucleus released two to three fast-moving neutrons that could act as detonators. The neutrons from one atom could split two others. The neutrons from these two split four more. The four could cause the detonation of eight. The eight — sixteen. With an ever-increasing number of fast-moving neutrons flinging themselves about, splitting atoms at an exponential rate, scientists could create what was called a chain reaction — and generate enormous quantities of energy.

Which prompted the obvious question: To what purpose? Some conceived of harnessing the energy release to fuel factories and homes. Oth-

ers were drawn to — or feared — its use as an explosive. Within a week of
Hahn's discovery, American physicist J. Robert Oppenheimer sketched a
crude bomb on his blackboard.

Fermi, who had immigrated to the United States, trembled at the
thought of what might come. Staring out the window of his office at Columbia University, he watched students bustling down the New York
sidewalks, the streets crowded with traffic. He turned to his office mate,
drew his hands together as if holding a soccer ball, and harked back to
the words of Rutherford. "A little bomb like that," he said solemnly before
looking back outside, "and it would all disappear." Given the aggression
shown by Nazi Germany by the end of summer 1939, such a bomb, if it
could be built, might be needed in a world on the precipice of war. Plans
to obtain it were rapidly put together on both sides.

By annexing Austria and occupying Czechoslovakia, Adolf Hitler had
managed to pursue his goals without a fight until September 1, 1939, when
at 4:45 a.m. his 103rd Artillery Regiment sent its first "iron greetings" into
Poland. Panzer tanks swept across the border and bombers shot eastward
overhead. The German *Blitzkrieg* had begun and, Hitler promised, bombs
would be met with bombs.

Britain and France responded with a declaration of war. On September 3, Winston Churchill, First Lord of the Admiralty, rose in the House
of Commons and said, "This is not a question of fighting for Danzig or
fighting for Poland. We are fighting to save the whole world from the pes-
tilence of Nazi tyranny and in defense of all that is most sacred to man."

Less than two weeks later, on September 16, German scientist Kurt
Diebner sat in his office at the headquarters of Berlin's Army Ordnance
Research Department, Hardenbergstraße 10, waiting for the eight German physicists he had ordered to report for duty a few days before. "It's
about bombs," he told the recruit who drafted the list of attendees.

Thirty-four years of age, Diebner was a loyal Nazi Party member with
a presence as modest and retreating as his hairline. His suit fit too tightly
over his short, thin frame, and he wore round schoolboy spectacles that
constantly threatened to slip off his nose. In meetings, his words came out
halting and unsure. But despite his appearance and manner of speech, he
was an ambitious and eager man.

Born into a working-class family outside the industrial city of Naumburg, Diebner got himself into university by dint of hard work and clever-ness. First at Innsbruck, then Halle, he studied physics. While some of the other students dined out and had the means to care about the cut of their suits, he lived a threadbare existence. Drawn to the experimental side of physics, he worked diligently in the laboratory, his aim being to find a po-sition as a university lecturer — and to achieve the salary and prestige that came with it. While a student at the University of Halle, Diebner joined an esteemed fencing club, an important rung on the social ladder, and earned several scars on his face from duels.

Diebner gained his PhD in atomic physics in 1931. In 1934, the year Hitler became the führer of Germany, he joined the Army Ordnance Re-search Department, where he was tasked with developing hollow-shaped explosives. For years he pushed his boss to allow him to create an atomic research division instead. Such work, he was told, was "malarkey," with no practical use. But rapid advances in the field in 1939 made it clear that atomic physics was anything but malarkey, and Diebner was finally given the mandate to form a team.

When those among the best and brightest in German science arrived at Hardenbergstraße that mid-September day, they carried suitcases, not sure of where they were going to be sent. When they saw that it was Dieb-ner who greeted them, they shook his hand enthusiastically, knowing that at least they were not to be delivered to the front. They assembled in a conference room and were told that German spies had discovered that the United States, France, and Great Britain were pursuing projects in nu-clear fission. This much was already well known to the attendees. They had all read, and some had contributed to, the rush of international jour-nal articles on the subject. Now that war had been declared, the curtain on this open theater of science had fallen. Diebner informed them that they had been called together to decide whether or not it was possible, in prac-tice, to harness the atom's energy for the production of weapons or elec-tricity.

One of the men in the room was already dedicated to the former goal. In April, Paul Harteck, a physical chemist at the University of Hamburg, had sent a letter to the Reich Ministry of War explaining recent develop-ments in nuclear physics. In his estimation, he wrote, they held the "pos-sibility for the creation of explosives whose effect would excel by a million times those presently in use . . . The country which first makes use of [this explosive] would, in relation to the others, possess a well-nigh irretrievable advantage." Harteck believed the assembled group should pursue any such advantage.

Otto Hahn, on the other hand, was distraught that his discovery was now being developed into a weapon to kill. He tried to extinguish any enthusiasm for the endeavor by pointing to the many technical challenges involved in engineering an explosive or designing a machine to produce energy.

He noted from recent studies that it was the atoms of the rare uranium isotope U-235 (atomic weight 235: 92 protons, 143 neutrons) that fissioned most readily. Meanwhile, its more common cousin, U-238 (92 protons, 146 neutrons), tended to absorb neutrons that struck its nucleus, stealing their potential to foster a chain reaction. And unless fast-moving neutrons released from a split atom were properly slowed, the probability of U-235 fissioning was small. Natural uranium was made up of only seven parts U-235 for every thousand parts of U-238, and no method to separate the two isotopes existed. Furthermore, they would need to find an efficient moderator for U-235. Given all this, and likely other unseen challenges, Hahn believed that attempting to harness the atom for use in the current war was a fool's errand.

The debate continued for hours, until the scientists finally reached a consensus: "If there is only a trace of a chance this can be done, then we have to do it."

Ten days later, on September 26, Diebner called another meeting of his "Uranium Club." This time Werner Heisenberg attended. Heisenberg was considered the leading light of German theoretical physics, particularly after Hitler's rise had forced Albert Einstein and other Jewish physicists to flee the country. Initially, Diebner had resisted his inclusion in the group, because he wanted experimenters, not theoreticians, and because Heisenberg had called Diebner's academic research "amateurish." But those Diebner did recruit urged him to reconsider: Heisenberg had won his Nobel Prize at the tender age of thirty-one, and he was too brilliant to leave out.

Heisenberg proved to be a useful addition to the club. By the end of that meeting, the group had its orders. Some, like Harteck, would investigate how to extract sufficient quantities of U-235 from natural uranium. Others, like Heisenberg, would hash out chain reaction theory, both for constructing explosives and generating power. Still others would experiment with the best moderators.

Heisenberg made quick work with the theory. By late October he'd started on a pair of breakthrough papers. If they separated the U-235 isotope and compressed sufficient quantities into a ball, the fast-moving neutrons would set off an immediate chain reaction, resulting in an explosion "greater than the strongest available explosives by several powers of ten." Isotope separation, Heisenberg declared, was "the only way to produce explosives," and the challenges of such separation were legion. But constructing a "machine" that used uranium and a moderator to generate a steady level of power was an attainable goal. After the machine went critical, the number of chain reactions would stabilize and it would sustain itself. The amount of U-235 was still key: they would need an enormous quantity of natural uranium in its processed purified form — uranium oxide — to provide suitable amounts of the rare fissile isotope.

On the subject of moderators, Heisenberg dismissed plain water as an option. Its hydrogen atoms slowed the neutrons enough to promote the fissioning of U-235, but they also captured them at too high a rate. This left two known candidates: graphite, which was a crystalline form of carbon, and heavy water. In graphite, the carbon atoms acted as the moderator; in heavy water, it was deuterium. Both should prove effective in slowing neutrons down sufficiently and reducing to a minimum the number of neutrons parasitically absorbed.

Once they had enough uranium and an effective moderator, Heisenberg concluded, it was simply a question of calculating the machine's most efficient size (quantity of uranium and moderator), arrangement (mixed together or layered), and shape (cylindrical or spherical). His initial figures indicated that a sphere filled with at least a ton each of uranium and the chosen moderator separated in layers would be optimal. It was going to be big, but it would work.

Heisenberg gave Diebner the direction he needed to move forward — and the Nobel laureate's reputation contributed to persuading others to follow this path. Experiments would continue to separate U-235, but most of the effort was now focused on building the uranium machine. If they were successful, they would prove at last the importance — and utility — of atomic physics. Constructing a bomb would follow.

In recognition of the Uranium Club's work, Diebner was named head of the Kaiser Wilhelm Institute of Physics in Berlin, a body of preeminent reputation and the country's most advanced laboratory. Heisenberg was appointed to the board as scientific adviser, to placate those who were upset at having Diebner, a physicist of no renown, directing the august institute.

By year's end Diebner had dozens of scientists under his watch across Germany refining the uranium-machine theory and building the first small experimental designs. Progress had been made outfitting laboratories and ordering uranium oxide and other key materials.

Although the issue needed further study, the scientists' calculations indicated that heavy water was the best presently known moderator. The Uranium Club would require a steady, robust supply of the precious liquid. Unfortunately, the world's sole producer, Norsk Hydro's Vemork plant, was far away in an inaccessible valley in Norway, a country whose neutral status in the war made it an unreliable partner. The plant had also only recently restarted heavy water production in November 1939 and could supply little more than ten kilograms a month. Diebner considered building a full-scale heavy water plant in Germany, though it would cost tens of millions of marks and consume a hundred thousand tons of coal for each single ton of heavy water. Before he made any such move, however, he and Heisenberg agreed that they needed to make sure heavy water was a viable moderator. For those experiments, twenty-five kilograms should do. Diebner had a representative of the German conglomerate IG Farben, which owned 25 percent of Norsk Hydro, put in the order, to conceal the involvement of Army Ordnance.

By January 1940, with more physicists in his group asking for their own supply, prospective orders had grown to one hundred kilograms a month, every month. Norsk Hydro wanted to know the purpose of such a large order, but with experiments using heavy water now labeled SH-200, a high-level military secret, the IG Farben representative offered only silence.

Not long after, the Norwegians did find out, from Jacques Allier, what that purpose was: the potential development of an atomic bomb.

When Allier visited Vemork on March 5, 1940, he presented himself merely as an official with Banque de Paris. Axel Aubert led the meeting with the plant's lead engineer, Jomar Brun. From previously unsold supplies and the restart of production, the plant had a total of 185 kilograms at hand. All of it, Aubert told Brun, needed to be transported secretly to Oslo by truck. Brun asked to know why, just as he had asked why when Aubert had quietly told him earlier in the year to explore a fivefold increase in production to 50 kilograms a month. As before, Aubert declined to answer any questions and instructed that not a word of this special order was to be mentioned to anyone.

After settling these arrangements and finding a Rjukan welder to make twenty-six stainless-steel flasks that would fit neatly into suitcases, Allier returned to Oslo with Aubert to conclude their negotiations and prepare for spiriting the flasks out of Norway. The Norsk Hydro director general offered to give France the heavy water on loan, with no price attached, and told Allier that Norsk Hydro would provide France with first claim on what was produced in the future. Impressed by this generosity — and the alacrity with which Aubert had moved — Allier opened up about the intended use of the heavy water by Frédéric Joliot-Curie and his team.

On March 9, two trucks departed Vemork down the steep, ice-slick road. Brun rode in the first truck. At a nondescript house in Oslo, they unloaded the twenty-six flasks and entrusted them to Allier's care. The house was owned by the French government and was a stone's throw from an Abwehr safe house, but sometimes it was best to hide in plain sight.

To smuggle out the supply, Allier had grand visions of a submarine sneaking into Oslofjord and spiriting it away, but instead he settled for the old "bait and switch," supported by the three French spies he had recruited in Stockholm. Through several ticket agents and under various assumed names, they booked flights on two planes leaving Oslo's Fornebu airport at roughly the same time on the morning of March 12. One was headed to Amsterdam, the other to Perth, in Scotland. In case anything went wrong, they also bought seats on the same flights for two consecutive days after that.

At dawn on March 12, a frigid and cloudless morning, Allier and a fellow spy, Fernand Mosse, took a taxi three miles south of the city center to Fornebu. Dressed as businessmen, they made a big show of their upcoming trip to Amsterdam in front of the gate agents and baggage handlers who took their several large, heavy suitcases. Soon enough, they were crossing the tarmac toward the Junkers Ju-52 aircraft designated for their flight. Adjacent to their plane was an identical one, destined for Perth.

Once they were sure their baggage was loaded onto the Amsterdam-bound plane and its propellers began to spin, they readied to board. At that moment, a taxi drove onto the tarmac. Its passenger, Jehan Knall-De-mars, another of Allier's team, had pleaded with the gate agent to let him through in the taxi so he could make his plane to Amsterdam. The spy had the taxi park between the two Junkers, out of view of the Fornebu terminal. From the trunk, he unloaded several suitcases, together containing thirteen of the heavy-water flasks. These were hauled into the baggage hold of the Scottish-bound plane, which Allier and Mosse boarded, instead of the one to Amsterdam. Knall-Demars left in the taxi, hiding in the back as he passed through the gate.

Minutes later, the plane to Amsterdam barreled down the runway and lifted into the sky. As it headed south over the Skagerrak, the strait of sea between Norway and Denmark, a pair of Luftwaffe fighters drew alongside. They ordered its pilots to divert their course to Hamburg. When the plane landed in Germany, Abwehr agents busted open the cargo hold. Rummaging through the suitcases, they found a few that were particularly heavy. Inside them? Granite rubble.

Meanwhile, Allier and Mosse landed safely in Scotland with their stash. The following day, Knall-Demars arrived with the other thirteen flasks.

By March 18, all twenty-six flasks were stored in the old stone-arched cellars of the Collège de France in Paris. The first battle of heavy water was won. The next, however, was all too shortly to begin.

2

The Professor

T**O TRONDHEIM** they came, in the dark, early hours of April 9, 1940, cutting into the Norwegian fjord at twenty-five knots. A northerly, snow-flecked gale swept across the steel decks of the German cruiser *Admiral Hipper* and the four destroyers at its stern. They approached the three forts guarding the entrance to the former Viking capital, all crews at action stations.

A Norwegian patrol signaled the intruding ships to identify themselves. In English, the *Admiral Hipper*'s captain returned that they were the HMS *Revenge,* there with orders from the British government to "proceed toward Trondheim. No unfriendly intentions." As the patrol shined a spotlight across the water, it was blinded by searchlights from the *Admiral Hipper,* which suddenly accelerated to maximum speed, blowing smoke to obscure its whereabouts. Signals and warning rockets lit up the night. Inside the Norwegian forts, alarms rang and orders were given to fire when ready on the invading ships.

Fifteen minutes passed. Within the port's batteries, the ammunition had to be loaded, then the electrical firing mechanisms failed. By the time the inexperienced Norwegian soldiers were prepared to respond, the *Admiral Hipper* was already steaming past the first fort. At the second fort, the bugler who should have sounded the alarm had fallen asleep at his post, and the men were late to their guns. The moment they opened fire, their searchlights malfunctioned and they could not see their targets.

At 4:25 a.m. the small armada set anchor in Trondheim's harbor. Cutters began ferrying two infantry companies from the warships to the shore. All was sleepy in town as German soldiers spread out from the port into the defenseless streets. The Nazi invasion of Norway had begun.

In an auditorium at the Norwegian Institute of Technology (NTH), a twenty-minute walk from Trondheim's harbor, students, teachers, and a handful of other community members had gathered. Word of the invasion had reached Leif Tronstad before the break of day, and while his children slept he had rushed to the institute. From the scarce reports he and the others had received, all of Norway looked to be under attack. Bergen, Stavanger, Kristiansand, and Narvik had fallen alongside Trondheim, but Oslo was rumored to be holding out. The panicked assembly debated what they should do. One among them, a firebrand named Knut Haukelid who was visiting friends in the city, wanted to fight with whatever weapons they could find. Others preached caution; their small country, with its limited military, stood little chance against German might.

When Tronstad spoke, he held everyone's attention. At thirty-seven, he was the university's youngest full professor, and a favorite in its classrooms. Of medium height, he had blue eyes, a hatchet-sharp nose, and ash-blond hair parted neatly on the side.

Tronstad informed those assembled that, as a reserve officer in the Army Ordnance Corps, he had standing orders to travel to Oslo if war broke out. He advised those with military experience to do the same. As for the others, he said, each man needed to follow his own conscience on what action to take, but he reminded them that their country needed them. With that, he said his goodbyes.

Tronstad had feared this would happen — that Norway's "sleeping government" would leave the country unprepared to mount a defense in the event of an invasion. Since the day Hitler invaded Poland seven months earlier, it was plain to Tronstad that Norway would not be allowed to maintain the neutral stand it had held during the Great War. The fight between the Allies and the Nazis in mainland Europe had stalled, and the two sides had circled Norway for months, waiting to see which would make the first move toward bringing it within their sphere of influence.

There were good reasons for their interest, chiefly because Norway's long coastline offered potential naval bases to dominate the North Sea. All the while, the Norwegian government hoped measured diplomacy would trump the need for force. Only a few days before, Hitler's "Special Envoy," Curt Bräuer, had thrown a party for over two hundred Norwegian political and military guests. Midway through, the lights were dimmed and a film, *Baptism of Fire,* was screened. To a soundtrack of Richard Wagner, bombers obliterated Warsaw while a voiceover explained that this was the consequence for those who made friends with the British and French. The film concluded with a map of Britain bursting into flames and the flash of a Nazi swastika.

Hurrying home, Tronstad found the city around him quickly occupied by German soldiers. They marched in columns through the streets. They established machine-gun nests and mortar positions on bridges and key spots throughout the city, and called out warnings in German to not resist. Tronstad ignored them. Finally, he reached his two-story house on the leafy outskirts of the city center. He told his wife, Bassa, that they were not safe in Trondheim. He would take her and the children to a tourist mountain lodge at Kongsvoll, a hundred miles to the south, then he would travel on to Oslo to join the army.

Together they woke up their daughter, Sidsel, age seven, and two-year-old son, Leif Jr., and helped them dress and pack. Fifteen minutes later they piled into their Opel Super 6, a luxury German-made car. As they headed south over the Nidelva River bridge, two ash-colored bombers flew overhead.

"What kind of plane is that?" Sidsel asked.

"It is a German plane," Tronstad said, his first explanation of their hurried departure. "I'm afraid the war's come to our country."

They reached Oppdal, a small town and the end of the road in wintertime, that afternoon. The news they heard there was not good. The call to mobilize had finally come through, but the "iron ring" of defense around Oslo had proven to be a fiction: the Germans controlled the capital. King Haakon VII, a tall, austere man who rose to the throne in 1905, had fled the city, as had the Norwegian government. On the radio, Vidkun Quisling, the head of the fascist Nasjonal Samling Party, announced that he had assumed governmental powers. Tronstad knew now that it was no use, his traveling to the capital. He took his family to the Kongsvoll lodge and left the next morning to join the local army command.

The day after the invasion, the sleeping Norwegian government finally roared awake. Bräuer stated his demands to King Haakon. The king brought these to Prime Minister Johan Nygaardsvold and his ministers, informing them that personally he could accept neither Quisling nor surrender but that as he was a constitutional monarch, it was not in his power to decide. If the government chose to make peace with the invaders, however, he would immediately abdicate. The government sided with their king, who then broadcast a message to his people over the radio: they should struggle and fight until the invaders were thrown from Norway's shores.

Pockets of resistance formed throughout the country, but backed by British and French reinforcements, the Norwegians put the greatest effort into two places: the naval stronghold Narvik and two long valleys between Oslo and Trondheim. If the Germans controlled that region, which separated the north–south corridor from the capital, then they would hold the heart of Norway.

It was here, with German panzer battalions backed by the Luftwaffe pushing north from Oslo, that Tronstad fought his war. He was charged with preventing German troops from crossing the defensive Norwegian line. Headquartered in the Kongsvoll lodge, he created a network of men to report on enemy activity and saw to the plowing of yard-high furrows in the snow on the surrounding lakes to prevent German transport planes from landing.

Over three weeks, aided mostly by the British, the unpracticed and outgunned Norwegians fought ferociously in the two valleys, but they were steadily pushed back toward Trondheim. Believing the battle over, the British began evacuating at the end of April. On May 1, when the surrender order came through to his district, Tronstad and several others hid eighty-one boxes of ammunition in the mountains in the hope of a future fight. That afternoon, while Tronstad was away from his family, a German cavalry unit navigating the hilly terrain reached the Kongsvoll lodge.

In the pouring rain, the men, covered in mud and grime, marched up to the front steps of the residence. The commanding officer ordered Bassa Tronstad to clear the barn for his men and their horses. "We have cows there," she returned in fluent German. "If we take them out, we won't have any milk." The officer, surprised at the rebuff in his native language, said that his men would then sleep in the lodge.

"As long as they're clean," she replied.

Before the unit left, the commanding officer stood before Bassa and threatened, "If anything happens around here, you will be shot immediately."

When Tronstad came back to the lodge, he collected his family and returned to Trondheim. The battle still raged in Narvik, but that too was soon over, and King Haakon and the government fled to Britain by ship. The Germans now ruled Norway. Their presence was a violation of everything Tronstad held dear, and their occupation robbed him of the life he had built from nothing.

Three months before Leif Tronstad was born, his father, Hans Larsen, died of a heart attack. His mother, Josefine, supported her four sons by running a small kiosk and by serving as a maid at private dinner parties hosted by the wealthier families in their neighborhood outside Oslo. Growing up, Leif was either studying, at the running track, or working. After secondary school and a couple of years employed at an electricity company, he headed to Kristiania Technical School to study chemistry. From the list of national records he set, it's clear that he majored in athletics as well. Even with a factory job and his participation in sports, he graduated with the highest marks on record. In 1924, after fulfilling his military service and shoring up his finances, he enrolled at NTH.

In Trondheim, he was finally able to focus on science alone, and he excelled to an even greater degree. He spent most of his waking hours poring over his books. "I work as a slave these days," he wrote to his mother in his first year. "No need to send money. I live modestly, Spartan, and careful. Only bread and the cheapest kind of butter." He graduated with top honors, which earned him recognition from King Haakon himself. Leif Tronstad, who had taken his mother's maiden name out of respect for her years of sacrifice, was now a rising star.

All this time, he had courted his childhood sweetheart, Edla, whose nickname since the time she was a toddler was Bassa. She was a few years his junior; they had grown up on the same hill in Sandvika, a neighborhood overlooking the Oslofjord, several miles west of Norway's capital city. While away studying, he sent love letters to his "little angel" and "beautiful treasure." One day, he promised, "we will have each other forever and be the happiest people under the sun." At Christmas, unable "to wait until we get old with love," he asked her to marry him.

Following their wedding in 1928, the newlyweds moved to Berlin, where, as part of Tronstad's three-year doctoral grant from NTH, he worked as a researcher at the Kaiser Wilhelm Institute of Chemistry. Keen with languages, Bassa worked as his secretary and translator. Together, they got their first taste of Nazism. Emblazoned across the city were posters with close-up photographs of Hitler and headlines calling to avenge the Treaty of Versailles.

After this stint, they moved back to Trondheim, where Tronstad continued his research, lectured on atomic chemistry, and earned his doctorate with a well-received work about measuring the oxidized surfaces of iron and steel. More overseas study followed at the famed Cavendish Laboratory at Cambridge University in Britain, where he furthered his electrochemical work and got to know the university's leading lights, including Ernest Rutherford.

In summer 1932 the Tronstads returned to Norway; soon after, their daughter Sidsel was born. In his diary Tronstad wrote, "A great day today ... Gosh, is she pretty!"

Talented not just in the lab but also in theoretical work, Tronstad found many opportunities open to him. Since his student days, he had wondered whether he should work in industry or academia. The former paid well, and after a life spent saving every krone, that was attractive. Academia paid far less, but a professorship would provide the opportunity to teach and research pure science without being enslaved by the bottom line. In the end he told Bassa that, while he wanted to be a professor, he would leave the decision to her. "If you like, I can make as much money as you want," he said. She gave him her blessing to return to the university. He was soon a full professor at NTH, published scores of research papers, and dazzled his students by igniting chemical compounds at the touch of a feather. He bought a nice house a short walk from the university and a car to drive out to his mountain cabin, where the family could ski and hike. He taught his daughter to ice-skate, and in August 1937 he welcomed a son, Leif Jr., who liked to be pushed around the yard in a wheelbarrow.

During these prewar years, Tronstad also worked as a consultant to several Norwegian companies. He advised them on the manufacture of steel, rubber, nitrogen, and aluminum, and, for Norsk Hydro, heavy water.

On November 11, 1940, seven months after the Germans invaded, Tronstad visited Vemork at the request of Jomar Brun. Tronstad's former classmate, narrow of frame, with round spectacles that matched his softly featured oval face, had said he'd wanted help in making improvements at the hydrogen plant. In truth Brun was also looking for insight into why the Germans were so interested in heavy water, and what he should do about it.

Shortly after Rjukan was taken, a German general visiting Vemork had ordered production at the plant — and deliveries to Berlin — increased at a rapid rate. He'd revealed nothing about the intended use of the heavy water. Brun had already implemented the expansion of high-concentration cells from seven to nine. This upgrade, as well as the laundry list of improvements suggested by Tronstad over the course of his two-day tour, including adding nickel-plated anodes to the cells to prevent rusting and increasing their amperage, would enable the plant to supply the Germans with 1.5 kilograms a day of heavy water, an increase of five times its former production level.

In private, after their work at the plant was finished, Tronstad and Brun wrangled with the question of why the Germans needed so much heavy water. As an avid reader of science journals, Tronstad knew there was interest in the substance as a moderator in the new field of fission research. Further, he had collaborated with Ernest Rutherford in 1935 in an attempt to produce tritium (an even rarer isotope of hydrogen) at Vemork. They failed, yet Tronstad understood well Rutherford's theory that tritium and deuterium could be used to obtain a fusion reaction with a potentially enormous energy release. Nonetheless, Tronstad dismissed the idea that heavy water could be applied to any great military use. Instead, he and Brun conjectured that the Germans might want deuterium for some kind of poison gas, but they doubted this application would yield results. Whatever the reason, Tronstad decided, if the Nazis were interested in heavy water, then so too should he be.

As for Brun, they agreed that he should remain at the plant and do whatever was necessary to keep his job, including making changes that would raise production levels. This was the only way he could stay informed of new developments. Anything significant, and he should alert Tronstad.

After his visit to Rjukan, Tronstad returned to Trondheim, where he had resumed his teaching and studies at NTH. He channeled most of his prodigious energy, however, into his activities with the underground resistance, working particularly closely with several bands of university students who were pushing back against the German hold on the country. Some published illegal newspapers. Others operated at a higher level, con-

necting with the British Secret Intelligence Service (SIS), the foreign intelligence agency. Under the moniker Skylark B, these students sent coded
wireless-radio transmissions from the forests outside the city to London,
reporting on troop movements and naval activity.

Codenamed the Mailman, Tronstad gave them any technical help they
needed and provided intelligence of his own. With his industrial connections, he gathered information on Norwegian firms and how they were
helping the Germans. Norsk Hydro was but one focus of many. The resistance was a precarious web that he feared might be unraveled at any time,
yet he continued spinning new threads within it.

In March 1941 Brun reported that German officials, accompanied
by scientists, had once again descended on Vemork. They now wanted
a nine-stage cascade before the high-concentration stage — thousands
more electrolyzer cells. They demanded fifteen hundred kilograms of
heavy water a year, and they were making Brun "personally responsible
for a perfect running of the plant." Soon after, Alf Løken, a chemistry student and member of Skylark B, approached Tronstad. SIS wanted them
to provide everything they had on Vemork's heavy water. Tronstad told
Løken all he knew. This was sent by wireless to London.

The Tronstad family spent several weeks that summer in their mountain cabin. Since the occupation, Tronstad had tried to keep things as normal as possible for his wife and children. But he was nervous. The Gestapo
had tracked down Skylark B's wireless transmitter outside Trondheim and
had tortured a student who was involved in its use. His confession led to
several other arrests, including that of Løken. Others in Skylark B fled.
Around this time, Brun came to the city with urgent news. The Germans
now wanted to increase heavy water production to five thousand kilograms a year, and Paul Harteck, whom Tronstad knew from Cambridge,
was on his way to advise on new methods to obtain such levels. Realizing the importance of conveying this information to the British, but with
Skylark B in jeopardy, Tronstad found a courier — a man planning to escape by boat to Scotland the following week. He gave him precise details
on Vemork, including its monthly production figures. The courier wrote
these details down on some cigarette paper.

The Gestapo was close to breaking up Trondheim's resistance. On the
morning of September 9, a student visited Tronstad for advice on securing Skylark's radio equipment. The student was arrested that same afternoon. A week later, Tronstad's courier was seized at the wharf's edge. For-

tunately he was able to swallow the cigarette paper before being hauled away. On September 20, one of the arrested Skylark members, fearing he might break under torture, managed to pass on a message to a friend through the barred windows of the Gestapo prison: "The Mailman must disappear."

Tronstad warned Bassa that they might need to flee, and two days later, when informed of yet another arrest, he went straight to her. It was their thirteenth wedding anniversary. "We must go," he said. They made their way to the train station with some hastily packed suitcases; there was a 7:15 p.m. train to Oslo. Bassa and the children boarded first. Tronstad looked left and right, fearing the Gestapo might be waiting for him on the platform. Once safe on the train, he settled down in the sleeping compartment with his family and wrote the first entry into the small, black diary he would keep throughout the war. "Family, house and worldly goods have to be set aside for Norway's sake."

At 10:15 a.m. they arrived in Oslo and took another train to Sandvika. They climbed up the hill, past the small kiosk where Tronstad's mother used to work. Farther up, they reached Bassa's childhood home. He instructed Bassa to tell anyone who asked that he was in Rjukan. Then they embraced. "I'm not afraid of anything," he told Bassa. "That is exactly what I'm afraid of," she responded.

Outside the house overlooking the fjord, he kneeled beside his children. "Take care of your little brother," he told nine-year-old Sidsel. Then he turned to four-year-old Leif: "You must be good for your mother while I'm gone." He promised to bring back a little gift for each of them. Sidsel asked for a watch; Leif, a motorized go-cart. "Be kind to each other," Tronstad said before hurrying away, overcome with emotion. He was on his way down the hill when he heard young Leif call out, "Where are you going?"

Tronstad swung around. "Notodden."

"You'd better hurry up." Leif pointed to the train moving into the Sandvika station below. Tronstad started to run down the hill, but the truth was he was not catching that train. Rather, he would wait on the platform to catch one heading in the opposite direction.

In Oslo he collected fake identity papers, and the next morning he borrowed a bicycle and rode twenty miles north to Sandungen, where his brother-in-law worked as a forester. For a day, Tronstad marked trees for felling and wondered if he should just wait out the war as a "hermit" in the forest. But he dismissed the thought. That night, he wrote a letter to Bassa. Falsely dated the night they'd left Trondheim, the letter spoke of his inability to remain at a university where those allied with the Germans received "preferential treatment," while he was slandered. The letter continued: he was now in Sweden, but he had not done, nor would he do, anything "dishonest or illegal." The letter was meant to provide Bassa with cover in the event she was interrogated by the Gestapo.

The next morning, September 26, two members of the resistance network picked him up in a truck. They drove toward neutral Sweden, a hundred miles away. An hour's hike from the border, they got out and set off on foot through the woods. They crossed into Sweden at 5:00 p.m., and a couple of hours later Tronstad was arrested at a Swedish military post that had an understanding with the network. They gave him a steak, beer, and some coffee.

A month passed before Tronstad gained passage to Britain, from where he hoped to continue his fight to free Norway. His passport, stamped in Stockholm, was "Good for a single journey." Aboard a bomber converted to a transport plane, he crossed the North Sea at high altitude, breathing through an oxygen mask. After seven hours in darkness, the plane, having fought headwinds the whole way, finally landed in Scotland.

On October 21, Tronstad arrived at King's Cross station in London. As arranged by SIS, a room was booked for him at the St. Ermin's Hotel in the heart of Westminster, a stone's throw from the spy agency headquarters. London, a city he knew well from his student days, was a war zone. Soldiers crowded the streets, and a floating armada of gray barrage balloons darkened the sky, defending against German bombers.

In Stockholm, Tronstad had read about the Blitz in the newspapers. Since September 7 the previous year, Hitler had sent his planes into the heart of London to break its fighting spirit. Incendiaries sent fires sweeping across rooftops. Explosives ripped apart houses and buildings. Many thousands of people had been killed in the attacks and countless more wounded or left homeless. But it was one thing to read about the devastation and another to see it. Although the attacks had largely relented by May 1941, the streets were still strewn with the rubble of bombed-out buildings, and the people he passed had a joyless, though determined, mien.

When Tronstad finally went to bed, British fighter planes patrolling the skies thundered overhead. Such was his exhaustion from the journey that he slept easily through the din.

St. Ermin's, a horseshoe-shaped Victorian hotel, was the perfect location for clandestine meetings, with its grand, busy foyer and many nooks and crannies. On his first Sunday in London, Tronstad sat down across from Commander Eric Welsh, head of the Norwegian branch of SIS. Welsh ran Skylark B and had choreographed Tronstad's journey to London.

The spy was short and overweight, with a huge dome-shaped head. Slovenly dressed, he chain-smoked; cigarette ash dusted the front of his shirt. He may not have looked like a hero, but in the previous world war, Welsh had won medals for gallantry as a minesweeper. He became an officer in Naval Intelligence, focusing on scientists and their work, before being transferred to SIS. His wife was Norwegian, a relative of the composer Edvard Grieg, and his knowledge of the language earned him a position with a Norwegian paint company that serviced a lot of the country's industrial firms. In fact, Tronstad soon learned, Welsh knew Jomar Brun and Vemork quite well. He had sold the corrosion-proof tiles for the flooring in the high-concentration plant.

Welsh knew many things, a lot of them from an informant inside Germany codenamed the Griffin. This was the scientist Paul Rosbaud, who, as an adviser for the scientific publisher Springer Verlag, was close to Hahn, Heisenberg, and other leading German physicists. He provided early reports of the Nazi atomic program, but since Army Ordnance had taken over, the intelligence had slowed. Welsh also had the inside track on Britain's development of an atomic bomb, under what was known first as the MAUD (Military Application of Uranium Detonation) Committee, then as the Directorate of Tube Alloys.

Welsh spoke in a ramble and garbled his words, which made the little of import he shared with Tronstad almost unintelligible. Regardless, Welsh was not there to reveal his own secrets but to learn what his new arrival had to offer. Tronstad was open from the start. Vemork was now producing four kilograms of heavy water a day for the Germans — with more to come. The two spent the whole morning together, not all of it in discussion about heavy water. Tronstad knew from other sources that the Germans had secured uranium oxide from Norway, and there were also his contacts with former students and colleagues in Oslo, Berlin, Cambridge, and Stockholm, which might bring in more intelligence. Welsh expressed his hope that Tronstad would lend his own expertise to aid the war effort. No firm next steps were set, but Tronstad made it clear that he was eager to be of help.

And indeed, Tronstad's involvement in the resistance continued to grow. One day he met with a Skylark member who had escaped to London. Another day, he was at the Thatched House Club, a gentleman's club and den for members of the British secret services. He ate a dinner of oysters and pigeon with those tasked with commando operations in Norway. The next day he spent at Norway House, off Trafalgar Square, the center for the exiled community, where he was introduced to top government officials — "celebrities," he wrote dismissively in his diary. Next he met with leaders of Milorg, the umbrella organization of military cells of which he had been a part in Trondheim. They were pushing to be officially incorporated into the Norwegian Ministry of Defense. Then he had an appointment with the E-Office, Norwegian intelligence, then one with the defense minister, then an audience with the crown prince. Throughout, Tronstad looked for a way to be of most use to his country.

Time and again over his first six weeks in London, Tronstad was called into conferences with the British scientific community associated with Tube Alloys, all brought about through Welsh, who let it be known that Tronstad was a man "acquainted with the particular subject which I think interests you." Tronstad knew some of the scientists from his Cambridge days, but others were new connections. He was introduced to Harold Urey, who had discovered deuterium and who was on a fact-finding mission in London for the United States' own program. Any doubt that remained for Tronstad about the German need for heavy water was dispelled. It was to be used in atomic research, potentially for a bomb. The only unknown was where the Germans stood in the race to build it first.

During these weeks, Tronstad participated in several strategy sessions on how to stop the supply at Vemork. Wallace Akers, a chemist well known to Tronstad and now head of Tube Alloys, brought him in to discuss a British Air Ministry plan to bomb the plant. Another plan, codenamed Clairvoyant, targeted six hydroelectric plants in southern Norway, Vemork included, with six teams of saboteurs operating in tandem. Tronstad argued against both. The plant was a poor target for nighttime bombers, and Clairvoyant was too ambitious. Instead, he suggested a targeted operation by Norwegian agents or sabotage from the inside.

Tronstad was unsure how, or if, his advice was taken — he was very much the outsider — but it seemed that little came from these sessions and that the urgency from the British side to do something about Vemork waned. By the end of December, after participating in a six-week Norwegian Army training course in Scotland, he was still looking for his place in the struggle to liberate his country. Many wanted him to focus on scientific work for the war effort, but he wanted to return to the action. As he wrote late one evening, "I want to be close to those fighting on the frontlines for Norway's cause."

3

Bonzo

BEFORE DAWN ON December 2, 1941, Knut Haukelid was roused by dogs barking outside his room. There was a chill in the air, and frost clouded the windows overlooking the huge wooded estate of Stodham Park, fifty miles southwest of London. Quickly, Haukelid dressed in his new British uniform, its starched collar surpassed in stiffness only by his standard-issue boots.

Outside, he stood with roughly two dozen other Norwegians who had volunteered to attend special commando training to fight for their country. They came from every walk of life — rich, poor, and in between; from city, town, and backwoods country. A few had never handled a gun before; others were marksmen. Some were boys, barely eighteen. Most were in their twenties, and a few were old men — all of thirty, like Haukelid. Before the war, they had been students, fishermen, police officers, bankers, factory workers, and lost types, looking for their place in the world — again, like Haukelid.

At first glance, he looked like many of the others. Although his twin sister, Sigrid, was a Hollywood movie star, known as the "siren of the fjords," there was nothing particularly handsome about him. He had fair hair, blue eyes, and a medium build, with slightly hunched shoulders. At five ten, he was just above average height. But there was something in his look, in the way his face went from the crinkle of a smile to hard intensity in an instant, that was unforgettable.

To their grizzled Irish sergeant major, who went by the nickname Tom Mix, after the American western star, Haukelid and the others were all the same: men to be instructed on how to kill, to sabotage, and to survive, any way and any how. He taught them only one rule: "Never give your enemy half a chance." This suited Haukelid fine. From the time he was a boy, rules had always rubbed him the wrong way.

At 6:00 a.m. the squad of new recruits started their day with what Mix liked to call "hardening of the feet": a fast march across the extensive estate. After an hour of this, the recruits' feet were more blistered than hardened. A short breakfast was followed by weapons instruction. "This is your friend," Mix said, twirling his pistol around his finger. "The only friend you can rely on. Treat him properly, and he will take care of you." Then he took them out to a grove in the woods and taught them how to stand — knees bent, two hands on the grip of the pistol — and how to fire: two shots quickly in a row to make sure the enemy was down. If circumstances allowed, "Aim low. A bullet in the stomach, and your German will squirm for twelve hours before dying." Haukelid had grown up hunting, but this was something very different.

After two hours of shooting, they spent another in the gym. They somersaulted forward and backward. They jumped off high ledges and rolled forward into a standing position. They pummeled punching bags. They wrestled, learning how to take down and disarm an enemy with their bare hands. A long knife hanging from his hip, Mix wove into his lessons stories of his own fighting in World War I and his time policing in the Far East. The other instructors were the same, one sharing, "We killed so many Germans, we had to rise up on our tiptoes to look over the heaps of them."

A break for coffee, then the squad had signals training, learning how to send and receive in Morse code. This was followed by lunch, then a two-hour class in simple demolition. "Never smoke while working with explosives," Mix said, a lit cigarette perched between his lips, again offering the point that rules existed to be broken. They blew up logs and sent rocks skyward. Ears ringing, they moved on to orienteering, navigating the estate with maps and compasses, then field craft, stalking targets and scouting routes through the woods. From 5:00 to 8:00 p.m. they were free to relax and have a meal before the night exercises began. These consisted of more weapons, more explosives, more unarmed combat — now executed in the pitch darkness.

Through day after day of this schedule, his boots and collar softening with each training session, Haukelid hardened into a fighter. Though reasonably fit at the start, he became fitter still. On occasion, he would be invited into a room with an officer or a psychiatrist to be asked if the training was too much, too hard, if he wanted to quit. This kind of work wasn't for everybody, they said. It was for him. Firing two shots in rapid succession became a reflex, and his aim grew lethal to the range's paper targets. He learned how to time throwing a hand grenade ("One can go from here to London before it explodes," Mix said) and not to be sparing with them. He gained expertise in hand-to-hand combat and in the use of a knife. He grew skilled at demolitions, able to light a ten-second fuse with steady hands.

Three weeks later, Haukelid was told that this — all this — had only been the preliminary training. The instructor report on him stated: "He is a cool and calculating type, who should give a very good account of himself in a tight corner . . . Has no fear." Haukelid received marks of "very good" for field craft, weapons, explosives, and map reading. His signaling was simply "good."

On December 20, thirteen days after Japan bombed Pearl Harbor in a surprise attack, bringing the United States, the country of his birth, into the war, Haukelid boarded a train to Scotland for further instruction. He had a promise to keep, a promise made by his mother to the Gestapo before her son had left for Britain.

In 1905, after breaking from its union with Sweden, the new Norwegian government sought to create a constitutional monarchy. The democratically elected parliament would be the highest authority, with the king the ceremonial head of state. They settled on Prince Carl of Denmark, who was descended from Norwegian kings, as the best candidate, but he declared that he would fill the role only if the people voted for him. He was duly elected, and thus King Haakon VII rose to the throne.

That same year, Bjørgulf Haukelid immigrated to the United States. He settled in Flatbush, Brooklyn, far away from the remote mountain lodge for hikers and cross-country skiers in Haukeliseter, fifty miles west of Rjukan, that his father wanted him to take over one day. The construction of New York's subway provided ample opportunity for a civil engineer. Bjørgulf married a nurse, Sigrid Gurie, who had emigrated from Oslo, and on May 17, 1911 — Constitution Day in Norway — they had twins: Knut and Sigrid. The only thing closer to Knut than his sister was a small teddy bear he kept perpetually clutched in his arms. Its name was Bonzo.

The following April, the family boarded a steamer back to Norway. Knut's parents missed their homeland and wanted their children raised as Norwegian. With the recent industrial boom, there was no better time.

On the way across the Atlantic, the RMS *Titanic* signaled a distress call. The steamer carrying the Haukelids diverted from its route to help rescue survivors, but they were too far away and arrived too late.

In Oslo, Bjørgulf launched a successful engineering firm and bought a house in which to raise the children. Knut was a rascal from the start. Dyslexic and restless, he hated school. Sitting still in those hard chairs all day, listening to the drone of the teachers, was torture for him. Because he had a slight stutter, being asked to speak in class only tightened the screws. And so he got through the day by pulling pranks. Once he released a snake in the middle of class, earning one of his many expulsions. Knut preferred the outdoors, but in the city there was not much opportunity to get outside. What was more, his mother, not wanting him to get dirty, made him wear a sack with a hole cut out of the top when he played in the backyard. Theirs was a house with many rules and restrictions. His twin sister, who wanted to be a painter and live in Paris, struggled against their parents as well.

The one place Knut was able to run free was the Haukeliseter lodge. On weekends and in summertime, he skied, fished, camped, and hunted with his grandfather, Knut Sr., in the mountains and lakes of Telemark. He was told the old tales of trolls inhabiting and protecting the lands of Norway, and he believed them. His faith in these creatures lent even more magic to the woods he loved.

Always with an eye for adventure, Haukelid left for the United States just shy of his eighteenth birthday. At the same time, his sister left for Paris. He enrolled at Massachusetts State College but never graduated. A devotee of John Steinbeck and Ernest Hemingway, he took to the road. In the Midwest, he found a job on a farm. He liked the work and the open plains but bucked against some of the farm's puritan ways. At dinnertime he had to wait out the long prayer given by the farm owner before he could eat. One night, Haukelid offered to say grace. He sang a short Norwegian song — nothing to do with God — and dug into his meal as if he hadn't eaten in weeks.

After a few years, he came back to Norway. His father found him a well-paid job at Oslo's biggest bank, but Knut turned it down. One could earn more money, he told his father, fishing for trout. And off he went.

After several months of fishing, he moved again — this time to Berlin. He studied engineering, learned German, and contemplated his future. In 1936 he witnessed Hitler's propaganda parade at the Olympics. One night, when confronted by a drunk Nazi Party member who was spouting one vile statement after another, Haukelid dropped him with a punch.

At last, he returned again to Oslo. His sister, discovered in Paris by Samuel Goldwyn, was now starring in films with Gary Cooper. Haukelid finally buckled under his father's wish for him to get serious with his career and his life. He took a job with his father's firm, importing engineering equipment from the United States, and fell in love with a young woman named Bodil, a physical therapist who treated him for back pain brought on by his outdoor adventures.

In early April 1940, Haukelid had just finished building a jetty in Narvik and took a few days off to go cod fishing before stopping in Trondheim. While there, the Nazis invaded. After listening to Professor Tronstad at NTH remark on the need to take a stand, Haukelid and a few students commandeered a freight train and drove it almost halfway to Oslo until they found the tracks closed. They abandoned the train and took a bus to the nearest army mobilization point outside Lillehammer, but it had no weapons to offer. At the mobilization site, they learned that the Nazis had taken Oslo. The news moved many of Haukelid's companions to tears. Then came word that the Germans had demanded the king abdicate and the government step aside. When Haakon VII refused, Luftwaffe bombers had tried to kill him in his woodland retreat. Haukelid, who believed he had seen the bombers on their way to assassinate the king, had found his leader, and his purpose, and this brought him to tears also.

He finally tracked down a regiment battling the Germans. Its commander, a colonel, gave him a Krag rifle and thirty rounds of ammunition, then sent him into war. Over the next three weeks, despite having no military experience, Haukelid fought. His battalion ambushed a line of German panzers at a mountain pass, wiping them out with Molotov cocktails and a single cannon, but apart from that one success they experienced one pushback after another.

After his regiment's surrender, Haukelid tried to reach the fighting in the two valleys that ran between Oslo and Trondheim, but his countrymen were already in retreat. He then traveled to the capital and went to his parents' home, a spacious apartment at Kirkeveien 74. His father was away; only his mother was there to welcome him. Knut went into the bedroom where some of his belongings were stored and closed the door. "What are you doing?" his mother asked, coming into the room.

"Getting some things," Haukelid said.

"You need to get out and fight," she told him.

That was exactly his plan. From the closet, Haukelid pulled out the boots and skis he was there to collect. The plan was to head north with his close friend Sverre Midtskau to the strategic port town of Narvik. On his way, Haukelid grew sick and began spitting blood. By the time he recovered from his stomach ailment, the battle in Narvik was over as well. In his surrender, before being sent to a German prison camp, Norway's top general made a plea to all Norwegians: "Remain true and prepared" for the future fight.

The general's words stayed with Haukelid as he went into the mountains with Bodil, as far from the occupying Germans as he could get. He spent the summer earning enough money fishing to sustain them for the struggle ahead. Then he got word that Midtskau was in Oslo and wanted to see him. Haukelid left for the capital. It turned out that Midtskau had been to Britain, where he had received wireless training before being sent back to Norway aboard a British submarine. The others who came with him went north to start Skylark B in Trondheim, while Midtskau was tasked with launching Skylark A in the Oslo area. For that he needed Haukelid.

For months they moved from hut to hut in the woods outside Oslo, sending radio signals to Britain but hearing nothing in return. Through a range of contacts in the city, they gathered intelligence on the German command in the capital, everyone from Reichskommissar Josef Terboven, who served as Hitler's dictatorial representative in Norway, to General Nikolaus von Falkenhorst, who oversaw the German military forces, to SS Lieutenant Colonel Heinrich Fehlis, who ran the security services. Unable to make contact with London, Midtskau traveled to Britain by fishing boat to pass on the intelligence and retrieve new equipment. During his parachute drop back in, the radio set was damaged beyond repair. They continued with their efforts nonetheless and even hatched a plot to kidnap Quisling. Haukelid and his friend were daring and brave; they were also amateurish and terribly ineffective.

In early 1941 the Gestapo arrested Midtskau and another in their group, Max Manus. Both managed to escape. Haukelid was detained soon after, but the Norwegian police let him go. Despite these setbacks, the group continued with its spying.

Reichskommissar Terboven, the thin, bespectacled former bank clerk who had supported Hitler since his early days, moved quickly to consolidate Nazi rule. He removed any Norwegians not loyal to the "new order" from positions of influence: judges, clergy, administrators, journalists, business heads, policemen, local municipal leaders, and teachers alike. The Norwegian parliament was shut down, its members dismissed. The parliament building in the heart of Oslo now flew a Nazi flag and housed Terboven's administration. The SS installed themselves in nearby Victoria Terrasse, a stately government building that stretched over a city block.

The Nazis' presence extended well beyond Oslo. Travel after curfew or beyond a certain place without an identity card or pass was prohibited. Radios were banned. Anyone in violation was subject to imprisonment — or whatever punishment the Nazis chose, since it was the Nazis, not the police, who enforced the law. Nothing was published in Norway without the censor's stamp of approval. New schoolbooks were printed to teach students that Hitler was Norway's savior. Strict rationing of coal, gas, food, milk, and clothing left families scraping by. People were reduced to making shoes from fish skins and clothes from old newspaper. All the while, the Germans confiscated whatever they wanted for themselves, from the finest cuts of meat to the products of Norwegian industry, even taking the best houses.

Some Norwegians supported the new order. Many others merely did what they were told. But there were others still who pushed back against the Nazis, whether by refusing to provide a seat for a German soldier on the tram or by organizing cells of resistance to meet force with force. Haukelid was open to such violent resistance, but he didn't know where or how to act.

Then, in September 1941, workers throughout Oslo went on strike against the strict rationing of milk. With the invasion of Russia slowing into a brutal siege, Hitler wanted his occupied territories managed with a firm hand; Terboven was under pressure to act. Furthermore, SS General Reinhard Heydrich, the chief of German police and security forces, was in Oslo at the time, and the Reichskommissar dared not look weak in front of him. Threatening to bring anyone who would threaten his hold on Norway "to their knees," Terboven instituted martial law. Hundreds were arrested, and the security chief, Fehlis, ordered the execution of the two strike leaders.

At the same time, an intensified Gestapo hunt for underground cells led to the breakup of Skylark B in Trondheim. The Nazis quickly tracked down all links to Oslo. Midtskau and several others were rounded up. Haukelid fled to Sweden, but not before marrying his sweetheart Bodil in an informal ceremony.

Soon after, Gestapo captain Siegfried Fehmer, looking for new leads, descended on the Kirkeveien 74 apartment owned by Haukelid's parents. The former lawyer was the Gestapo's lead investigator of the resistance in Oslo. Unlike many of his fellow officers, he rarely wore his stiff gray uniform and black peaked cap, preferring instead plain suits, and employing a handsome smile to disarm interrogation subjects and recruit informants. Brought to Norway by Fehlis, who had helped train him as a bloodhound, Fehmer tried to assimilate with the locals, at least by taking Norwegian women to bed and learning the language — the better to do his job. Blond, six feet tall, intelligent, with a sharp memory, he was widely known in the resistance as the Commissar.

Fehmer found only Haukelid's mother, Sigrid, and new wife, Bodil, in the apartment. He led the two women down to a waiting car, which, he warned, would bring them to Møllergata 19, the Gestapo prison. On the way, Fehmer asked Sigrid if she knew where her son was. Sigrid slapped the German officer hard across the face. Fehmer asked again. Then again. "He is in the mountains," Sigrid finally said. Thankfully she had managed to destroy her son's American passport before the Gestapo arrived.

"No," Fehmer said. "He is in England. Our contact in Sweden tells us that he has already been taken across the North Sea. What do you think he is doing there?"

"You will find out when he comes back!" she promised.

Fehmer brought her and Bodil to Møllergata for further questioning, but he got nothing more from them and they were released.

In fact, Haukelid had yet to leave for England, and he received word in Stockholm of the arrests of his mother and wife. He returned to Oslo, but the security clampdown made resistance activity impossible, so he once again crossed the border and finally left for Britain by plane. In London he met first with Eric Welsh, his handler, who wanted him to return immediately to Norway as a spy. Haukelid had other ideas. He wanted military training, and he wanted to fight. In his view, it was the only way his country would be freed. Welsh sent him to Norway House.

There, in an attic-floor office overlooking Trafalgar Square, Haukelid met Martin Linge. Wearing the gray-green uniform of a Norwegian Army captain, Linge had a winning smile and a firm handshake, and like every one of Linge's potential recruits, Haukelid was charmed by the officer straightaway. A former actor, Linge had been attached to a British unit when it landed near Trondheim in April 1940.

At first, Linge spoke of how things were in Norway, of places they both knew. There was no doubt he was trying to read Haukelid: what he wanted, why, and if he was capable of achieving it. He explained that there was a lot of work to do if Haukelid had a mind for it. He would be joining a small company, numbering only a couple of hundred men. It was overseen by the British but was made up only of Norwegians. Military experience was helpful, but not necessary, because he would be practicing unconventional warfare, often behind the lines. It was the kind of warfare where one man in the right place could make a big difference. The training would be tough, even brutal, and the operations even more so. The specifics of these were not made clear to Haukelid, but he wanted in nonetheless.

"Are you married?" Linge asked.

Haukelid answered yes. Most recruits answered no, which Linge tended to follow up with the riposte, "That's good. Then we won't have to send flowers to your widow." Liking what he saw in Haukelid, Linge welcomed him to Norwegian Independent Company No. 1. Within a week, he was issued his boots and British uniform.

After a long train ride, then a ferry across Loch Morar on the cold, windswept coastline of western Scotland, Haukelid arrived at the next stage of his training. Some instructors at Meoble, an old hunting lodge, were Norwegians who had undergone the course already and now led the way, translating directions from their British commanders when needed.

Marches through Stodham Park were replaced with scrambles through thick brush, fording ice-cold rivers, and rappelling down steep ravines — or crossing them on ropes. With both British and foreign weapons, the men practiced instinctive shooting (shooting without the use of sights) and learned close-quarter firing. They stalked the shadowed pine woods, firing at pop-up targets. They moved through building mockups, clearing the rooms. In demolitions, they graduated from blowing up logs to destroying railroad cars and factories. They crafted charges and incendiary devices of all sizes. Haukelid was amazed at what a small charge placed in the perfect spot at the perfect time could do: it could stop an army, render a weapons plant useless.

They became faster at sending and receiving messages in Morse. They practiced how to kill silently with a knife, making sure the blade slipped into flesh, avoiding bone. They were taught how to break into safes, how to use poison, how to incapacitate someone with chloroform. They learned how to follow a route to a target by memory alone, without maps or compass. They studied how to camouflage themselves in the field, how to crawl through a marsh and reach their enemy undetected, how to take them down without a sound — without even a weapon.

"This is war, not sport," their instructors reminded them. "So forget the Queensberry rules; forget the term 'foul methods' . . . these methods help you to kill quickly." Haukelid learned that a sharp blow with the side of his hand could paralyze, break bones, or kill. Weak points on the human body included the back of the neck, just to the left or right of the spine; between the Adam's apple and the bridge of the nose; the temple; and the kidneys. He practiced these blows on mannequins, over and over, until his hand felt like it would break.

Even the occasional night off, like Christmas or New Year's Eve, was instructive. Their teachers brought Haukelid and the others in his squad to a pub, where they plied their charges with strong ales and whiskey. At first it seemed like a good time, but they learned later that they were observed all evening to see who drank too much, who made a spectacle of himself, or, worst of all, who spoke out of turn. One had always to be on one's guard.

All the lessons were then woven together in staged raids, both in the light of day and in darkness. At Meoble, the instructors called these "schemes." The men were given a target — a railway line, military barracks, airport, or factory. They devised an attack plan: the route, whose responsibility it was to scout, to do recon, to provide cover, and to hit the target. Then they executed the plan, carrying real weapons and anything else they would need. When possible, they placed live charges. Some of these raids took hours. Others took them away from Meoble for several days.

The instructors and some of the other recruits played the role of the Gestapo, eager to thwart the raid, ready to be as rough as they needed to be.

It was a merciless regime and, like before, Haukelid was asked regularly if he wanted out. He refused. They may not have been learning traditional warfare, with fixed lines and flanking maneuvers, but David was fighting Goliath, and Haukelid knew that "gangster school" (what they were told the Germans called places like Meoble) was exactly where he needed to be. The chief of Meoble thought the same, informing his bosses in London that Haukelid was "a really sound man and cunning. Has done well. Has no fear. Another excellent student who would do well in almost any special job."

Before leaving Meoble, Haukelid discovered that the Norwegian company was not unique. Rather, it was part of an expansive organization called the Special Operations Executive (SOE), with country sections across Europe and beyond.

Founded soon after Winston Churchill was elected prime minister, its directive was to "set Europe ablaze" with commando missions against the Nazis. Hugh Dalton, head of the Ministry of Economic Warfare and the SOE's first leader, formulated the basis of its purpose. "We must organize movements in enemy-occupied territory comparable to the Sinn Féin movement in Ireland, to the Chinese guerrillas now operating against Japan, to the Spanish Irregulars who played a notable part in Wellington's campaign . . . We must use many different methods, including industrial and military sabotage, labor agitation and strikes, continuous propaganda, terrorist acts against traitors and German leaders, boycotts and riot." Culling staff and methods from SIS's D section (D for destruction) and the Army's similar MI (Military Intelligence) unit, its masterminds, who started the organization from three rooms at St. Ermin's Hotel, referred to themselves as the Ministry of Ungentlemanly Warfare.

On January 14, 1942, Haukelid arrived at Special Training School (STS) 26, in the Scottish Highlands near Aviemore, the home of Norwegian Independent Company No. 1. He was eager to be sent into action. Roughly 150 Norwegians lived in three hunting lodges (Glenmore, Forest Lodge, and Drumintoul) amid cragged granite mountain peaks, steep pine-forested valleys, and endless stretches of moors. The place, nicknamed simply Twenty-Six, reminded Haukelid almost too much of his homeland, but that meant it was ideal terrain to prepare for missions. He was soon to realize that there were none in the offing.

While Haukelid was training at Meoble, the British had launched two major combined air, land, and sea operations: one against the Lofoten Islands in northern Norway, the other on a pair of coastal towns, Måløy and Vågsøy, in western Norway. Taken together, the attacks aimed to weaken the German hold on the coastline and stop production of fish oil, used to make TNT. On December 27, 1941, British destroyers bombarded Måløy, silencing its coastal batteries and leveling the German troop barracks. The Royal Air Force dropped smoke bombs to blind German defenders to a beachhead assault by six hundred British forces. They were joined by Captain Linge and twenty of his men. The Måløy garrison surrendered soon after the attack began, and Vågsøy was taken with similar ease. Tragically, Linge was killed in the assault, hit by a sniper as he rushed the German headquarters.

In the Lofotens, seventy-seven Norwegian Independent Company men had accompanied the British landing force. They made a swift advance. By nightfall, several German garrisons and Nazi officials had been captured. The people of the island chain, who had come out in support when they were made to understand that the British were there to stay, aided the assault. The next day, the threat of a furious German counterassault by air and sea spooked the operation's commander. He called for an immediate retreat. When the Allied force, including the Norwegian commandos, withdrew from the archipelago, the local population spat at them, knowing the reprisals they would face at the hands of the Nazis.

When Haukelid first came to STS 26, he found a company in despair. The Lofoten debacle and the subsequent realization that the British had not even informed the exiled Norwegian government of the operation caused a furor. The death of Captain Linge broke their hearts. Many threatened to join the regular army unless their missions were aligned with the aims of their country's leadership. The men whispered that they were considered by the exiled Norwegian government nothing more than "false beards" and a "half-trained bandit gang" who caused more trouble than they did good. A dozen members openly mutinied; they were disciplined and sent away.

There was nothing for Haukelid to do but wait and see how the situation unfolded. He was happy to at least be surrounded by Norwegians like himself who had risked everything to come to Britain and train to fight. Though Haukelid had arrived by plane, most of the others had come via the so-called Shetland Bus, an armada of Norwegian fishing boats that traveled across the stormy, unpredictable North Sea to the Shetland Islands. One in his company had even made the journey alone, in a rowing boat. Others, like Jens-Anton Poulsson, a lanky twenty-three-year-old soldier from Rjukan, whose tobacco pipe was never far from his mouth, had traveled for over six months around the world — from Norway, to Sweden, Finland, Russia, Turkey, Egypt, India, South Africa, and Canada — to finally arrive on Britain's shores.

For two weeks Haukelid trained with his new company and roamed the snowbound countryside. There was never enough food at STS 26, and though hunting was technically against the rules, bagging a stag for extra nourishment was not exactly discouraged. Haukelid, who had taken the nickname Bonzo, after his childhood teddy bear, loved the sport it provided.

But the pall over his company remained. Then, on January 31, several visitors came up from London by night train. The company showed off their shooting and raid techniques, and then hosted a dinner at one of the lodges. Oscar Torp, the Norwegian defense minister in exile, and Major General Colin Gubbins, second in command of the SOE, were the guests of honor. Torp gave a rousing speech, promising a new era of cooperation between the exiled Norwegian government and the British. Their aim in Norway was twofold: the long-term goal of building up Milorg, which was now under Army command, in the anticipation of a future Allied invasion of Norway; and the short-term goal of performing sabotage operations and assisting in raids to weaken German military and economic strength. The Independent Company would be the tip of that spear, and no operation would occur without Norwegian consent. When he finished, the men cheered and pounded on the tables.

Then Torp and Gubbins introduced the two officers who would command them. The first was Lieutenant Colonel John Wilson, the new chief of the SOE Norwegian section. Wilson told them that he had "Viking blood" in his veins but that over the generations it had thinned, like his graying crown of hair. He had a short but upright bearing, a quiet, stern voice, and determined manner. Wilson, who was leader of the International Scouting Movement prior to the war, had helped design and run the SOE's training schools. Next to him, dressed in uniform and cap,

stood Leif Tronstad, whom Haukelid had briefly met on the fateful morning of the German invasion. Coordinating closely with Wilson, Captain
Tronstad would oversee the company's training, planning, and execution
of operations.

The next day, the guests toured the camp and met with the men. That
evening, Torp led a memorial service for Captain Linge. They sang, and
read from Scripture, and a Norwegian chaplain recited a poem written in
1895, before their country's independence from Sweden:

> We want a country that is saved and free,
> and which does not have to answer for its freedom.
> We want a country, which is mine and yours . . .
> If we do not have this country yet,
> We will win it, you and I.

In time, the Norwegian company assembled and led by the man they
eulogized would be unofficially renamed in his honor: Kompani Linge.
Captain Leif Tronstad would now help direct the Norwegian soldiers in
their fight for freedom.

4

The Dam-Keeper's Son

O N THURSDAY, March 12, 1942, Einar Skinnarland found himself sitting on an operating table at a hospital in Kristiansand, a port town in southern Norway. The doctor was insisting that he be anesthetized for the surgery on his left leg. A week before, he had fallen and dislocated his kneecap. He had forced it back into place, but blood pooling behind his knee had caused dangerous swelling. No anesthesia, Skinnarland said. He was leaving Kristiansand soon, and he feared the drugs would hamper his recovery. The doctor warned yet again that the operation would be exceedingly painful. "No," came the answer. The doctor shook his head, no doubt thinking his patient must be insane.

Skinnarland rested back on the table as the nurses secured his leg. Then the doctor cut into it with a scalpel. Skinnarland — red-haired, all broad shoulders and wiry muscle — did his best to steel himself. His face, usually set with an easy, bright smile, seized up. He gritted his teeth, and his narrowly set eyes seemed to narrow further still. As the doctor probed into his open flesh, Skinnarland endured the procedure with barely more than a murmur. After draining the blood and fluid from behind the kneecap, the doctor made sure the bone was set properly and then stitched his patient up. He would need a few nights to recover in the hospital. Again, Skinnarland refused. A tightly bound bandage would have to do. A couple of hours later, walking stick in hand, he hobbled down the stairs and into a taxi.

Over at the port, across from the seamen's hostel, he entered the ship chandlery. The proprietor, a cheery bachelor who always had a good story to tell, brought Skinnarland into the back room. There he tried to rest, but the pain radiating from his leg was too intense. Instead, he focused on oiling his revolvers. He planned to persuade the crew of the *Galtesund*, a 620-ton coastal steamer that was shortly due into port, to change their course and bring him and his friends across the North Sea to Aberdeen. Until then, he would endure, as members of his family had always done.

For centuries, Skinnarlands had lived along the banks of the serpentine Lake Møs. It was unforgiving backcountry, impenetrable for six months of the year on anything but skis. Running east from Lake Møs in a fast current, the Måna River followed a course through the Vestfjord Valley, its waters dropping almost seven hundred meters along the way — several times precipitously over falls.

In the sixteenth century, a small village of wooden houses developed on the banks of the Måna. Its few hundred inhabitants lived in the shadows, the sun hidden behind the high valley walls for most of the year. Visitors were unwelcome; indeed, the locals were rumored to be "shameless bodies of the Devil whose chief delight is to kill bishops, priests, bailiffs and superiors — and who possess a large share of all original sin." Hyperbole notwithstanding, it was a forbidding place, but the beauty of the area and of its falls, particularly the thunderous Rjukanfossen, which dropped 340 feet, brought sightseers from great distances away. Painters would stand just out of reach of its spray and try to capture its magnificence, their landscapes often including the 5,600-foot-tall peak of Mount Gausta towering in the background.

At the turn of the twentieth century, the painters and tourists had to make way for the bustle of industrialists. In 1905 Norwegian engineer Sam Eyde and physicist Kristian Birkeland invented a way to take the nitrogen occurring naturally in air and use it to manufacture fertilizer. Their process required large amounts of electricity. The rumbling Rjukanfossen was ideal. Norsk Hydro was formed, and soon the Vestfjord Valley was attracting engineers and construction workers. They dammed Lake Møs, blasted tunnels into the hillsides for pipelines, and built a series of power stations and factories. Near the site of the old village, they also built a new town: Rjukan. There were hundreds of houses and worker dormitories, a grid of lighted streets, a railway and station, a church, firehouse, police force, schools, hospital, and even a community garden and a dance hall.

Civilization may have come to the Vestfjord Valley, but in 1918, when

Einar Skinnarland was born, the eighth child of the dam-keeper Hans Skinnarland and his wife, Elen, life up at Lake Møs remained much the same as it had always been. The Skinnarlands lived thirteen miles west of Rjukan, in a large wood-planked cabin beside the dam. Despite the developments below, from November through the early summer months, the area remained reachable only on skis. The family planned for the long winters, storing whatever food, clothes, and household items they might need. If something broke, it was up to them to see it fixed.

Growing up, Einar spent as much time on skis as he did on his feet. He skied to haul wood and perform other chores around the dam. He skied to the neighboring farms to play with friends. He skied over the frozen lake to the boarding school on Hovden Island, ten miles from his house, where he attended elementary school, and then he skied to middle school in Rjukan. More than that, he skied for fun. His older brothers were championship racers and ski jumpers, as evidenced by the line of small but prized silver trophies on the shelf in the room he shared with them. For them, it was pure enjoyment to ski a marathon's distance to Mount Gausta, climb for three hours to reach the summit, race down at breakneck speed in five minutes, and then return home for supper.

Although they lived on the edge of civilization, the Skinnarlands had their share of ambition. Hans and Elen bought parcels of land around Lake Møs and had opened a general store by the dam. Among their children, there was a nurse, a hotel owner, an assistant dam-keeper, and a pair of engineers. Considered the brightest of the bunch and the "golden boy" in his mother's eyes, Einar earned high marks in school and went on to study general engineering at a local technical college. From there he headed to Oslo to do his military service at the esteemed Engineering Corp Officers School. He wanted to build big things, dams and the like. Then war broke out.

On April 9, 1940, Elen Skinnarland put down the phone and wept. The Germans were invading, and she knew that her sons would soon be off to war. In Oslo, Einar was already in the thick of it, witnessing Luftwaffe bombers screaming overhead, toward the airport. He mobilized, but like many soldiers in the Norwegian Army experienced only a month of retreat and defeat, marked with sporadic fighting. His brothers Torstein and Olav saw more action, participating in the heated resistance around Rjukan that made it one of the last towns in southern Norway to surrender. They came home wearied and traumatized.

When Einar returned to Lake Møs in the middle of May, German troops had moved into his brother's hotel by the dam, turning his sibling into their servant and appropriating his stores of food and beer. The soldiers made a mess; judging by the scuff marks on the walls, they must have slept in their boots.

In some ways, life returned to normal. Einar took a job with Norsk Hydro as a dam-construction supervisor, and he lived again at his childhood home. He dated a young woman named Gudveig from Bergen. They had met through his childhood friend and Norsk Hydro colleague Olav Skogen, who also was from Rjukan. Skogen was seeing Gudveig's sister. The four skied together and took bicycle rides through the countryside.

But there was another side to Skinnarland and Skogen. Since the invasion, they had set themselves to undermining the Germans in any way they could as part of the area's nascent Milorg cell. Skogen was one of its leaders, and Skinnarland stored arms and organized a group of ten reliable men around Lake Møs. He traveled back and forth from Oslo and obtained a mimeograph machine for use by an illegal newspaper. He connected with the Oslo Milorg cell about setting up a wireless radio near his house and even dug a room out from underneath the floor of a mountain cabin in which to maintain it. But no set was found for him.

In the fall of 1941 a local Rjukan boy arrested by the Gestapo turned informant in the hope of better treatment. He told agents that Skinnarland was hiding weapons in a cabin on the edge of Lake Møs. They arrested Skinnarland but were able to find only a common radio, albeit illegal, at the cabin. Nonetheless, he was held for ten days in Rjukan jail. Upon his release, he was more determined than ever to aid the cause.

At the start of 1942, Skinnarland again traveled to Oslo, this time to ask Milorg's signals chief if he could be sent to Britain to train as a radio operator and bring back a radio set with him. He was provided with fake papers under the name Einar Hansen. In early March he told his family that he was heading into the mountains on an extended hunting trip. Instead he left for Flekkefjord, a port town west of Kristiansand, where a British boat was scheduled to pick him up. It was while traveling to Kristiansand that he'd injured his leg.

In Flekkefjord, he met with two other members of the resistance: curly-headed Odd Starheim and his sidekick, Andreas Fasting. The son of a ship's captain and one in his own right, young Starheim was already a legend in resistance circles. Soon after the German invasion he had commandeered a small boat, which he named *The Viking*, and had gone to Scotland, where he became one of Martin Linge's first recruits. He returned to Norway and, over the past year and a half, had developed an extensive intelligence network in the southern part of the country that informed on German naval movements and fortifications. Now on the run from the Gestapo, Starheim needed to escape the country, and he was charged by the underground to take Skinnarland with him. On meeting in Flekkefjord, Starheim told Skinnarland that their transport had been canceled because of storms on the North Sea. They needed to find another way. Given the hunt for Starheim, crossing into Sweden was out of the question. It was Starheim who came up with the idea of hijacking a steamer. Over games of chess, the three hatched their plan to hijack the *Galtesund*.

But if Skinnarland was to be of any use in overtaking the vessel, he would first need to have his leg attended to by a doctor.

Two days after his surgery, on March 14, Skinnarland boarded the *Galtesund* in Kristiansand for the trip to Flekkefjord. On the overnight journey, with walking stick in hand, he scouted out the steamer's passengers, crew, cargo, and coal supply. There was only a single passenger, a man traveling to Stavanger, farther up the coast, to attend a christening. No Germans. Twenty-two crew. Knudsen, the captain, was a burly old sea dog who was unlikely to look kindly on someone attempting to take his ship. The *Galtesund* held a typical cargo for these coastal waters, including a couple of crates of Norwegian tobacco. The coal bunkers contained sixty tons, more than enough to reach Scotland. When the steamer arrived in Flekkefjord the next morning, Skinnarland disembarked to meet up with his fellow hijackers.

A few hours later, he walked slowly back toward the harbor. On his suit lapel, as before, he wore a Nasjonal Samling badge — just another Norwegian in line with the new order, like the local girls on the arms of German soldiers, promenading in the brisk early evening.

The revolvers tucked inside his suitcase told a different story.

He crossed through a small cobblestone square lined with bright green and yellow wooden houses and then moved onto the quay, where stevedores were loading the last cargo onto the *Galtesund*. Its white-painted hull settled more deeply into the water.

Boarding the steamer, Skinnarland made his way to the cabin in which he had stayed the previous night. Half an hour later, Starheim and Fasting arrived at the quay, then proceeded to the deck. At 5:00 p.m., on the third and final ring of the ship's bells, the *Galtesund* shook to life under their feet, its propellers cutting through the black water. Belching black smoke from its funnels, the steamer chugged away from the dock.

Once the vessel was clear of the sheltered harbor, Skinnarland left his cabin, his two loaded revolvers tucked into his belt and several lengths of rope around his waist. He was nervous. Very nervous. Starheim was practiced at this kind of action, but Skinnarland had no training and feared he would make a mess of things.

At 6:20 p.m. he met Starheim outside the saloon. They were about to open the door, to enter and seize the captain, when the ship's second mate came up from behind. He asked Starheim to produce his ticket. Skinnarland took his hand off the saloon-door handle, not sure what to do. "I'm afraid I've arrived too late," Starheim said innocently. "So I thought I would pay onboard."

The second mate led him off to pay the fare. Minutes later, Skinnarland and Starheim were back at the saloon. They eased open the door, Starheim first, Skinnarland second, each with two drawn revolvers. "Hands up," Starheim ordered. Only the captain and the passenger heading to Stavanger were inside. Neither responded until Starheim repeated himself. Then he said, "We're officers of the Norwegian Navy, and we're not alone; at this instant my men are seizing the engine room. I'm now assuming command of this ship."

The captain protested, but Starheim cut him off. Soon Skinnarland had the captain and passenger tied up with ropes. The two hijackers then headed to the bridge. The first mate tried to make it out a side door, but Skinnarland, more sure of himself now, prevented the escape. The pilot let go of the helm, but Starheim ordered him to maintain course. Below deck, Fasting secured the engine room with a pair of stokers he had recruited to the cause, and before long the *Galtesund* was theirs, without a shot having been fired.

Now they had to survive the voyage to Aberdeen. Starheim wanted to set a course directly west, but they were nearing a coastal fortress manned by the Germans, and the pilot warned them that they should proceed on their planned course at least until the sun set. If they were not seen heading to the next scheduled stop, the Germans might suspect something. After dark, they would have until the break of the following day before their absence was noted.

Starheim followed his advice, and once darkness fell the pilot finally headed away from the coastline. Shortly after, as arranged by Starheim when they were still in Flekkefjord, a compatriot sent a wireless message to London. "We have captured a coast ship of 600 tons . . . We make for Aberdeen. Please give aircraft escort because we expect attack from German aircraft tomorrow morning."

Through the night, he and Skinnarland remained in the bridge, drinking coffee to stay awake. At dawn, as a fog rolled over the sea, they heard the rumble of a plane in the distance. Fearing it might be a German search aircraft, they prepared for the worst. Everyone put on safety vests and the lifeboats were prepared. Starheim scoured the sky with binoculars while also trying to discern whether the plane's engine sounded British or German. In a momentary break in the fog, he spotted the silver and black Nazi cross on the tail. Then the fog blanketed them again — saving their lives.

At 2:00 p.m. the sky finally cleared, and the hijackers paced the bridge, knowing that the ship was again an open target. Then Fasting spotted another plane on the horizon. The red, white, and blue of the RAF brought cheers and sighs of relief. The seaplane circled over the steamer, and Starheim had the second mate signal: "*Galtesund* making for Aberdeen and wants pilot." A moment passed, and the plane signaled back: "Congratulations." They had an escort. The following morning, an armed trawler from Aberdeen led them through the minefields to port.

To Skinnarland's surprise, he was sent immediately to London by overnight train.

Upon exiting the Baker Street Tube station, Leif Tronstad headed into Chiltern Court, a block of flats that rose above the London Underground tracks below. The brass plate beside the arched door read INTER-SERVICES RESEARCH BUREAU. It was the kind of ominous name that would have been appreciated by H. G. Wells, one of the building's first residents and author of *The World Set Free*, a novel published in 1914 that predicted atomic bombs "killing and scorching all they overtook." The "Bureau" was actually a cover name for the SOE, and Chiltern Court one of several buildings along Baker Street that housed its staff.

Tronstad walked down the long hallway, then let himself in to the Norwegian section's offices. Colonel Wilson had asked him to come meet one of the *Galtesund* hijackers.

Since his first uncertain weeks in London, Tronstad had made himself an indispensable actor in the Norwegian struggle against Germany. At an officer's training course in Scotland, he earned his captain's rank, some sore muscles, and a close connection with the men who would fight on the ground. He authored an exhaustive report for the Ministry of Economic Warfare, which oversaw the SOE, on the Norwegian industries supplying the Nazi war machine, and this had helped identify prime targets. He sat on two high-level government committees to chart out the future of the Norwegian resistance and its partnership with Britain. He recognized the importance of Milorg, and this secured him the friendship of Oscar Torp and Wilhelm Hansteen, the Norwegian Defense Minister and Army commander-in-chief, respectively. By spring 1942, as leader of the Norwegian high command's Section IV, tasked with intelligence, espionage, and sabotage, Tronstad was at the center of the secret war against the Nazis in his homeland.

Although at a remove from the struggle on the ground, Tronstad maintained good intelligence on the breadth of the Nazi horrors. After the failed Lofoten raid, there had been massive reinforcements of German troops, bringing the number to 250,000 and unleashing untold privation on the people who were forced to feed and house them. Gestapo torturers were known to have stomped on the torsos of prisoners, locking them in dark cells for weeks — and some of those prisoners had taken their own lives rather than betray what they knew. Close friends in the Trondheim resistance had been murdered in cold blood. Tronstad tried to keep his spirits up, firm in his belief that Norway would soon win its freedom. At the start of the new year, he wrote in his diary, "No sacrifice will be too great to achieve this goal."

His own family was not untouched by the suffering. For months Tronstad had heard no news. Then he learned that soon after Bassa and the children had returned to Trondheim, the Germans seized their house and financial assets. Forced into the streets, the family traveled to Oslo and, with the help of a family friend, set up house in Høvik, seven miles west of the city center. The Gestapo investigator, Fehmer, took Bassa in for interrogation, keeping her overnight at Victoria Terrasse. He grilled her over her husband's disappearance, and she responded with the letter Tronstad had written to her explaining why he had left. In the morning, Fehmer let

her return home but instructed her to report regularly to him, thus keeping up the intimidation. Tronstad prayed each day that he would see her
and his children again soon and that they would understand why he had
needed to be away from them at this difficult time.

Tronstad greeted the new arrival from Norway waiting for him in Wilson's office. If what Wilson and Starheim reported about Einar Skinnarland was true, then this construction supervisor from Lake Møs was their
man, and he would be returning immediately to Norway.

"There are two questions we want to ask you," Tronstad said to Skinnarland. "To begin with, do you think that you have been missed by now?"

"I don't think so," Skinnarland replied. "I let it be known that I was going into the hills on a hunting trip."

Tronstad looked to Wilson, who followed with the next question. In
three months of working together, the pair had quickly grown to trust
and understand each other. "Are you prepared with what training we can
give you to parachute back to the Hardangervidda before you're missed?
We want to be kept posted about the situation at Vemork." Skinnarland
nodded. "I'm afraid," Wilson continued, "that you're tackling a very tough
assignment. There's not an hour to lose. And I can't promise you any immediate help. Others — some of whom you may know — will follow later.
Quite frankly, we're greatly worried about what's going on at Norsk Hydro."

Wilson and Tronstad detailed the increase in production at the plant.
They needed to know everything Skinnarland could find out about German activities at Vemork, any intended use of its heavy water, as well as
details on the plant's security and staff. Skinnarland once again committed to the assignment.

Now that he had positioned himself to hear about what operations were
planned for Norway, Tronstad knew that a bombing run on Vemork had
been dismissed by the British Air Ministry the previous December. The
SOE's Clairvoyant operation, which aimed to cripple the supply of power
to industries of German importance, had gone further into the planning
stages, with Vemork one of the operation's six targets. A team was assembled, led by Rjukan local Poulsson, to be dropped into the area to set
out lights in the Vestfjord Valley. These lights would lead bombers to the
plant at night. Poulsson also had instructions to sink the ferries on Lake
Tinnsjø that served as the main transportation link between Rjukan and

the outside world. Tronstad argued against such drastic, far-ranging, and dangerous attacks on critical elements of Norwegian infrastructure, and ultimately, a few weeks before, the operation had been winnowed down to hitting a single hydroelectric plant on the western coast, where aluminum was produced.

Before Skinnarland's arrival, Tronstad and Wilson had decided to send Poulsson and his team to the area in April to launch a guerrilla organization that would sabotage targets and establish independent wireless contact directly with London. Skinnarland could prepare for their arrival and, given his extensive contacts at Norsk Hydro, and the fact that nobody knew he had gone to Britain, return to work and act as an intelligence agent on the inside.

After their meeting, Tronstad took his new recruit out for a drink at a local pub. He was impressed by Skinnarland's willingness to serve, no matter what was asked of him — and a lot was being asked of him — with no explanation of why Vemork's heavy water might be of such importance. Norway needed more "young heroes" like him, Tronstad wrote later that night in his diary.

On the clear, moonlit night of Saturday, March 28, barely two weeks after Skinnarland left Norwegian waters, a British Whitley bomber, known as the "flying barn door" for its boxy rectangular fuselage and broad wings, neared the Norwegian coastline. Its twin Rolls-Royce engines made for an awful racket, and the aircraft rattled and shook as it crawled through the sky toward its target point north of Lake Møs. Inside, a freshly enlisted sergeant in Kompani Linge, Einar Skinnarland, awaited his drop. His white padded jumpsuit, a sleeping bag, and a thermos of tea kept him warm in the frigid air. Beside him was a large steel tube packed tightly with two Sten guns, fourteen Luger pistols, 640 rounds of ammunition, and twenty fighting knives. A smaller container held some spare clothes, fake papers, 20,000 Norwegian kroner, a pair of pistols, a camera with enough film for eight hundred pictures, and a few other items, which were listed in the inventory as "presents for members of his organization."

They crossed into Norway without incident: no flak from the German batteries, and the two Whitley gunners quiet at their posts. In an hour, maybe less, Skinnarland would have to plunge through the belly of the fuselage into the darkness and hope his silk parachute opened above him.

He was terrified.

Over the course of the previous week, he had been given the sparest of instruction. Most SOE agents sent behind enemy lines underwent at least ten weeks of intensive training, from preliminary courses at Stodham Park, to paramilitary work in Meoble, to "finishing school," where they were taught everything there was to know about being a secret commando. Skinnarland spent two days at STS 52, the specialized school for radio operators, where he was given a crash course in using a wireless set and sending and receiving messages in code. From there he went to STS 51, an airport outside Manchester ringed by a large park, for parachute training. This usually took place over the course of seven days; Skinnarland got three. At the entrance to STS 51 two large posters hung on the wall. The first warned troops arriving in enemy territory not to think of themselves as heroes. The other advised that it did not matter how one jumped from a plane, Mother Nature would see that one ended up on land. "How you are received by her is entirely up to you." On the first day, he practiced forward and backward somersaults on mats, then he was hoisted up on a trapeze to practice the same while falling from a height.

He was told by the master sergeant that with a parachute one did not float gently to the earth like a feather. One hit the ground like a stone. Paratroopers unable to quickly transfer vertical motion to horizontal motion broke bones — usually a leg but sometimes also an arm or a few ribs. The school's infirmary was filled with proof of this, should any be required.

On his second day, Skinnarland stood on a wooden platform several hundred feet high, which was attached to a stationary balloon; some likened it to a "dinghy in the high seas." A parachute strapped to his back, he dropped through a hole in the center of the platform. A seventy-five-foot, stomach-in-the-throat free fall was halted by the abrupt opening of the parachute. Skinnarland landed hard, doing his already bum left leg no favors. On his final day, he successfully jumped — twice — from an actual plane. "He showed great keenness," his instructor wrote in his notes, though "his training was rushed."

Now, as the Whitley approached the drop point, Skinnarland ran through what he had been taught: "Feet together and launch off gently. Relax every muscle when you hit the silk and —" It was one thing to parachute in daylight, the landing zone flat and clear to the eye. It was another to fall into a midnight blackness over territory pocked with rocks, precipices, and frozen — but perhaps not frozen enough — lakes. Skinnarland was tough, but he was not fearless.

At 11:44 p.m. his drop dispatcher, Sergeant Fox, called him to action stations. Fox sent the large tube container of supplies out first. Then Skinnarland edged up to the hole, reluctant to dangle his legs over oblivion. Fox raised his arm and then lowered it, signaling Skinnarland to go. Skinnarland hesitated, then shouted out that they were not in the right zone, but his words were lost in the growl of the bomber's engines. Fox signaled him again. Skinnarland hesitated still. For twenty minutes, the plane circled around the drop site, Skinnarland unsure whether they were over the right spot, unsure of himself. At last Fox approached his reluctant charge and yelled that they lacked the fuel to fly much longer.

"We're going back," Fox said.

"No, I'm jumping," Skinnarland responded, though he remained still.

In the small container by his feet, among the presents designated for his organization, was a silver-plated spoon with a painting of the Houses of Parliament on its handle. If he survived the jump, he planned on giving it to his mother.

Finally, a few minutes after midnight, Skinnarland took a deep breath and dropped into the dark sky.

5

Open Road

THE TUBE ALLOYS COMMITTEE met on April 23, 1942, at Old Queen Street in London, gathering in a seventeenth-century townhouse with tall windows and a fine view of the early spring in St. James's Park below. As usual, the scientists, led by Wallace Akers, former research director at Norsk Hydro's rival Imperial Chemical Industries (ICI), had a lot to talk about: experimental work on fusing a bomb, cooperation with the Americans, the expansion of a model isotope-separation plant, and further orders of uranium oxide. As at every meeting, they discussed the Germans, but this time with heightened urgency.

Since 1939 the British scientific establishment had feared the Nazis would obtain an atomic bomb. Hitler's invasion of Poland, then his boast that he would soon "employ a weapon" for which there was no answer, made the danger that much more imminent, prompting Sir Henry Tizard, head of the Air Ministry's research department, to investigate the production of a British bomb. As the government's chief scientific adviser, charged with developing new technologies like radar, Tizard's word carried a lot of weight. And so the inquiry began.

Two young physicists, Otto Frisch and Rudolf Peierls, both Jewish refugees from Germany, put the British firmly on their path. On March 19, 1940, their report, "On the Construction of a Super Bomb," landed on Tizard's desk like a thunderclap. Frisch and Peierls detailed how little more than one pound of pure U-235 — divided into two (or more) parts that were then smashed together at a high velocity — would initiate an explosion that would "destroy life in a wide area . . . probably the center of a big city . . . at a temperature comparable to the interior of the sun." Then they raised the specter that German scientists might soon "be in possession of this weapon." The only way to counter this threat, they concluded, was for Britain to have the technology as well.

The following month, the British government launched the MAUD Committee. Exploratory research began with some of its preeminent scientists, as well as Peierls and Frisch. Foremost, they proposed building a plant to separate rare U-235 from its cousin U-238. This plant would cost as much as a battleship. In July 1941 the group delivered a road map for an atomic bomb program. Tizard remained skeptical, particularly as to its cost. He thought it best the Americans handle everything. But the project now had a champion in Lord Cherwell, the Oxford physicist who Churchill said could "decipher the signals from the experts on the far horizons and explain to me in lucid homely terms what the issues were." On August 27, 1941, Cherwell recommended moving forward with producing the first bomb within two years. The experts gave the project 10–1 odds of success, but Cherwell told the prime minister that he would bet little more than "even money." He continued, "It would be unforgivable if we let the Germans develop a process ahead of us by means of which they could defeat us in war or reverse the verdict after they had been defeated."

Churchill wrote to his War Cabinet, asserting, "Although personally I am quite content with the existing explosives, I feel we must not stand in the path of improvement." The cabinet agreed, promising "no time, labour, material or money should be spared in pushing forward the development of this weapon." Thus the Directorate of Tube Alloys was formed.

Throughout this period, fears over the German bomb persisted. From far and wide came whispers, rumors, threats, and fact — which, mixed together, made for the typically confusing brew that governments called "intelligence." Two German pilots were overheard on a tram speaking about "new bombs" that were "very dangerous" and had the power of an earthquake. One German émigré physicist warned that there was pressure from high within the Nazi government to build a bomb and that the Allies "must hurry." Another warned that the Wehrmacht had taken over the Kaiser Wilhelm Institute of Physics. A military attaché in Stockholm reported, "A tale has again reached me that the Germans are well under way with the manufacture of an uranium bomb of enormous power, which will blast everything, and through the power of one bomb a whole town can be leveled." Other reports chronicled a mysterious September

1941 meeting where Werner Heisenberg admitted to Niels Bohr, who was living in Nazi-occupied Denmark, that a bomb could be made, "and we're working on it."

The effort to obtain further intelligence was largely the territory of Eric Welsh. From the Griffin the British had some insight into machinations in Berlin. Peierls and Frisch also provided hints of German interest in a bomb by culling lists of German physicists and dissecting any papers they published on atomic research. Again, these added ingredients to the brew, but nothing definitive.

The best intelligence the British received came through German activity at Vemork. As early as April 1940 Jacques Allier had alerted his British allies to Nazi efforts in uranium research using heavy water from the plant. After the occupation of Paris two months later, the Norwegian supply of 185 kilograms under Joliot-Curie's care was secreted out of the country aboard a British ship before the Germans could seize it. The twenty-six flasks were hidden in Windsor Castle until the Tube Alloys scientists began their own experiments with the material.

The continued German focus on heavy water into 1942 was of even greater importance because of a newly discovered element called plutonium. The British were centering their efforts on isotope separation to produce enough pure U-235 for their bomb. But British and American scientists knew this was not the only path toward an explosive. In mid-1940, before the curtain of censorship fell completely, a widely published paper in the *Physical Review* revealed that when uranium (atomic number 92) is bombarded by neutrons, some split the rare isotope U-235 nuclei but others are absorbed by the much more prevalent U-238 nuclei, transmuting it into the isotope U-239. This unstable isotope decayed by what was called beta emission, which had the effect of increasing the number of protons in the nucleus by one while reducing the number of neutrons by the same amount. This new element, neptunium (atomic number 93), readily decayed once again, creating yet another element, but a stable one: plutonium (atomic number 94).

As further Allied experiments showed, but now in classified reports, plutonium was fissile, similar to U-235, and could be used as explosive material. However, unlike U-235, plutonium was chemically different from uranium, and thus the two were more easily — and much more cost-effectively — separated. If one managed to engineer a self-sustaining reactor with uranium and a moderator, it would breed enough plutonium that could then be extracted to help build a bomb. This kind of reactor, Allied physicists theorized, required between three and six tons of heavy water. Through Leif Tronstad, the British knew the Germans were attempting to produce similarly large quantities of heavy water at Vemork.

At the April 23 Tube Alloys meeting, Akers and his group of scientists discussed the findings of a new SOE source in Norway (a quick-working Skinnarland). According to his coded messages, heavy water production was up to 120 kilograms a month — and increasing. Something, the men at Old Queen Street decided, must be done, and soon. In the minutes of their meeting sent to Churchill's War Cabinet, they stated, "Since recent experiments have confirmed that element 94 would be as good as U-235 for military purposes, and since this element is best prepared in systems involving the use of heavy water, the Committee recommends that an attempt should, if possible, be made to stop the Norsk Hydro production." If Vemork had not been a significant target before, it was definitely one now.

In the weeks that followed, Tronstad suddenly found himself preoccupied with heavy water. On May 1 Wilson sent him a note, asking him to determine where and to whom in Germany Norsk Hydro delivered its supply from Vemork. The same day, he consulted with Akers on the construction of a British heavy water plant. Soon after, he sat down again with Eric Welsh, who wanted him to set up a spy network, both inside and outside Vemork.

Skinnarland was already providing good intelligence, but it simply wasn't enough. The Tube Alloys Committee speculated that the Nazis were pursuing a reactor to produce plutonium, a sure road to a weapon, and everything must be known about their activities. Germany was largely closed to intelligence work, but Nazi scientists traveled to Vemork, Oslo, and Stockholm — places where Tronstad had close contacts.

On May 11 Tronstad wrote two letters, the urgency in his tone clear. The first was to "the Master": Jomar Brun. Tronstad requested detailed sketches, diagrams, and photographs of the Vemork plant, as well as production figures — anything Brun could discover about the German use of "our juice," and the specific address where it was sent "so we can give our regards to the people there." Tronstad wrote: "You can take this as your war effort! It must be pursued with all means." He signed the letter Mikkel, "the Fox," his new alias on coming to Britain.

Tronstad addressed his second to Harald Wergeland, his former student and now a University of Oslo professor. Wergeland had studied under Heisenberg and was close to several other German scientists. Addressing him as "My Dear Young Friend," Tronstad wrote, in thinly veiled code, "We must know if the Germans have managed to tame the very smallest creatures." He wanted to know everything Wergeland had learned from his recently reported trip to meet with Niels Bohr. Tronstad instructed him to insinuate himself into the Nasjonal Samling Party, then to attempt a research trip to Germany to "soak up everything possible as quickly as possible."

When these letters were brought to Welsh, Tronstad demanded that SIS and the SOE be sure to keep strictly separate communication lines with Skinnarland and with Brun at Vemork. No connection between the two men must be suspected. They did not want another atrocity like the one that had recently happened at Telavåg.

Almost a month before, the two British services had both used the coastal island village southwest of Bergen as an entry point for their operations. On April 17, two Kompani Linge men arrived by boat into Telavåg to organize resistance groups and conduct sabotage operations. The Gestapo was already on alert after some SIS agents had been seen in the area days before. Some loose talk and gifts of British flour gave away the presence of the Linge men. They were cornered in a house, and a shootout followed. One of them died, the other was captured, and a Nazi SS officer was killed. Reichskommissar Terboven was incensed at the brazen action and by the discovery of arms depots in the village. Norway was under *his* control, and he would see that this much was clear to all its people by making an example of Telavåg.

He ordered the village razed and personally watched it happen: every house and building burned down, every boat in the harbor sunk, every animal killed, and the entire community — men, women, and children alike — sent to various concentration camps. Terboven also had eighteen citizens who were caught trying to escape executed by firing squad. He was dutifully following Goebbels's recommendation: "If they can not learn to love us, they should at least fear us."

Tronstad feared the same would happen to those he had tasked with spying for him. Such thoughts weighed heavily on him, especially since he slept safely at night in a house beside the expansive Hampstead Heath while some of his countrymen were in constant jeopardy.

· · ·

If Lillian Syverstad was stopped by a German soldier, and there were always German soldiers patrolling Rjukan, she would simply say that she was on her way to visit her sister, Maggen, who was married to the assistant dam-keeper Torstein Skinnarland up on Lake Møs. Lillian, a pretty eighteen-year-old, worked in the town's bookshop, and she could smile and charm her way out of most situations. That particular June day, she carried a folded note, which had been passed to her by a friend in town. It was likely from her brother Gunnar, a laboratory assistant at Vemork, but Lillian never knew — and never looked to see — what the messages were. It was for the best. She could not confess what she did not know. Once she reached the dam, she made sure nobody was watching, then left the note behind a simple round stone. Soon afterward, Lillian's childhood friend, Einar Skinnarland, retrieved the note, and yet another scrap of intelligence on Vemork was collected.

On the night of his parachute drop into Norway, a strong northwest wind had blown Skinnarland into a rocky hillside. He hit the ground hard, his spine compressing like an accordion. When he stood and gathered his parachute, he felt something pop in his lower back, but there was nothing to be done about it. He quickly found the small package that had been sent out after him, but the tube container with the weapons was nowhere in sight. Finding it in the dark was hopeless.

From the silhouette of Mount Gausta, Skinnarland reckoned he was at least ten miles northwest of Lake Møs. His back shooting with pain, his knee still tender, he made his way through the craggy mountains toward home. He arrived just before dawn. Only his brother Torstein knew where he had been. The rest of his family greeted him warmly and asked how his long hunting trip had gone. He begged off any questions about his stiff back, and there was little time for a reunion.

Skinnarland set off on skis to meet his friend and confidant Olav Skogen, operating out of a mountain lodge halfway to Rjukan. The local Milorg leader, with a shock of dark, curly hair swept high over his broad forehead, guessed from the outset where Skinnarland had been. Hearing about the drop, Skogen stressed the need to find the container of weapons he'd been supplied as quickly as possible. The two set off for the drop site but found nothing. The container was likely hidden in a drift of snow and would never be recovered.

Skinnarland returned to his job building a dam at Kalhovd, eighteen miles north of Lake Møs. There was little time to see Gudveig, who lived in Bergen, but now and again he sent her letters, saying nothing of his undercover work. He began gathering intelligence on Vemork, partly through his own reconnaissance but mostly through his contacts inside the plant.

Skinnarland did not know it, but Jomar Brun was also providing a wealth of information about Vemork to London. On October 3, 1941, Paul Harteck and another German scientist had traveled to Rjukan with Consul Erhard Schöpke. A Nazi zealot with the medal-bedecked uniform to match, Schöpke was responsible for exploiting the country's industry for the Third Reich as a member of the Wehrwirtschaftsstab Norwegen (War Economy Staff for Norway). Accompanying them was Bjarne Eriksen; Norsk Hydro's former lawyer had replaced Axel Aubert as director general on his retirement. Despite the company's earlier support of the Allies, Eriksen had essentially taken the position since the occupation that the firm's survival was more important than patriotism — or, as some of its workers said more bluntly, "Long live Hydro . . . To hell with Norway."

After Brun gave a tour of the heavy water plant, the men gathered in the stately administration headquarters in Rjukan. They enjoyed a nice dinner and then, around a warm fire, smoking cigars and sipping whiskey, Schöpke got to the business at hand. Production numbers were off and something needed to be done. In the twenty months since the Germans invaded, Vemork had delivered 390 kilograms of heavy water, despite German orders of 1,500 kilograms a year. Not only was Norsk Hydro falling far short of the first target, but the company needed to more than triple production, to 5,000 kilograms a year.

Then Harteck spoke. There were two ways to increase production to such a level. First, they could expand the electrolysis plant as it now operated, adding more electrolyzers to the initial cascade and doubling the size of the high-concentration plant. Second, they could test and institute a new technology (catalytic exchange) that held out the promise of not wasting the deuterium burned off as gas during electrolysis. In the end, they decided to do both.

Brun was instructed to develop a plan to institute these ideas. Unwilling to help, and still unsure of why the heavy water was needed, Brun dragged his feet. In the months before, he had actively slowed production by contaminating the high-concentration cells with cod-liver oil, making the heavy water foam. He could not continue interrupting production this way without exposure, however. In January 1942 he was ordered to Berlin. At the Army Ordnance Department in Hardenbergstraße 10, he endured more meetings and more demands for increased production, chiefly from one Dr. Kurt Diebner. When Brun asked Diebner what was the purpose of such quantities of heavy water, Diebner answered they would be sent to the "quinine factory" to make a refreshing tonic. Brun was not amused.

Over the next several days, Brun worked with Harteck to formulate the way forward at Vemork. The plant revisions needed to be implemented immediately, no matter the cost. With the improvements, Harteck figured they could produce at least eight kilograms a day. Further, Brun was told, Norsk Hydro needed to consider constructing heavy water plants at two other hydroelectric power stations: Såheim (in Rjukan proper) and in Notodden. They were smaller power plants, with more limited electrolysis facilities, but together they could add another six kilograms a day, bringing total production to five thousand kilograms a year, possibly more.

Some of the scientists Brun met in Berlin sympathized with Norway's plight and spoke quietly of their disgust with Hitler. They warned Brun to be careful of what he said to whom, particularly to the more ardent Nazis such as Diebner. Even so, none of them would reveal what they needed the heavy water for, although they assured him it was not for the war. It was a disturbing trip. Too often Brun heard the phrase "Heil Hitler." Signs on the trams warned of enemy ears. There were only spare amounts of food at the Kaiser Wilhelm Institute's canteen, and the offices were barely heated. One day, Brun passed several young Jewish girls on a corner who had yellow stars sewn onto their shirts.

On returning home, Brun had no choice but to begin instituting the expansion. Workers doubled the number of high-concentration cells to eighteen. The nine-stage cascade, already huge, grew to include over forty-three thousand cells. In addition, Brun started a pilot test of the catalytic exchange process, and plans were developed for Såheim and Notodden.

In May, Terboven visited Vemork, then Harteck returned to refine his new exchange method before it was finally implemented. Production was up—from an average of 80 kilograms a month in December 1941 to 130 kilograms in June 1942—and still rising. Såheim and Notodden were soon to be operational as well.

All this information and more Skinnarland and Brun, operating independently, sent in coded messages through a system of couriers to Oslo, then on to Sweden. Some were secreted inside toothpaste tubes. Others were taped to the backs of the messengers bringing them across the border. Then they were sent by plane to London, where they made their way to Chiltern Court.

One report from Skinnarland detailed the meager security at the plant, stating that the Germans "depend too much on the surrounding natural defenses. The night guard is usually Georg Nyhus, a middle-aged, decent fellow. A neighbor. He should not be hurt. His only job is to check the permits of workers through his window, after they have been cleared by the two sentries on the bridge."

Another from Brun, labeled "High Concentration Plant for Enrichment of Glucose," included an inventory of the plant down to the lead pipes, sand seals, rubber connections, and flanges of the high-concentration cells. Microphotographs of building blueprints, detailed drawings of equipment, and production figures followed. It was everything one might need to build a heavy water plant, or indeed to destroy the only one in existence.

On June 4, 1942, Kurt Diebner watched a line of guests — some in fine suits, others in uniform — file into the lecture room at Harnack House, headquarters of the Kaiser Wilhelm Institute. In addition to scientists involved in atomic research, the attendees included General Emil Leeb, Admiral Karl Witzell, and Field Marshal Erhard Milch, respectively the armaments chiefs of the German army, navy, and air force. The hawk-eyed Albert Speer, the newly appointed minister for armaments and war production, had called them together to decide the future of the atomic program. Even though this had been launched on Diebner's initiative, it was Heisenberg who took to the stage to present their findings, sidelining Diebner.

In the first two years of the Uranium Club's existence, they had made steady advances in atomic science. Thanks to the Nazis' military successes, they had an easy channel to Vemork's heavy water, a substance made all the more important by the final determination that it was both superior to graphite as a moderator (based on a mistaken calculation by one of its chief scientists, Walther Bothe) and easier to obtain in a highly purified form than the carbon product. They also had access to tons of Belgian uranium ore and a cyclotron in Paris commandeered from Frédéric Joliot-Curie to experiment with subatomic-particle collision.

Over seventy scientists, distributed over a number of institutes, were pursuing basic but necessary research on everything from the energies of fission products, to several methods of U-235 isotope separation, to the construction of a uranium machine, to, finally, the likelihood of a new fissile material created from such a machine: element 94 (what the Americans were calling plutonium). They had broken ground on a new laboratory next to the Kaiser Wilhelm Institute of Physics (dubbed the Virus House to discourage unwelcome visitors). All looked bright for the future.

Of all the developments, the most exciting related to the construction of a uranium machine. In Leipzig in September 1941, with assistance from Heisenberg, Professor Robert Döpel built a small spherical machine with two concentric layers of uranium oxide and heavy water, a beryllium neutron source at its core. They submerged the sphere in a vat of water and awaited the test results. The increase in neutrons was small, but it was there, evidence that the machine was successfully splitting U-235 atoms. With more layers of heavy water and higher-grade uranium, Heisenberg knew "in his bones" that they would have a self-sustaining pile. From that point forward, he said, there was "an open road ahead of us, leading to the atomic bomb." Diebner agreed. In his mind, success was now a matter of shifting all their basic research into an industrial program.

But two months later, with the Russians counterattacking on the Eastern Front and Hitler calling on Germany to focus on meeting the short-term demands of the war, Erich Schumann, Diebner's supervisor and no fan of "atomic malarkey" from the start, called for another review of the research. On December 16, Schumann recommended to his generals that responsibility for the program be handed over to the Reich Research Council, the civilian-run, industry-centered body for basic and applied scientific research. A weapon, even if achievable, was still too far off, Schumann determined. The generals punted a decision to a second conference in February 1942. By then, one of the council's lead scientists, Abraham Esau, was already being put forward as a potential new head of any overall group. A pioneer in wireless telegraphy, professor of physics, and a man of considerable influence in the Reich, Esau had been stripped of fission research by the army when Diebner was made head of the Uranium Club in 1939. Now he looked poised to have his revenge.

For that second Army Ordnance conference, held on February 26, Diebner made his case in an exhaustive 131-page report. "In the present situation, preparations should be made for [harnessing] atomic energy . . .

all the more in that this problem is also being worked on intensively in the enemy nations, especially in America." With five tons of uranium metal and heavy water, and a self-sustaining machine, "a bomb of the greatest effectiveness" using between ten and one hundred kilograms of plutonium was in sight. Diebner offered a step-by-step plan to reach this goal. He simply needed the manpower, supplies, and capital to achieve it.

The very same day, the Reich Research Council held its own meeting on atomic physics with Hahn, Heisenberg, and Esau as lead speakers. Attendees at the conference came away impressed at the technology's future potential. Goebbels wrote in his diary, "Research in the field of demolishing atoms is so advanced that its results can perhaps be used for waging this war. Here tiny efforts result in such immense destructive effects that one looks forward with horror at the future course of this war."

But the army generals remained unconvinced. Diebner could not promise success with certainty. Unable to justify such expense and effort without the guarantee of a weapon within a year, the army forfeited control of the Uranium Club to the Reich Research Council. A further blow for Diebner came when Heisenberg was chosen to be the new director of the Kaiser Wilhelm Institute of Physics, and Diebner was forced to vacate his offices there.

Four months after the shakeup, Diebner still believed there was a chance for an industrial-scale program. Speer, who had maneuvered the project within his sphere of influence, had called the June 1942 meeting at Harnack House to decide how much backing he should give it.

Heisenberg took the stage to present the scientists' findings. Tall, blue-eyed, with a sweep of straw-colored hair, he commanded the room in a way Diebner never could. He began by giving a theoretical overview of atomic science, then dove down into the specifics of isotope separation, uranium machines, and the production of plutonium. Then he hit his audience with the potential of the technical application of this science. With a uranium machine, they could "power ships, possibly even aircraft, with the greatest imaginable range." With plutonium, they could produce explosives that "will be a million times more effective than all previous explosives." One general, who had visions of dropping bombs on New York, wanted to know how big such bombs would be. Heisenberg cupped his hands and said, "About the size of a pineapple."

Heisenberg then shifted to downplaying expectations. They were still at the basic research phase, he said. More theory needed to be developed, more experiments performed. There were many obstacles standing in their way as well, including supplies of heavy water. One day, far in the future, a bomb might "turn the tide of war," he concluded; but first they needed a working reactor, and that was a long way off.

Speer then asked how much money their project needed. Heisenberg proposed a sum of 350,000 marks — in effect, nothing at all. There were other programs, namely V-1/V-2 flying bombs, whose scientists had demanded billions of marks and tens of thousands of workers in order to complete their projects and see them put to use in the war. Speer was flabbergasted, Diebner furious that Heisenberg would ask for such a trifling sum. If one was certain of a pineapple-sized bomb deciding the war, and if one was bent on producing it, much more would be needed to bring it to fruition.

After the meeting, Speer steered his backing toward programs like flying bombs that he believed would most benefit Germany's immediate war strategy. Basic research on harnessing atomic power would continue. The Reich Research Council would lead the project, and Army Ordnance would help fund it. Scientists would have ready access to supplies, and they would remain exempt from military service to do their work. But unless there was reason to reconsider, all this research was focused on future potential. There would be no massive project to see it made of any use in this war.

Diebner was undeterred. He returned to his research at Army Ordnance's Kummersdorf Testing Facility in Gottow, fifteen miles southwest of Berlin. With a host of young physicists he had recruited in the first days of the program, he quietly continued building his own uranium machine with a very different design from the ones that others had engineered. Hard and fast, he pressed on toward a bomb.

Prime Minister Winston Churchill chewed on a cigar as he stared out at the Atlantic's moonlit waters on June 17, 1942. Seated in the cockpit of a Boeing Clipper flying boat, he had many matters weighing on his thoughts as he flew toward the United States. The whole of continental Europe remained under Hitler's heel, and though British and American bombers pummeled Germany night after night, only a cross-Channel invasion could liberate the Continent. But Churchill knew that Allied forces were far from ready for such an attack. He must convince the U.S. president, Franklin D. Roosevelt, to delay an invasion — one of the two key purposes for his twenty-eight-hour journey across the Atlantic. The other was to discuss atomic bombs.

After a glass of champagne and a restless nap, Churchill fastened his seat belt for landing beside his trusted chief pilot, Rogers. The Clipper glided past the Washington Monument and landed on the Potomac. In the capital he met with General Marshall, and the next morning Churchill flew up to Hyde Park, New York, the site of Roosevelt's family estate. The American president greeted him on the tarmac. Physically, the two men were a study in contrasts: the short, feisty British bulldog beside the tall, smooth American lion. But Churchill and Roosevelt were both intellectuals as well as cunning politicians, and they shared the terrible weight of leading their people through a great war. They were also good friends.

Driving his blue Ford Phaeton, which featured special hand-controlled levers to accommodate his physical disability, Roosevelt whisked his guest off on a hair-raising tour along the Hudson River bluffs. For two hours they spoke of the war, and Churchill was encouraged by how much more they settled zipping across the estate than they would have done on opposite sides of a crowded conference table.

Earlier that week, Roosevelt had read through a report that set out a plan for a massive U.S. Army program to build atomic bombs. His chief scientific adviser, Vannevar Bush, who'd founded Raytheon, was spearheading the effort, with an estimated cost of $500 million.

The program's genesis mirrored in many ways the British Tube Alloys program. In August 1939 Albert Einstein, in contact with a group of scientists who had recently emigrated from Europe, sent a letter to Roosevelt warning of the need to exploit the explosive potential of fission before the Germans did. A "Uranium Committee" of leading physicists was formed, and over the next two years, with limited funds but a lot of ambition, they conducted research, welcomed insight from the British, and concluded that a devastating new weapon was indeed possible — and that there were several potential routes to success. By summer 1942, word had come from Europe that the Germans might have already realized a working nuclear reactor — something the United States had yet to achieve. As one scientist wrote Bush, urging a decisive American effort, "Nobody can tell now whether we shall be ready before German bombs wipe out American cities."

With a handwritten note that simply read "Ok. V.B.," Roosevelt approved the development of the program, which was codenamed the Manhattan Project. In typical American bigger-is-better fashion, its leaders decided that every route to a bomb should be pursued, including using heavy water reactors, plutonium, and U-235 isotope separation.

On June 20, Roosevelt and Churchill held a meeting in a small, dark study that faced the front porch of the Hyde Park mansion. Once Roosevelt's children had used the space as a classroom, but now it was his quiet hideaway, with bookshelves and nautical prints on the walls and a huge oak desk that took up most of the space.

Churchill sat, a globe beside his feet, and got straight to the point. "What if the enemy should get an atomic bomb before we did!" he later wrote, recounting the meeting. "However skeptical one might feel about the assertions of scientists, much disputed among themselves and expressed in jargon incomprehensible to laymen, we could not run the mortal risk of being outstripped in this awful sphere." Their two countries needed to "pool our information, work together on equal terms, and share the results, if any, equally between us." If Churchill expected a debate, he didn't get one. Roosevelt agreed wholeheartedly with the proposal, and given the ongoing Nazi bombing raids on Britain, they decided that the United States should be the center of activity.

They also discussed the German focus on producing heavy water — "a sinister term, eerie, unnatural," Churchill later said. A few days after the prime minister flew back to London, the War Cabinet put forward plans, of the highest priority, for a raid on Vemork.

Part II

6

Commando Order

JENS-ANTON POULSSON, a Rjukan native, wanted a mission — and if his commanders would not give him one, he would come up with his own. He had nearly circumnavigated the globe to come to Britain to join Kompani Linge, and since his arrival in Britain in October 1941 he had heard a lot of plans but seen no execution. His best prospect had been to lead one of the six teams in Operation Clairvoyant, his task specifically to guide nighttime bombers toward the Vemork power station by setting out lights in the Vestfjord Valley. But then that operation was abandoned and, as he wrote in his diary, "the greatest opportunity of my whole life" slipped away.

Thus in late February 1942 Poulsson traveled down from Scotland to pitch his bosses in London a new plan. Meeting with a member of Colonel Wilson's staff, Poulsson proposed the idea of a small team that would organize resistance cells around Telemark and prepare to sabotage railway lines. He drafted the details in a report, then returned to Scotland while the plan was considered.

Weeks passed and no answer came. He was sent on a training scheme to attack an airport. Then in early April he got orders to go to STS 31, the "finishing school" at Beaulieu, a forested estate in southern England. Over the next three weeks, he received training in espionage and living an underground life. He learned how to develop a cover ("Your story will be mainly true"); shadow a target; recruit informants ("A few drinks may be helpful"); build up an underground cell; establish a covert headquarters; and thwart counterespionage efforts, including losing a tail ("Lead

him through a long deserted street and then plunge into a crowd"), staying alert ("A familiar voice or face suggests an agent is being followed"),
and manufacturing a good alibi. His instructors taught him how to sur-
veil a target, to merge into the background on a street, to burgle a house,
to open handcuffs, to read a room for a quick escape. He became skilled
in leaving hidden messages, in microphotographs, ciphers, and invisible
inks. It was all very different from the kind of warfare he'd imagined.

He studied the enemy, everything from its organizations, uniforms,
and regulations to its detective measures, wireless-interception abilities,
and interrogation techniques. If he was ever to find himself under questioning, his lecturers said, "Create the impression of an averagely stupid,
honest citizen." The school's commander, Major Woolrych, told his students, "Remember: the best agents are never caught. But some agents . . .
they are inclined to relax their precautions. That is the moment to be-
ware of. Never relax. Never fool yourself by thinking the enemy are asleep.
They may be watching you all the time, so watch your step."

Poulsson graduated from STS 31 with a somewhat mixed instructor report: "Much more intelligent than he would appear at first sight as he has
a very retiring disposition. He has, however, a thorough understanding of
the work . . . Could make a good second-in-command." Even though he
was tall—six two—with a mop of curly dark hair, a lean face, and bright
blue eyes, Poulsson was a retreating presence in a crowded room. He preferred to remain at the back, clouded in pipe smoke. "A good second-in-
command" was, however, far from the truth.

Poulsson had been born in Rjukan, where it was said that one was
raised in either sun or shadow. Norsk Hydro's top brass lived in grand
houses on the sunny northern hillside of the Vestfjord Valley, while the
rank and file found themselves living deep below, down by the river. As
Jens-Anton's father was a chief engineer at Norsk Hydro in town, he grew
up in the light. His family had a storied history—nobility, ship captains,
high-ranking army officers, English knights—and owned almost ten
thousand acres of land in the Vidda, including an island on Lake Møs.

Named after his father and grandfather, Jens-Anton was the sixth of
seven children. He had blond curls and the habit of smoothing them down
with one hand, and he was shy in company, his nose usually pressed into
his sketchbook or an adventure tale. He devoured stories of war, polar ex-
peditions, and survival in the wild, but although he was a strong reader,
he didn't much care for school. His interests lay in the outdoors. Pouls-

son received a shotgun for his eleventh birthday and soon after killed his
first grouse. With his best friend and neighbor, Claus Helberg, he spent
his early teens wandering the Vidda, skiing, fishing, hunting, and hiking.
A quiet, calm authority, Poulsson was the unspoken leader of his group of
friends.

He never doubted what he wanted to be in the future: a military officer. A straight arrow, he liked rules and regimens. At his school in Rjukan,
there were two classes of children his age, one wild and unruly, the other
well-behaved. Wanting to tame the former, the principal moved Poulsson
and Helberg into the disruptive classroom, and within a few months discipline had been restored. At fifteen, Poulsson spent a summer at a military camp. He was given his own Krag-Jørgensen carbine and learned
to march in step. At twenty he joined the Army's Second Division NCO
school. He was there when the Germans invaded. Within five days his
battalion, which was deployed solely in a defensive position, surrendered
and retreated to Sweden. "The saddest day of my life," Poulsson wrote.
They hadn't even put up a fight.

After a long billet in Sweden, he returned to Norway and holed up
outside Rjukan for several months, bristling to do something. Unable to
obtain passage to Britain by boat, he skied back to neutral Sweden, and
from there he journeyed around the world. In Turkey he witnessed "mud
and stone huts and beaten oilcans for roofs." In Cairo he found "flies and
street vendors the biggest plagues." On heavy seas to Bombay, he experienced "stomach aches and head aches." In India, the camp was "populated
by large amount of lice, not nice bedfellows." Still, it was an adventure,
and an eye-opening one for a young man who had never before traveled
outside his homeland. During the six-month journey, he worried at times
if he had what it took to be a good soldier. One night he wrote in his diary, "One never knows one's own reactions the first time one comes under
fire."

After concluding his spy training, Poulsson returned to STS 26 and
learned that his proposal to build up resistance cells around Rjukan had
been accepted. At last, he would find out what kind of soldier he was. Operation Grouse was due to depart in a few weeks, Poulsson at its head. He
and his team were to survive the harshest of winter conditions out in the
wild, like the alpine bird for which their mission was named, while waiting for the green light for operations.

One unlucky delay followed another, and soon the long Norwegian summer days made the launch of the mission too dangerous. Parachute drops into Norway were limited to a very narrow window. For half the year, there was too much light at night for planes to cross over the countryside unseen by the Germans. For the other half, particularly during the long winter, drops needed to occur around the full moon, when the darkness was cut by just enough natural light that pilots could navigate by landmarks — and parachutists could spot a safe place to land.

With the operation now delayed until at least late September, Poulsson wondered whether he might be better off rejoining the regular army. Others in the company, like Knut Haukelid, felt the same, even though the Norwegian Army soldiers who had made it to Britain were similarly frustrated with inaction. Reassured by their Kompani Linge commanders that they would soon get their chance, they remained.

In the meantime, Poulsson finalized his small team: Arne Kjelstrup, a short, broad-chested plumber born but not raised in Rjukan, who carried a bullet in his hip from fighting the Germans during the invasion. He had accompanied Poulsson on his round-the-world journey to join Kompani Linge. Knut Haugland was a slightly built twenty-four-year-old with a thick shock of fair hair and a thin, boyish face that belied his exacting intelligence. A carpenter's son from Rjukan, he had become a first-class radio operator. And Knut Haukelid, whom Poulsson often went out stag hunting with in the Highlands, knew what it took to survive and operate in the Vidda.

While waiting for their orders to come through, Haukelid stumbled and shot himself in the foot during a training exercise in the countryside. Doctors told the crestfallen commando he would not be "fit for duty" until at least October. Poulsson quickly decided on his replacement: Claus Helberg, his childhood friend. Now leaner, taller, and fitter than most, and with a mischievous twinkle in his eye, Helberg had found his own way to Britain in the early spring to join Kompani Linge. He would need parachute training, Poulsson knew, but there was time for that.

Throughout August, Poulsson and the others prepared for their operation, gathering enough supplies to fill eight tubular containers, which would be dropped with them. The inventory list was two pages long, supplies weighing almost seven hundred pounds: ski gear, boots, gaiters, windbreakers, woolen undergarments, sleeping bags, cooking utensils, tools, cigarettes, candles, tents, kerosene, rucksacks, maps, frostbite ointment, a wireless set and two six-volt rechargeable batteries to power it, guns, ammo, and food. No one was more exacting in his requirements than radioman Knut Haugland. Often to the rankling of the British quartermasters, he specified the exact type of batteries and other radio equipment needed for the operation. That was his way.

On August 29, a hot, sultry day interspersed with thunderstorms, Poulsson traveled to London to meet with Colonel Wilson and Leif Tronstad at Chiltern Court to finalize their plans. The Grouse team would drop near Lake Langesjå, ten miles north-northwest of Rjukan, with Einar Skinnarland on the ground to guide the plane in. Haugland knew Skinnarland well from the local Rjukan resistance, and all of the team were well acquainted with the Skinnarland family. (Einar's brother Torstein was a ski-jumping legend in town.) If for any reason it was not possible for Skinnarland to act as guide, they would blind drop and head to Lake Møs on their own. Wilson and Tronstad laid out their operating instructions, the focus on forming "small independent groups" to prepare for operations against future targets. These included German communications, bridges, and roads. Vemork was not mentioned. As far as Poulsson knew, this target was no longer on the table since the shutdown of Clairvoyant.

Two days later, the Grouse team left for STS 61 at Gaynes Hall, near Tempsford airport outside Cambridge. The distinguished mansion had once been the home of Oliver Cromwell, but now served as the SOE launch point for foreign agents headed overseas. The Grouse team would continue to train here, and wait.

That same day, August 31, Leif Tronstad sat in a smoke-filled room on Old Queen Street, the Tube Alloys headquarters, and raised the prospect of Grouse leading an attack on Vemork. Seated around the table with him were Colonel Robert Neville, the chief planner of Combined Operations, Wallace Akers, and Akers's former ICI assistant, Michael Perrin, a key member now of the British atomic program.

When Lord Louis Mountbatten took over Combined Operations in October 1941, the command he inherited, charged with missions that brought together naval, air, and land forces, was in a state of shambles. And indeed, since then the operations of the forty-two-year-old royal-blooded British naval hero had, at best, a checkered record. Stories of the disastrous mid-August beachhead assault at Dieppe were only just beginning to recede from newspaper headlines.

Since Churchill's return from America, the War Cabinet had tasked

Mountbatten with investigating a possible operation targeting Vemork. Neville, his chief planner and a Royal Marine, looked like he could take on the task single-handedly.

The four men considered several potential courses of action to stop the production of heavy water at Vemork: (1) an attack from within by Norsk Hydro men, (2) infiltration by Poulsson and his team, (3) a six-man SOE attack party to blow up the pipelines (mirroring an early Clairvoyant plan), (4) a Combined Operations raid of between twenty-five and fifty men to destroy the pipelines and the plant, and (5) an RAF bombing.

Tronstad argued against an air attack: with all the hydrogen and ammonia produced in the area, the town of Rjukan might be wiped out in a devastating explosion, and it was unlikely any bombs would penetrate deep enough into the plant to destroy the high-concentration stages located in the basement. As for recruiting saboteurs who already worked at the plant — an inside job — he did not believe they could find enough trustworthy people at Vemork to pull it off. Instead, Tronstad wanted his Grouse team at the forefront of a direct attack. They knew the area, and according to the most recent intelligence, there was only limited security at the plant. With an additional six-man sabotage team to carry out the demolition, the group would have good odds of success.

Neville was unsure — German defenses might be stronger than reported. He favored British sappers (combat engineers) executing the attack, with the Grouse team acting as guides. Fifty soldiers could overcome any resistance, and with their strength in numbers, they could perform a larger attack on the plant, making certain it was removed as a threat. The trouble would be getting the men out and away from Norway. Neville recognized that this challenge made the sappers very likely a "suicide squad."

The four men knew Mountbatten would make the final decision, but it looked like the Grouse team would indeed have a role to play in the Vemork plan.

Tronstad was desperate to be part of any operation on the ground as well. Yes, he was contributing to the war effort. He had his own intelligence network. He recruited Norwegian scientists to aid the British defense industry. He advised on potential chemical attacks. He helped steer the strategy, training, and operation of Kompani Linge. But at times he felt like he was fighting a paper war, of reports and conferences. He wanted away from this "abnormal life." He felt that others were suffering the burdens of the conflict while he remained in London. Many of his close friends were dead; the Gestapo had evicted his family from their home and hounded his wife for information on his whereabouts. Brun and Skinnarland were risking their lives every day spying for his country. Tronstad wanted to do the same.

After celebrating his thirty-ninth birthday that March, he had quit smoking and begun exercising diligently. In June, he went through parachute training at STS 51. Each evening, he tried to get in a "little commando work" in the expansive park, Hampstead Heath, near his house.

Believing himself prepared for any mission, he pitched to Major General Gubbins, the SOE chief, his own involvement in Grouse. But Gubbins told Tronstad that his place was in London. The Allies could not risk losing his insight and leadership. Coming to an uneasy peace with staying behind, Tronstad threw himself into his Kompani Linge command.

His resolve was strengthened by the news out of his homeland. Across Norway, average citizens were actively resisting the Germans any way they could. Earlier in the year, teachers had gone on strike, refusing Nazi demands to teach the new order to their pupils. Terboven had ordered the arrest of the most recalcitrant teachers — five hundred in number — sending them to a concentration camp in the Arctic seaport of Kirkenes. The journey took sixteen days, the prisoners crowded inside the cargo hold of an old wooden steamer, with little food or water and no toilets. They were forced to work twelve hours a day on the docks, alongside Soviet prisoners of war, and were ill fed, poorly housed, and beaten on a whim. Some died. Others went mad. Still, they resisted.

"War makes the mind very hard," Tronstad wrote in his diary, thinking of the latest news of their hardship. "Becoming a sensitive person again will not be easy."

Throughout September, as Knut Haukelid watched the rains sweeping across Scotland and nursed his injured foot, he wished passionately that he had been able to join the Grouse team. From the team's letters, however, it sounded like they were as stuck as he was. In one, headed "Somewhere in England," Poulsson wrote, "If you think we have left, then you are damned wrong ... A week's waiting for fine weather which never comes. Otherwise it is all right here — the house full of FANYs [field army nurses]." Then, on September 9, "There is a red light today and we hope for the best. We are now ready to start."

Haukelid awaited word that they had dropped safely. Once they connected with Tronstad by wireless and were securely in place in Telemark, the plan was for him to join them with another Linge member. If only for that damn foot . . .

At the end of September, another letter arrived. "Of course we came back. Motor trouble." The following day brought yet another note from the Grouse team. "Another unsuccessful attempt. Fog in the North Sea. Devil take the lot! But tails up."

Then silence. Nothing. Surely they were gone now, landed in the Vidda, without him.

General Nikolaus von Falkenhorst, commander of German military forces in occupied Norway, strode through Vemork's grounds on October 1, impressed by its natural defenses but conscious that they were insufficient to protect the plant from British bandits. There needed to be floodlights, more guards, more patrols, barracks for his troops, potentially an antiaircraft battery. Mines must be laid in the surrounding hillsides and alongside the penstocks running down into the power station. The fences around the grounds had to be raised and topped with rings of barbed wire. The narrow bridge leading to the plant required a reinforced gate.

With a face that looked like it had been chipped from stone, Falkenhorst was a soldier of the old school. He came from a noble German family, had fought in World War I, and won several promotions before his country again found itself embroiled in war. During the advance on Poland, he shone. When Hitler needed a commander to take Norway, Falkenhorst was recommended, in part because of his brief stint in Finland in 1918.

The Führer had given him only a few hours to return with a plan. Falkenhorst, who knew little of Norway, sketched out the attack based in part on what he learned from a Baedeker travel guide found at a local bookshop. His success with the invasion had not brought another command in the continuing German advance. Instead he found himself stuck in Norway, guarding the country like a common sentry. He kept himself on decent terms with Terboven and the SS but savored none of their brutality in keeping the occupied country in check. However, there was no doubt that if given an order from Hitler, he would follow it, no matter what.

After his inspection of Vemork was complete, Falkenhorst gathered its directors, engineers, workers, and guards. He explained that only eleven days before, the power station at Glomfjord had been blown up in a British commando raid, halting the aluminum works that depended on it. Grabbing one of the guards from behind, Falkenhorst demonstrated to his audience how fast and ruthless these commandos could be in an attack. He warned that they might arrive in town as ordinary passengers on the train or bus but that they would come "equipped with automatic weapons with silencers, chloroform, hand grenades, and knuckle-dusters." Vemork, he concluded, must be prepared.

The price of failure—or for those who aided a sabotage operation—was soon after made clear. On October 5 men in British uniform raided an iron-ore mine outside Trondheim with what German intelligence believed was clear help from the Norwegian resistance (in fact, it was an operation concocted by Tronstad and executed by Kompani Linge). The next day, the city woke up to find posters declaring a state of emergency; the Reichskommissar Terboven arrived by overnight train, accompanied by SS Lieutenant Colonel Heinrich Fehlis and scores of his Gestapo. After the RAF's bombing of their Victoria Terrasse headquarters two weeks before, the SS was eager for blood.

In the town square, Terboven gave a speech. "I have sincerely, and in good faith, had this country and its people's best interests at heart . . . I have waited magnanimously, and for a long time, but I have now realized that I am forced to take severe measures. When we National Socialists first realize that we have to intervene, we do not follow the democratic method, hanging the little fish, while the big ones swim away. Instead we get hold of big ones, those who want to remain in the background . . . This evening, the population will be made aware of this principle." Terboven and the SS picked out ten prominent local citizens—a lawyer, newspaper editor, theater director, bank manager, and shipbroker among them—"to atone for several sabotage acts." Later, Fehlis's execution squad shot them in the back of the head.

The Swedish border was effectively closed, and Fehlis led an exhaustive hunt for resistance members—indeed, for anyone holding contraband (radios, arms, or large sums of money). His troops searched tens of thousands of people, vehicles, houses, and farms. In the end, they arrested ninety-one individuals as well as every male Jew over fifteen years of age. Some of these prisoners were executed as well.

Terboven intensified efforts to prevent any future raids and to break the will of the Norwegian people. New border regulations, ration cards, and travel permits were instituted. The list of violations punishable by death now included providing shelter to enemies of the state and attempting to leave the country. Across Norway, thousands were arrested, often indiscriminately. Prison transports to Germany increased. Informants were pressed for names of those in the resistance. Torture intensified. If a known resistance member couldn't be found, the Gestapo took his or her parents or siblings instead.

In mid-October Hitler delivered a secret order, the *Kommandobefehl*, to his generals across Europe, including Falkenhorst, to further punish the Allies for their commando attacks: "Henceforth all enemy troops encountered in so-called commando raids in Europe or in Africa, are to be annihilated to the last man. This is to be carried out whether they be soldiers in uniform, or demolition groups, armed or unarmed; and whether in combat or seeking to escape . . . If such men appear to be about to surrender, no quarter should be given to them — on general principle." The order clearly violated the written and unwritten codes of war.

7

Make a Good Job

WHEN COLONEL WILSON summoned Poulsson and his radio operator Haugland to London on October 12, the Grouse team members could not help but fear that their mission was going to be postponed yet again — or canceled altogether. Their pilot had called off the first flight attempt because of dense fog. On the second, their Halifax airplane was already over Norway when one of its engines burned up and they were forced to turn back. They had almost needed to parachute out over Scotland, but the pilot dropped their heavy gear, sufficiently lightening the plane for an emergency landing. By the time they were ready to try again, the narrow window each month, where there was enough moonlight, had passed. "Finally" — as many Kompani Linge members said when their planes were about to go wheels up on an operation — might now be "Never."

At Chiltern Court, the colonel got straight to the point. Rather than fomenting resistance in Rjukan, Grouse would now be the advance operation on a British Army action against Vemork. First, they would recon a landing area for either a parachute drop or Horsa glider. Second, they were to act as a reception committee, putting out lights to direct the aircraft as well as operating a homing radio beacon. Third, on the night of attack, they would guide between twenty-five and thirty Royal Engineers to the target. As for facilitating their escape, or the specific purpose of blowing up Vemork and its heavy water plant, Wilson told them nothing. They knew better than to ask.

Underscoring the top-secret nature of the mission, Wilson informed

Poulsson and Haugland that Helberg and Kjelstrup were not to be told of the mission change until the team arrived in Norway. Further, they were to blind drop into the Vidda. For security reasons, Einar Skinnarland would no longer receive them, nor were they to have any contact with him or his family. If their paths never crossed, the Germans would not be able to tie the two together. The Grouse team was to leave at the first available date of the next moon phase, October 18 at best, and the operation would take place the following month, giving them time to prepare. Then Wilson led them into a room full of maps and reconnaissance photos. He wanted them to pick out a safe spot within the vicinity of Vemork for the sappers to land. Some suggestion was made about a mountain clearing, but Poulsson worried that the British soldiers would have trouble navigating through the snow and harsh terrain. In November, an early snowstorm would demand practiced skiers.

Come back with an answer, Wilson told them.

The two Rjukan men didn't need a map, as they had traversed every inch of the surrounding area on skis and by foot. Together, Poulsson and Haugland came to the same conclusion: the Skoland marshes. The marshes were a wedge of unpopulated land at the eastern point of Lake Møs. They were southwest of the dam and next to a mountain road connecting Rjukan to Rauland that was closed during the winter months. The British Royal Engineers would have a clear run down the road to Vemork, only eight miles away.

The spot was particularly well known to Haugland. As a boy, he had hiked with his family along the pass toward Rauland, fishing for trout and picking cloudberries along the way. One time while they were camping there, his eldest brother, Ottar, came down with scarlet fever. Their father fitted a rucksack across his chest and hoisted his big, fourteen-year-old son onto his back, bringing him over the mountains to medical attention. It was then that young Knut understood the meaning of strength.

Poulsson and Haugland presented their selection to Wilson, and he took it to Combined Operations. Over the next few days, the two were briefed on what they were to communicate — and when — for the entirety of the operation, down to the codes and passwords. All their instructions were given verbally and memorized. "This is Piccadilly," Poulsson was to say on receiving the commandos. "I wanted Leicester Square," they should respond. It was unlikely that there would be any other British troops arriving by glider in November in the marshes, but Poulsson and Haugland did not question the need for passwords.

On the day of their departure for Scotland, Wilson brought them in for one final meeting. "This mission is exceedingly important," he said. "The Germans must be prevented from getting their hands on large quantities of heavy water. They use it for experiments, which, if they succeed, could result in an explosive that could wipe London off the map." The colonel must be a little overexcited, Poulsson and Haugland thought; no explosive could do such a thing. Perhaps he simply wanted to inspire them to the task at hand. Nevertheless, they assured him that they would do everything they could to see the operation was a success.

"Make a good job out of it," Wilson replied.

At STS 26's Drumintoul Lodge, in a side room with a small fireplace, Joachim Rønneberg set up his headquarters for Operation Fieldfare. The young second lieutenant pasted scaled maps of Norway on the walls and inventoried a list of winter survival equipment he would need, all to prepare for his recently approved mission to establish a resistance cell in the Romsdal Valley. From there, he and his fellow Kompani Linge member Birger Strømsheim would sabotage critical German supply rail lines.

For almost a year and a half, since first meeting Martin Linge at Norway House, Rønneberg had been waiting, like most of his compatriots, for an operation. Linge had convinced him that his first choice, the Navy, could do without him and that there was "valuable work" for him with the Norwegian Independent Company. Months of training followed. Some nights, Rønneberg had trouble sleeping, profoundly disturbed that he was essentially in a "vocational school for butchers." It was a life so different from the peaceful, quiet one in which he had been raised.

Rønneberg was from the prominent harbor town of Ålesund, on Norway's northwestern coast. For generations, his family had owned a collection of businesses centered around the fishing industry — everything from ships, barges, and wharf sheds to a rope factory and export business.

Growing up, Rønneberg felt most at home in the outdoors. He loved Alpine skiing and raced whenever he could. He often ventured into the mountains for days at a time. He especially liked orienteering. By the time he was a teenager, he was confident enough to head out on his own, finding his way with only a map and a compass. "If you were alone in the wild you never felt alone," he said. "You were not afraid. You knew within yourself what to do."

At twenty, Rønneberg intended to join the family business. When called up for national service, he volunteered to be a land surveyor's assistant instead of joining the military; thus, when war broke out, he was not mobilized. While he remained in Ålesund, plagued with guilt, his friends went into battle. Some died, including his closest friend, who was killed by German soldiers waving the white flag of surrender as a ruse to draw him near. Although Rønneberg did not join the underground, he detested the Nazis' presence in Ålesund, soldiers marching and singing in his streets, acting like they were a law unto themselves. He burned to do something. In March 1941 he arranged for ship transport to Britain.

Fearing his parents might try to stop him, he wrote a note to them that was delivered only after he had left: "If you only knew what it has cost me to put a good face on things. You will wonder why I did not come to dinner, why my bed is empty . . . You can seek solace in the fact that you are now sharing the same sacrifice as many families in our beloved country and also that I will never feel more free than on the day we cast off from Norwegian soil and plow the sea, bound for freedom's last hope. Live well then dear mother and father and brothers . . . We will meet again before too long. You will always be with me wherever I go in the world." He crossed the roiling North Sea in a forty-five-foot fishing boat, often manning the helm when the others in the crew were seasick.

When he joined the Norwegian company, Rønneberg was twenty-one years old. He had no military training and no experience in war or resistance work. Nonetheless, he excelled at the SOE's training schools, particularly with explosives and raid exercises. He also made peace with the brutality he was being taught, as he knew it would allow him to survive in any situation.

After finishing training, Rønneberg was sure he would be sent on a mission, but Linge called him to London and told him, "We're off to the tailor." There, he was suited up in a fresh uniform for his new role as an instructor of recruits. At first, this meant that he was primarily involved in translation and liaison activities (mastering his English by reading detective novels), but with each new squad that came through, he led more of the instruction. From Stodham Park he was sent to Meoble, then finally to Aviemore. He became an expert in demolitions and devised many of the sabotage schemes against bridges, railway stations, and military barracks that his students executed. At first he felt ill fitted for the role. He thought he should be on real missions, doing real work. But every time an operation leader asked for Rønneberg to be on his team, he was told by the SOE, "No, we need him here." Eventually he embraced his role, taking his trainees on practice marauding raids through Scotland. Instead of using explosives, they tagged intended targets with white chalk. If what the trainees learned under his direction helped make their missions successful, then he was making a difference. Meanwhile, he developed and pitched his own operations to his superiors, but none had yet been set in motion.

In fall 1942 the SOE signed off on Operation Fieldfare. Once Rønneberg's preparations were finished, and the moon phase was right, he and Strømsheim expected to be dropped into Norway to make a start.

"Number one, go!" the dispatcher yelled through the cold wind whipping into the Halifax. It was 11:36 p.m., October 18, 1942. With a surge of excitement and fear, Poulsson edged himself out of the open hatch on the plane's belly. He tipped forward, careful to keep his head clear of the opposite side of the hatch. And then in an instant he was falling, falling fast. The sixteen-foot line connected to his parachute pack and a steel cable in the Halifax went taut. The chute emerged from its pack like a butterfly from its cocoon. He continued to free-fall until the air swept into the silk chute and he was yanked sharply upwards by the straps tight over his shoulders. The sound of the plane's engines faded, and he was floating downward from a thousand feet.

The Vidda spread out beneath him in the clear moonlight: its snow-peaked mountains, isolated hills, lakes, rivers, and narrow ravines. It was a place both beautiful and terrible, and Poulsson knew it must be respected. At three thousand feet above sea level, it was exposed to unpredictable weather and high winds that could hurl a man off his feet. In the winter, a skier could be sunning himself on a rock one moment only to find a storm sweeping through the next, bringing blinding slivers of ice and snow and temperatures below minus-thirty degrees Celsius. Norwegian legend had it that it could grow cold enough, quickly enough, to freeze flames in a fire. According to simple fact, it could kill the unprepared in two hours.

The Germans had steered around the thirty-five-hundred-square-mile plateau when they attacked Norway, and even now the country's occupiers dared venture only far enough into it that they could get out by sunset. There were no roads, no permanent habitations in the huge expanse of land. Only skilled skiers and hikers could reach its scattering of hunting cabins. In the valleys, one could find birch trees, but many areas were simply frozen, lifeless hillsides of broken scree, one mile indistinguishable from the next.

As he scanned the landscape in preparation for landing, Poulsson found himself unable to identify the flat Løkkjes marshes, twenty miles west of Vemork, where they were expected to drop. Instead, there was only snow-patched hillsides of boulder and rocks — ideal for snapping one's neck.

He landed hard, but luckily without injury, and quickly got loose from his parachute before a gust of wind took him for a limb-breaking ride across the rough terrain. He called out to the other members of the Grouse team, who had followed him out of the plane. Kjelstrup and Haugland were in good shape, but Helberg walked gingerly, having come down against the edge of a boulder. On examination, the back of his thigh was swelling, but there was no fracture. He did not complain.

For the next few hours, they searched the hillsides for their eight containers of gear, most importantly their stove, tent, and sleeping bags. If a storm hit without those essential supplies, they would be in trouble. Though they located this equipment, it was too late to do anything more than take shelter from the wind and settle in for the night. They made camp, huddling together beside a boulder. It was cold, but bearable. They all wore long underwear and two pairs of wool socks. Then gabardine trousers, buttoned shirts, and thick sweaters. Over these went parkas and windproof pants. They also had wool caps and two pairs of gloves, as well as balaclavas and goggles, but there was no need for those now.

Poulsson dug into his pouch of tobacco and prepared his pipe, a ritual that somehow eased the nerves of the others. He lit the pipe, puffed a couple of times, and then addressed the Grouse team members who were as yet uninformed of the mission change. "There's a new order of the day," he told Helberg and Kjelstrup. No longer were they there to build up a network of resistance cells; instead, they were the advance team for a sabotage operation. On hearing the plan, Helberg thought it was a suicide mission for the British troops: How would they escape Norway? All four Norwegian commandos, however, were happy they would be in on a bigger job. As Haugland thought: *You don't jump out of a plane over your occupied country to contribute a little something.*

Divided into a pair of tents, using their parachutes as ground sheets, the four slept wrapped in their sleeping bags. They woke to a stunningly clear blue sky, the rugged hills surrounding them cast in sharp relief. They were home now, far from soggy Scotland; the air was crisp and dry. Examining the terrain, Poulsson determined they had landed on the edge of the Songa Valley, more than ten miles west of their intended drop point.

The men spent two backbreaking days collecting the rest of their supply containers, scattered about the area. On inspection, they found some serious problems. First, their British suppliers had neglected to pack enough kerosene for their small Primus stoves. These stoves would have allowed them to hike a straight course over the barren mountains to the Skoland marshes, where they would meet the Royal Engineers. But crossing this type of countryside was too great a risk without some source of heat, so they would have to travel through the valley, where there would be cabins for shelter and birch trees for firewood. That added a lot of distance and at least several days to their journey.

Second, their suppliers had made some critical mistakes with the radio equipment Haugland ordered. They had failed to include bamboo poles for rigging the antenna. They had also replaced the standard-issue Ford car batteries, used to power the wireless set and their homing beacon, with batteries weighing twice as much. Worse, these batteries were stamped MADE IN ENGLAND. The British connection would put anyone involved in recharging the batteries in serious danger if they were caught. They tried lashing together ski poles with parachute cord to form an antenna, but they failed to reach London by wireless. Now they faced a forty-five-mile trek across the Vidda, uncertain if they would be able to make radio contact with their handlers or arrive at the drop location in time to meet the British sappers.

8

Keen as Mustard

O N OCTOBER 20, Einar Skinnarland was working in the dark. On the edge of the frozen, dagger-shaped expanse of Lake Langesjå, he waited through yet another night for Grouse's arrival. The cold bit at his face, and from the winds that swept across the treeless, boulder-strewn terrain, he could tell a storm was coming. He had received news of the team's intended drop from a BBC evening broadcast that stated, "This is the latest news from London," instead of the usual, "This is the news from London." As with the two previous mornings, he returned home empty-handed, his face raw from the wind and the bitter temperatures.

Over the six months Skinnarland had spent living a double life, the fear of discovery had been constant. During the day, he worked for Norsk Hydro. At night, he continued to operate his intelligence network. He had several informants at Vemork, and he learned that one, who had been asking questions at the plant for him, had been brought in by Rjukan's police chief. Nothing came of it, but a loose word, a single mistake, by any of a number of people, including Skinnarland himself, and he would be lost. Despite these risks, he was too often, frustratingly, given less information than he needed from his handlers in London, whether because of communication breakdowns or because of need-to-know secrecy.

This was the job he had volunteered for, but how long could he keep inviting disaster for men who never showed up? He would head to Oslo in a few days, deliver his latest intelligence, and find out what he was to do next. Until then, he would continue to spend his nights at Lake Langesjå, and wait.

Jomar Brun picked up the ringing telephone. On the line was a University of Oslo student who introduced himself as Berg. He was calling from Rjukan. The Fox wanted to meet with the Master in person as soon as possible. Tronstad was in London. Berg asked if he could come to Vemork to discuss the matter with Brun, who lived on the same precipitous crag of rock as the plant itself.

"I'll see you in an hour," Brun replied.

After hanging up the phone, Brun rang the newly reinforced guard station on the suspension bridge and instructed the Germans manning the post to allow the visitor to pass. Shortly after, the student, a member of Milorg's "export department," arrived on foot. He told Brun to prepare to leave as soon as possible. Winston Churchill himself had asked that Brun be brought to London forthwith.

Brun insisted that his wife, Tomy, accompany him. Further, he needed some time to collect as much intelligence as he could before departing. Berg agreed, remaining in Vemork to help orchestrate the Bruns' travel to Oslo.

Two days later, on October 24, Jomar and Tomy Brun crossed the suspension bridge and traveled by bus to Rjukan. Brun had told his bosses that he had a doctor's appointment in the capital. He carried a heavy satchel of drawings, photographs, and documents, as well as two kilos of heavy water. With Berg, they took a ferry down Lake Tinnsjø, then a train to the capital. Berg led them to a safe house in the city and left. Their new handler, Gran, another Milorg exporter, was now in charge. Brun passed over his intelligence, and Gran told him it would be microphotographed and couriered separately.

Along with additional blueprints and drawings, Brun had much to reveal. Vemork was no longer fortified by its natural defenses alone. Sappers had strung barbed wire over the fences around the plant and its penstocks. They had started to set land mines, booby traps, and alarm tripwires. Austrian troops numbering close to a hundred and armed with automatic weapons were now billeted in Rjukan, Vemork, and beside the Lake Møs dam. The suspension bridge over to Vemork was guarded around the clock. Electrical wires were being run to the plant's rooftop to power searchlights, and antiaircraft guns were expected soon. The Gestapo was rumored to have moved into Rjukan, setting up at its best hotel.

Gran gave the Bruns new passports and travel passes, and asked them to hand over to him their engraved wedding rings, in case they were arrested along the way. The night before they left for Sweden, there was an urgent knock on the safe-house door. Gran had been followed by the Gestapo and had lost their tail only narrowly. The safe house was no longer safe. He led the Bruns on a hasty run through the dark streets to a luxury apartment behind the Royal Palace. The next day, they took a northbound train from the city with a new handler, stayed overnight at a farm by the border, and then walked across an unguarded bridge into Sweden.

At about the same time, Einar Skinnarland arrived in Oslo. With his own intelligence on Vemork's new defenses, he corroborated the reports from Brun. While in the capital, Skinnarland learned that he could stop his nightly vigil out at Lake Langesjå. The boys would not be coming. That was all he was told.

At 9:00 a.m. sharp the day after the Grouse team had left England, Lieutenant Colonel Mark Henniker entered a cold, corrugated-steel Nissen hut on the Bulford military base, eighty-five miles west of London on the Salisbury Plain. The thirty-six-year-old had a trim mustache and a face hardened by his war experiences in France. Inside the hut were men from two field companies of the Royal Engineers. Before they'd enlisted, these young men were mostly skilled tradesmen: mechanics, electricians, carpenters, cobblers, plumbers, and the like.

Henniker told the sappers that he knew they were as "keen as mustard" for action, but that he had to make himself clear. There was a big, very dangerous, very secret mission ahead. If they failed in its objective, the Germans might well win the war within six months. He needed volunteers but said that anyone who wanted to decline, whether because they were recently married or had a child on the way or were simply not up to it, would face no shame, no questions. Henniker had reluctantly taken charge of the training and planning of the Vemork mission from Combined Operations. He had concerns that the RAF was not up to the challenge of navigating to the drop site; and, as a career military man, he did not much like the idea of his sappers hitting a target in uniform, then switching to civilian clothes to effect their escape. They should fight their way out like the soldiers they were. An order, his seething general informed him, was an order.

Every single sapper stepped forward to volunteer. Their number included Wallis Jackson, a burly twenty-one-year-old, skilled with explosives and more than happy to knock some discipline into new recruits. He also spoke French and wrote gentle letters to his mother and three sisters back in Leeds. Beside Jackson stood Bill Bray, a former truck driver and taut line of a man, who was expecting his first child the following January.

Henniker instructed them to tell anyone who asked that they were preparing for an endurance contest against American paratroopers. In their first week at Bulford, Jackson, Bray, and a few dozen other Royal Engineers marched, practiced shooting, slept on straw mats, and suffered through lectures on how to keep their feet healthy (two pairs of dry socks). Then they were shipped off to northern Wales, where, weighted down with rucksacks, they were sent on long treks through the mountains of Snowdonia, up at dawn, done at dusk. Then again. They slept huddled closely together, a mound of men sharing not enough blankets. At the summit of one mountain, Bray collapsed from exhaustion. His company rustled him back on his feet and split up his kit between them. Those who fell and did not get back up were bounced from the mission. Any questions about why such conditioning was needed, what the mission entailed, and where it would take place were answered only with, "You aren't to know. Not yet."

Few had even heard of the mission's codename: Operation Freshman. Henniker was troubled by the name because it struck at the core of his doubts over the plan. Combined Operations had decided to bring his sappers to their target by plane-towed gliders instead of dropping them by parachute. Although the Germans had used an armada of these silent gliders in their invasion of Belgium, most notably landing them inside Fort Eben-Emael, once considered impregnable, this was the first time the British had attempted their use in an operation. Henniker wanted to see his men do their job and get out alive. He was not interested in breaking new ground.

Group Captain Tom Cooper, Henniker's counterpart in the RAF, had selected the latest model of glider for the mission: the Horsa Mark I. Measuring sixty-five feet from tip to tail and eighty-five feet from wing to wing, the Horsa was constructed of a solid wooden frame with an arched, plywood outer shell. It resembled a coffin, some said. It could carry a payload of four tons within its narrow fuselage — a jeep or an artillery gun, for example. There were collapsible wooden seats for twenty-eight soldiers.

The two-pilot cockpit contained a simple steering column and rudder bar, along with a compass and gauges for air speed and pressure, rate of climb/descent, and tow-cable angle (what pilots called the angle of the dangle).

It was the duration of the dangle that troubled Cooper. His 38 Wing command typically flew two-engine Whitleys, and the glider pilots were practiced at being towed behind this aircraft. But Norway was a four-hundred-mile flight across the North Sea, and only the four-engine Halifax would be able to haul a glider that distance—and back, if needed. His crews would need training to fly Halifaxes, with and without gliders, and he would need to borrow these heavy bombers from other commands. Naturally, the planes he was provided with were far from being the pride of the RAF. Even if everything were to go smoothly on the flight, Cooper still worried over the dangers his crews faced in Norway. The pilots would have to land the fragile gliders at night, in unknown territory prone to tempestuous weather and uncertain terrain.

Cooper voiced his concerns to Combined Operations, as did Colonel Wilson and Leif Tronstad, who were in charge of organizing the reception crew on the ground. The military planners weighed this risk, among others, against not destroying Vemork, and the decision was taken to use gliders.

Combined Operations had settled on its tactical plan over a series of meetings in September and October. A nighttime bombing raid by the RAF offered bad odds of hitting the target and would likely kill many innocent civilians. An attack by Kompani Linge saboteurs was turned down because a successful demolition would demand hundreds of pounds of explosives, too much for a small force to carry. Further, the Norwegians were not considered sufficiently trained to blow up the power station and heavy water plant. Thus the British sappers were selected.

Fifteen would suffice for the raid, but given that some might be killed before reaching the target, Combined Operations decided that two forces of fifteen would guarantee the task was done. After all, as the planners reported in one meeting, "In all probability there could only be one attempt at the Freshman objective and that must be successful."

The SOE suggested bringing in the sappers via Catalina flying boat, giving them a chance to hit the target and then escape by plane. But the steep approach to the lakes, plus the fact that they might be frozen, ruled this out. Arriving by parachute was also discounted: To drop the sappers near the plant, the planes would have to fly too low, too close to Rjukan, risking detection. Worse, too many sappers might be injured on landing in the rugged terrain, and they might be too spread out to find one another promptly. Gliders, released at ten thousand feet, would land all the men together with all the equipment required for the operation. To reduce the risk of the party being discovered, the planners deemed it imperative that the operation proceed on the night of the drop.

The plan took shape. The Grouse team would use lights to signal the two gliders, each carrying fifteen sappers, to land in the Skoland marshes. They would also employ the Eureka/Rebecca system, a new, untested technology that used radio signals to provide a homing beacon to planes. They would guide the sappers down the road on the north side of Vestfjord Valley to the plant, if possible by bicycle. At Vemork, the British troops would cross over the suspension bridge, neutralize the small number of guards, and place almost three hundred pounds of explosives to blow up the power station's generators and the hydrogen plant. Once away from the plant, they were to separate into groups of two to three men and change into civilian clothes. Then they were to hike a two-hundred-mile route, often through populated territory, to Sweden. They would be equipped with maps, ten days of rations, and a few Norwegian catchphrases if they came across patrols. ("I've just been out buying stores for my mother.")

The new intelligence that had come in from Skinnarland and Brun, revealing an influx of soldiers and new fortifications at Vemork, prompted efforts to adapt the Combined Operations plan to reflect the situation on the ground. Suggestions were made to increase the force to 250 to 300 men or to launch a large-scale daylight bombing attack, best suited to the U.S. Bomber Command. But such changes would delay the operation by months. And a bombing raid in daylight hours would kill even more civilians than would a nighttime attack.

Although the operation was clearly now more hazardous, Lord Mountbatten pushed Churchill to move forward with Operation Freshman as planned. "It is of great importance that [it] should take place at this time as greater difficulties will be experienced during subsequent moon periods," Mountbatten advised. Given the threat of a German bomb, Sir John Anderson, a War Cabinet member and the ultimate authority on Tube Alloys, confirmed yet again to the operation's planners that the mission was "of the highest priority."

At noon on November 2, Henniker brought his weary sappers by train to an unmarked station north of London. Awaiting them was a line of

Humber Snipe automobiles. With curtains drawn over the windows, the cars took them to STS 17, the SOE's industrial sabotage school in Brickendonbury Hall, a Jacobean manor that had most recently been a private preparatory school.

If the British had one advantage to counter the unknowns of the weather and the landing site, it was their intelligence on the target itself. There to meet Henniker's men for their stay at Brickendonbury Hall was Leif Tronstad. Thanks to the intelligence from Skinnarland and Brun, he was able to give the sappers blueprints of the buildings as well as photographs and drawings of the equipment to be targeted. He provided insight into where employees and security staff were at any given time, inside and outside Vemork. The sappers came to know virtually everything about the plant, down to the type of locks on the doors, the location of the keys, and the number of steps to reach the heavy water high-concentration plant on the basement floor. Major George Rheam, the British master of industrial sabotage, had even built a wooden mockup of the heavy water concentration cells for the sappers to practice on.

There was just one problem. It had been more than two weeks since the Grouse team had left Britain, and Tronstad had heard nothing from them. Without the four young Norwegians, without regular radio contact, Operation Freshman was a no-go.

Knut Haugland was cold, hungry, exhausted, and wet. His team members, skiing in a line behind him through the Songa Valley, suffered the same. Burdened by seventy pounds of equipment, Haugland often sank into the deep, heavy snow. The candle wax he had spread on his skis was proving useless; given the mild weather, the new snow clumped like gum to the bottoms, making the trek a snail's slog. The limited amount of actual ski wax they had been equipped with had to be kept in reserve for the night of the sabotage operation.

Haugland navigated around birch trees and the rugged, boulder-pocked terrain, sticking close to the banks of rivers and small lakes. In the full of winter, he could have easily skied straight across them, but on this morning of October 24, the ice was not yet completely frozen. In the few patches he thought they could cross, the surface water on the ice left their boots and socks drenched. After an advance of a few miles, Haugland and the others stopped, emptied their rucksacks, and took a short break to eat. Poulsson had rationed them each a quarter slab of pemmican, four crackers, a little spread of butter, a cut of cheese, a piece of chocolate, and a handful of oats and flour — for the day. The pemmican, a pressed mix of powdered dried game, melted fat, and dried fruit, was treasured above all. Altogether, they were probably burning twice as many calories as they were eating each day. Their rest over, the four returned to their morning starting point, this time with empty rucksacks, and retrieved the other half of their equipment and food, another seventy pounds each. This they hauled across the countryside, retracing their tracks in what at times felt like Sisyphean labor. A slight misdirection of their skis and they would sink to their waists.

Their intended departure from the drop site had been delayed by a day due to a terrible storm. Through the night of October 21, the four hunkered down inside their tents, surrounded by their eight containers, as a blizzard raged. They awakened to four feet of damp snow and a forty-five-mile trek ahead. In ideal conditions, they could have skied this distance in a couple of days. But even after leaving unessential supplies in a depot dug into the snow, they still had 560 pounds to carry, including one wireless set, two batteries, the Eureka beacon, a hand generator, field equipment, weapons, and food stores. Divided by four, this was 140 pounds for each of them, impossible to haul unless they split it into two journeys. With the double-backing, their journey would be 135 miles, and the condition of the snow was far from ideal.

At the end of their third day of trekking, having advanced only eight miles from their drop site, they came across an abandoned farmhouse beside Lake Songa, where they found some flour and frozen meat. They built a fire, melted snow in a pot, and then softened the meat in the boiling water. For the first time in almost a week, they feasted until their bellies were full. Set by the crackling fire, their wet socks, boots, and clothes steamed as they dried out. Better still, they found a welcome surprise in the cabin: an old wood-and-canvas sledge Poulsson's father had given him as a child. During the invasion, Norwegian forces had borrowed it, and it had somehow ended up at the farmhouse.

Over the next six days, the team made slow, steady progress east, their supplies split between their rucksacks and the sledge. No longer would they need to make the double journey. But it was still tough going. At one point, Poulsson fell through the surface of a half-frozen lake, and Kjelstrup, his body stretched flat on thin ice, had to pull him free with a pole. Poulsson also developed a pus-filled boil on his left hand, and had to carry his arm in a sling when not on skis. At night, the men continued to break into cabins for shelter, but none held the same booty as the farmhouse. They devoured their pemmican, sometimes cold, sometimes mixed with oats or flour in a hot gruel, but they were always left wanting more.

As one day followed the next, the four grew thin and their beards scraggly, their cheeks and lips blistered from the constant wind, cold, and toil. They were almost always wet, as their clothes never dried completely at night. Their skis, not wholly impermeable, grew heavy as logs. Their Canadian boots became so frayed, they had to take an awl and yarn to them each morning to keep them from falling apart altogether. Had it not been for all their hard training in Scotland, they would already have given up.

As they traveled, Haugland confiscated enough fishing rods from the cabins they sheltered in to construct a mast for the antenna. One night, attempting to fire up his wireless set, it short-circuited. Too late did he notice that the set was covered in condensation from being brought into the warm cabin out of the severe cold. Next time, he would wait for the set to dry out inside before starting it up.

By October 30, seven miles southwest of the Skoland marshes, Poulsson scribbled in his small diary, "We are fairly done in." They settled into a small hunter's hut at a place called Reinar and decided on a course forward. Helberg offered to return to the farmhouse alone to steal more of its stores. Poulsson and Kjelstrup would find the easiest path ahead to the marshes, and Haugland would try to reach London by wireless yet again.

Each of them understood that the radio operator's job was the most essential, and nobody felt the weight of this more than Haugland. The voice that carried through the darkness and distance, the voice that brought help, that rescued lives — that was what he was meant to be. He had always wanted to be a radio operator, an ambition fueled by a naval adventure novel he had read as a teenager, which featured a radioman saving his whole crew. After graduating high school in Rjukan, Haugland attended the Army's signals course in Oslo. Operating a radio set in Morse had been like learning to play the piano. At first he was all thumbs, slow and clumsy, but after practice, practice, practice, he found that he was a virtuoso, tapping out dots and dashes without thinking.

On graduation in 1939, he worked as a wireless operator on a three-thousand-ton merchant ship that traveled between Norway and Iceland. The war at sea was already underway, and Haugland listened to distant SOS distress signals from sinking ships, knowing that his own was too far away to be of any help. When Norway was invaded, Haugland found his skills in dire need, especially in the pitched battles at Narvik. A forward observer, he scrambled through the countryside, locating enemy troop and artillery positions and radioing them back to his commanders. He was constantly under fire, whether from mortars, incendiary bombs, or machine guns, and he witnessed the horrors they wrought. One night, as Haugland sheltered in a hastily dug trench, a man stumbled toward him, his chest riddled with bullets, one through his cheek. Other days, he passed bodies almost unrecognizable as human. Throughout, Haugland stayed at his wireless set. He was a fast radio operator, and he learned something a soldier only knows when tested: he was almost preternaturally cool under pressure. The worse things became, the more deliberate and calm he was.

On the call to lay down arms, he felt a terrible blur of emotions: anger, hopelessness, confusion, sadness. Hitler now controlled Norway, and he was supposed to just accept it. In Oslo, he got a job in a radio factory. He was soon volunteering to build sets and establish wireless stations for the resistance. Several times the Gestapo and the state police arrested him. Each time, he talked himself out of the bind. But when an informant definitively gave him away, Haugland had to escape. The Germans put a one-thousand-kroner bounty on his head.

Haugland joined Kompani Linge soon after he arrived in Britain. "Quiet, keen, hardworking and very intelligent," his instructors at Stodham Park reported. "Full of courage and painstaking. Expert at Morse," they said at Meoble. Haugland then attended STS 52, the specialized school for wireless operators, where his teachers said he should be instructing them. He ciphered codes and tapped in Morse faster than most, and he could build a radio from the sparest of material. He was deliberate in everything he did, causing some of his fellow students to quip that he was still making plans while the others finished the actual job. Nonetheless, his instructors reported, "Quick thinking and unfailing attention to detail . . . He has a well-reasoned course of action to meet each emergency."

Now, thirteen days into his first operation, Haugland had still not managed to get his radio to work, and it burned at him. Alone at the Reinar cabin, determined not to fail again, Haugland prepared his wireless set and antenna mast. This time, he had kept the set and its battery inside the cabin all night. When ready, he turned it on, hoping the shortwave signal would reach the English operators at Home Station, Grendon Hall, Northamptonshire.

Straightaway he got reception, and then, an instant later, nothing. The battery, all thirty "Made in England" pounds of it, was dead. Its charge should have lasted a month. Not yet defeated, Haugland tried to recharge it with a hand generator, but the crank would barely budge because of its short handle. When it did, the insulation on the generator's wires started to burn because it was not suited for such a large battery. There was no way around the fact that he needed a new one.

Eventually Haugland's teammates returned to the Reinar cabin. Helberg was worn out from a long trip through gusting winds to the farmhouse, and Poulsson and Kjelstrup had both narrowly escaped a deadly plunge into a half-frozen lake. Haugland shared with them the dismal news about the battery. The mood in the cabin was grim. In sixteen days, the moon would be in position for the operation against Vemork. Their team needed to be in place to reconnoiter the landing site, to provide weather reports and intelligence on the facility, and to receive the British sappers. They still had many miles to go, and they had no working radio.

After burying the battery, the men decided that one of them would go on ahead of the others to the Lake Møs dam, to seek help from Torstein Skinnarland. Helberg volunteered to go. That his countryman was up for the journey so soon after his trek to the farmhouse inspired Poulsson to write in his diary, "Helberg proved the old saying, 'A man who's a man goes on till he can do no more, then he goes twice as far.'" Even though the Grouse team had been ordered to not make contact with the Skinnarlands, Poulsson felt they had no choice. No radio: no mission.

At the dam, Helberg found Torstein Skinnarland. The two did not personally know each other, but Einar had told his brother that they might soon have some friendly visitors in the area. Helberg only explained they were on a "special mission." Torstein promised to provide a new battery, along with new boots and some food, but it would take time. Einar had some supplies like that hidden away, but Torstein did not know where they were. Helberg then departed.

With new snowfall and a drop in temperature, conditions on the ground improved, and the other Grouse team members made quick progress east over the next two days. Helberg met them at a river crossing on his return from the dam.

In the early hours of November 5, the Grouse team arrived at a cabin beside Lake Sand, three miles east of the Skoland marshes. Inside, the four — dirty, unshaven, and half-starved — collapsed from the strain. Their boots were in shambles, their sweaters shrunken and in tatters, and two of their rucksacks had been chewed up by dripping battery acid. They slept that night like the dead.

The next day, Haugland started constructing his aerial antenna again. He lashed together two towers made from fishing poles, set them fifteen feet apart, and strung insulated copper wire between them. Then he nailed together several boards and secured these to the cabin wall. He strung the antenna from his wireless set through the corner of the cabin window, up the boards, and into the wire strung between the two towers. While he worked, Poulsson and Kjelstrup skied to the marsh to inspect the landing site, and Helberg left for the dam, hoping that Torstein had been able to secure a new battery for them.

After dark on November 9, Haugland finally thought he had everything set to restart the radio — including a new battery supplied by Skinnarland. He had already coded his message, a jumble of seemingly random letters on the notepad in front of him. It needed to be short and quick because the Germans operated D/F (direction finding) radio stations that tuned in on broadcasts from Norway. If Haugland transmitted for too long, and two German stations were close enough, they might get a cross-bearing on their position. As a radio operator, Haugland knew that brevity and speed saved lives — his own among them.

The new battery feeding in a steady current, Haugland powered up the wireless set. With his three team members watching closely, and his hand trembling from the cold and excitement, he sent out his identity call sign. He immediately received an answer in return. They had contact with Home Station. Poulsson and the others whooped in celebration as they congratulated the radioman.

Haugland delivered his first message: "Happy landing in spite of stones everywhere. Sorry to keep you waiting. Snowstorm and fog forced us to go down valleys. Four feet of snow impossible with heavy equipment to cross mountains. Had to hurry on for reaching target area in time. Further information. Next message."

Grouse was in place for its mission and ready to guide the sappers to their target.

9

An Uncertain Fate

IN HIS FIFTH-FLOOR office in Kingston House South, bordering Hyde Park, Leif Tronstad faced a mountain of paper—most of the sheets stamped TOP SECRET in red ink. Reports, minutes of meetings, wireless messages, and letters—they came at him in piles. Working quickly, he penciled his remarks on some of the documents and dictated answers to his secretary, Gerd Vold Hurum, for others, but mostly he delivered his replies via the two secure phone lines in his office. This method was faster, prevented misunderstandings, and cut down on further paperwork, all of which had to be filed in the huge steel safe behind his desk.

Tronstad did not much care for the continuous deluge of paperwork, but he was thrilled at the sudden multitude of decoded messages from Grouse. A week before, the operators at Grendon Hall thought they had made contact with the team, although there was something amiss with the call sign. Tronstad sent a message: "We were very glad to hear you . . . We hope everything is all right with you." There was no response. Fearing the worst, and breaking security protocol, he dispatched a courier to connect with Einar Skinnarland to see if he could get in touch with the men. Before that could be done, wireless contact was finally made. "Battery run down so forced to make contact with Tante Kjersti [Einar Skinnarland] . . . Fixed six kilometers from glider landing place," read the message. "Enemy troops in area are Austrians. About ninety-two Germans in future. Operation still can be done with success. Telephone lines ought to be cut. Waiting for orders . . . Too much snow for cycling."

After conferring with Combined Operations, Tronstad and Wilson sent back a list of questions to be answered, ending their message, "Stick to it and good luck." Over the next few days, a steady stream of messages reported that Grouse's Eureka was in working order and that the landing site was in a perfect place, out of sight of the Germans and on "nice, flat ground approximately 700 yards long. No trees or stones." Further, they signaled, "Engineer Brun" had disappeared with his wife. Grouse obviously had good intelligence.

On November 12, Tronstad walked from his office to a South Kensington hotel, a short distance away, where he met up with Eric Welsh. Together, they knocked on the door of a room registered to a Dr. Sverre Hagen and his wife. Brun, who had been given this alias by the British spy agency, answered with Tomy at his side. After a warm greeting, the three men sat down to discuss Vemork. In late August, Kurt Diebner had personally visited Norsk Hydro headquarters and had made it clear that "all necessary measures" were to be taken for "the fastest execution of the work." Throughout September, the plant expansion had continued, and the first of the catalytic exchange units was installed in the sixth stage of the cascade. Heavy water plants were also being installed at Såheim and Notodden.

Brun also told Tronstad and Welsh about a revealing conversation he had had with Hans Suess, one of the scientists assisting Harteck in the Vemork upgrade, just a couple of weeks before his escape. Confident that Suess was anti-Nazi, Brun had asked him about the German interest in heavy water. Suess revealed that they needed five tons as a neutron moderator for a uranium machine but said that none of their research was of any "immediate application" in the war. Rather, it was a long-range project, most likely for peaceful purposes.

While Brun said that he believed Suess was being honest, the German physicist was either wishful in his thinking or uninformed about the breadth of the Nazi program. The fact that Diebner, an Army Ordnance scientist charged with making weapons, was so closely involved signaled that the project was unlikely limited to peaceful purposes. Vemork had delivered roughly fifteen hundred kilograms of heavy water to his "quinine factory." He needed three thousand kilograms more for his uranium machine. In a year, with full production, he would have it.

This could not be allowed to pass, Tronstad said. Plans were already in motion to blow up Vemork. He needed Brun to provide additional drawings and details in the next couple of days. These would be passed to the

British commandos preparing for the raid. Despite Tronstad's misgivings about destroying a plant he had helped to build, the more Tube Alloys told him about their own atomic research efforts, of "super bombs" that would equal "1,000 tons of TNT," the more Tronstad knew the Germans needed to be stopped.

While this mission developed, Tronstad was involved in scores of other operations, from the recently launched Bittern, which sought to assassinate certain Norwegian Nazis and informers, to the planning of Carhampton, a plan hatched by Odd Starheim to seize a convoy of merchant vessels. Many in Kompani Linge were also now engaged in missions similar to Grouse. They prepared sabotage operations, set up wireless transmission sites, and established underground resistance cells throughout Norway. The list of mission names read like an ornithologist's dream: Chaffinch, Cockerel, Crow, Feather, Hawk, Heron, Lark, Mallard, Partridge, Penguin, Pheasant, Raven, Swan, Thrush — among others.

There was much to worry about, but Tronstad was comforted at least by the news that his family was well. In October, Bassa had written to him: "The little boy is quite something. He is thrilling, beautiful and enchanting . . . Sidsel is a sweet and kind girl." As for herself, Bassa said, "Everything is fine and well," though she wondered, "Do you think the war will end before Christmas? We so long for peace." Then she asked him to send her news of his work and life and said how she looked forward to "the time we will see each other again." Her letter must have crossed paths with his own, as he had penned a note only days before. "I am well, as well as humanly possible away from you and the children." He told her how he spent long days doing "interesting work" that had nothing to do with his former profession, and then finished, "Live quiet and secluded. Don't be afraid for me, dear friend. I will manage and expect to find you all again . . . I miss you dearly . . . Trust in us. We will be coming soon. Sustain yourself until then."

On November 15, at 11:30 a.m., Tronstad joined Wilson, Henniker, and several others at Chiltern Court to go over the plan for Operation Freshman one last time. He delivered the latest messages from Grouse: they suggested the sappers bring snowshoes but said that even under bad conditions the march to Vemork should not take more than five hours. The team would cut the telephone line between the Lake Møs dam and Rjukan on the night of the operation. There were two guards at the door to the hydrogen plant, and the sappers would have no trouble overpowering them.

Finally, if reinforcements came from Rjukan, the suspension bridge could easily be held. Henniker said that he would need only two guides from Grouse on the approach to the plant and that he would provide them British battle dress (so that, if captured, they would not be implicated as Norwegians). Their role would end before the Royal Engineers went across the suspension bridge. The other Grouse men were to operate the wireless and destroy the Eureka beacon, a technology the British did not want to fall into German hands.

Overall, the mood of the meeting was positive. Toward its close, Tronstad presented diagrams of the plant. He said he understood the mission's importance but worried that destroying all the power station's generators would hurt the livelihoods of most of the Norwegians in Rjukan and eliminate the fertilizer supplies his country desperately needed. Instead, he sketched out a plan to save two of the twelve generators, which would have the same effect on heavy water production but keep Vemork alive as a hydrogen plant. Henniker offered to raise the point with his superiors, but time was running short for making changes to the plan.

Tronstad was sure he had been given a polite brushoff by these "handsome and undoubtedly brave soldiers," as he wrote in his diary that night. However, he was undeterred, and he prepared a report, which did help him achieve the exclusion of two generators from the planned demolition. In it, he wrote, "Good policy for destruction of plants in Norway is namely to do just as much damage as strictly necessary — to prevent the Germans from winning the war — and nothing more."

In a dark, windowless corner of the cabin on Lake Sand, Poulsson tended a huge pot of boiling sheep's-head stew on the wood-fire stove. He fancied himself a bit of a cook, so he had added some canned peas and whatever else he could find in their stores to improve the flavor. Not that the others would complain, he knew, as all of them were starving. The boys had even set a cloth over the table.

Earlier that day, Haugland and Helberg were on their way back to the cabin when they came across a sheep that had strayed from its flock and become caught in some rocks. They killed it, then carried it back to the cabin over their shoulders. After skinning and chopping it up, Poulsson prepared the stew.

Finally, Poulsson called the others to the table; their feast was ready, and the smell wafting from the pot made everyone's mouths water. Outside, the wind howled, driving snowflakes into the cabin through the breaks in the walls. They could barely see one another in the flickering light of the single candle.

On his way to the table with the huge pot in hand, Poulsson slipped on one of the reindeer pelts on the floor. The weight of the stew tipped him off balance. As he fell, the stew spilled across the floor, sheep's head and all. Everyone looked at it, then at their cook. Without a word, the four got down on their knees. They scooped up whatever they could onto their plates. In the end, there was nothing but picked-over bones and a string of jokes from Kjelstrup about having "hair in my soup." To a man, they still thought it delicious. They went to bed with full stomachs.

In the three days leading up to November 18, the start of the full moon phase, the Grouse team remained busy. They scouted the route down to the target. From a hidden position on the opposite side of Vemork, they monitored the guards on the bridge.

Helberg continued to venture back and forth from Lake Møs for supplies and batteries. One evening, on his way to meet Torstein Skinnarland, he found Einar Skinnarland instead. Helberg did not reveal the names of the others on the Grouse team, and Skinnarland knew better than to press. He simply offered his compatriots any additional help they might need in the days ahead, whether supplies or intelligence on Vemork. He quickly became part of the Grouse team.

Haugland was the most occupied, constantly coding and decoding messages to and from London. Some of his scheduled daily broadcasts took place after midnight, and he emerged from the shell of his sleeping bag only as far as he needed to tap out his Morse. Mostly he sent word about the weather: the direction and strength of the wind, the height of the clouds, and the visibility. Weather was everything now, and right up to the time of the operation it had been very unpredictable. One day it was clear. The next, there were high clouds, and the next they were low, blanketing the valley. Some nights it snowed lightly; others were merely freezing cold. Only the winds were constant. They came in squalls from shifting directions, and time and again knocked over the antenna mast, meaning that one of them had to climb up on top of the cabin to set it back in place.

November 17 was the third fine day in a row, and Haugland transmitted as much in his afternoon broadcast: "Larger lakes like Møs ice-free. Last three nights light and the sky absolutely clear. Temperature about minus five Celsius [23 degrees Fahrenheit]. Strong wind from the north has quieted down tonight. Beautiful weather."

That same day in Scotland was dreary and gray. At STS 26, Knut Haukelid knew nothing of Grouse's mission, nor of the fate of his friends. In early October, his doctor had finally cleared him for training and hard duty. Still, there was no word yet on the plans for him to drop into Norway to set up additional resistance cells, nor was he tapped for any other missions. Instead he was ordered to travel down to the south of England for an armored-fighting-vehicle course. If he could not be a commando operating secretly behind enemy lines, then perhaps he might go the opposite route in the Army, bringing the fight to the Germans in a roaring twenty-seven-ton hunk of steel. They would see him coming, but given that he would be commanding a tank crew armed with a heavy machine gun and an antitank pounder, let them.

Sardined inside the domed dorsal gun turret of a Halifax bomber barreling through the darkness over the North Sea, Mark Henniker was praying the crew could save them from disaster. Through his headset, he listened to their heated exchange.

"Which is it, Wilkie?"

"The one by the main spar," the pilot, Squad Leader Wilkinson, yelled to his rookie flight engineer, who was hesitating over which fuel valves to shut on and off. "For Christ's sake, be quick or we'll have an engine die on us."

"Is it the right-hand or left-hand?"

"The right-hand one, just behind the main spar. No, you bloody fool, not that one."

"Can't get it to budge, sir."

"Turn the thing the other bloody way. Christ! What happened?"

Henniker saw a bank of clouds course past. One of the port-side engines was dead, and the aircraft pitched from side to side. They were losing altitude quickly. Minutes before, he had been enjoying the exhilarating rush of streaking across the sky with the purr of the Halifax's engines in his ears. Clouds billowed under him, while above his head the stars looked altogether like diamonds on a velvet cloth. It was November 18, the start of the new lunar phase, and he had decided to accompany one of the two Halifaxes on a test run to the landing site in Norway to acquaint the plane and glider crews with the terrain.

In the panicked exchange, Henniker heard the navigator tell the pilot what to expect once they broke through the cloud base: "We should either see the Norwegian coastline at once, or we might charge into a mountain almost immediately."

"We shall turn for base," Wilkinson said. "I hope to God we get there!"

The Halifax banked sharply, and Henniker feared they would not make it the four hundred miles back to Scotland. If this "great monster" ditched into the sea, it would sink within ten minutes, faster if the petrol tanks were still full.

Two hours later, they landed at Skitten, a remote RAF airfield on the northeastern tip of Scotland. Group Captain Tom Cooper, who was piloting the other Halifax, returned soon after. Even with the clear night over the Vidda, nobody on his plane had spotted the landing site. He told Henniker that the mountains and valleys, lined up in indistinguishable bands of "dark and light," reminded him of tiger's stripes. Short of perfect weather and a keen eye to locate the Skoland marshes, the mission depended on the Eureka/Rebecca signals bringing the gliders to their spot.

Under a tight schedule, Cooper had done everything possible to ready his side of the operation. His 38 Wing pilots underwent tens of hours of Halifax training and had perfected homing in on landing sites by radio beacon. In their final nighttime practice, they had missed a direct flyover by only forty yards. The glider pilots had trained hard as well. Still, Cooper knew that even in daylight, a smooth, on-target landing was always a bit of a trick.

He had also tested in a freezer the communication link woven into the towline that connected the glider to the Halifax. It was critical that the link worked in the frigid air over Norway. He stressed to his superiors that his aircrews could have used more training and the Halifaxes more maintenance work. They must, as one of his commanders wrote, make do, as "it is too late to stop, and we must hope for the best." Churchill had already been notified that Operation Freshman was set to go.

The Skitten cooks managed to rustle up some bacon and eggs for the returning crews, then Henniker fell into bed for a few hours' rest. At 10:30 a.m. the next day, November 19, he received a weather report from the Grouse team: another cloudless day with light northwest winds. He met with Cooper to decide if the operation would go ahead. By evening, the men and planes would be ready. The only question now: Would the weather hold? Cooper had his own meteorologists at Skitten, but Combined Operations had also sent him a Norwegian one, Lieutenant Colonel Sverre Petterssen, to advise. Petterssen, a former MIT professor, spent the whole morning studying charts and the latest weather reports. Despite the message from Grouse, he was worried about the strong westerly currents over Scotland and the upper reaches of the North Sea. They might bring tough conditions by late evening. Petterssen advised waiting a couple of days, when an "outbreak of Arctic air" would bring ideal conditions. Henniker and Cooper considered the Norwegian meteorologist's forecast, but because Skitten's own forecast did not ring any alarm bells, decided that the operation would proceed at 6:30 p.m. Norwegian time.

Wallis Jackson, Bill Bray, and the twenty-eight other Royal Engineers were well primed for their operation. They ate sandwiches and smoked cigarettes outside the Nissen huts on the barren seaside Scottish moor that made up Skitten airfield. There was some banter, the false bravado of men about to head into action. Others handled their nerves in silence. There would be a final briefing, but they already knew what they needed to know: they were heading to Norway to blow up a power station and hydrogen plant.

One did not have to be an Oxford don to deduce the location and nature of their target. After they had returned to their Bulford base from Brickendonbury Hall, they were given Norwegian-labeled clothing and lessons in how to walk in snowshoes (practiced in tall grass), sure clues they were headed to the land of ice and snow. Other giveaways were a visit to a power station and lessons from a Norwegian scientist (Tronstad) about how an electrolysis cell worked. The sappers still had no idea what purpose was served by the "very expensive liquid" produced at Vemork. However, given the blanket of security everywhere they traveled, and the orders to strip off their uniform's badges and insignia, they knew it must be important.

In their final week at Bulford before traveling up to Scotland, they had rehearsed their attack and hiked for miles every day. When one of their team twisted his ankle, their lieutenant said to the medical officer, "I need 250 miles out of this man. Will he make 100 percent recovery to take this task on? I don't want his blood on my hands." The sapper was bounced from the mission, and every single one of his mates knew then that they had a hard road ahead of them. It was a tough choice between which was the most intimidating: a three-and-a-half-hour night flight over the North Sea in a wooden glider or a long escape by foot through occupied territory to Sweden.

Still, none backed out.

At Skitten, most of the sappers wrote home, some betraying their worry. To his mother, Wallis Jackson simply wrote, "Mamie, if you send my laundry and letters here it will be okay. Looking forward to my next leave. Bags of love, Wallie. P.S. Writing this in bed where it's warm." Bill Bray, who was suffering a cold and sore throat, wrote a quick letter to his wife, hours before they were to leave. "A few lines in haste to let you know I am just off on a raid. I can't say where but don't worry too much darling if you don't hear for a couple of weeks or so. But I shall be back for Christmas so get that chicken ordered up . . . Darling, remember I love you and adore you. Don't worry too much dear because I shall be back so bye bye and God bless you. From your ever loving hubby, Billy."

After the sappers finished their tea, Henniker and his two lieutenants, Alexander Allen and David Mehtven, gave them a short briefing. "Whatever happens," Henniker concluded, "someone must arrive at the objective to do the job. Detection is no excuse for halting." He wished them good luck and Godspeed. Then the men finished kitting up. They wore steel helmets and British Army uniforms with blue rollneck sweaters underneath. They each had a Sten gun, a rucksack filled with ten days of rations, a sleeping bag, explosives, and other equipment. Some of them carried silk maps with the target circled in blue and a false escape route to the west coast of Norway. After the operation, these would be dropped to throw off their pursuers.

Under a slight drizzle, the sappers strode out onto the runway where two black Horsa gliders stood behind their Halifaxes. Wilkinson was flying Halifax A, with Cooper onboard to supervise the overall flight. Arthur Parkinson, a twenty-six-year-old Royal Canadian Air Force pilot, was captaining Halifax B. While the Halifax crews went through their checks, the sappers boarded the gliders. Jackson was in Glider A, Bray in B. Most of them were in their early twenties, and one observer noted that they looked like schoolboys. The men took their positions in the fuselage and strapped on their safety harnesses. The floor beneath their boots was corrugated metal, its long channels designed to prevent slipping on the vomit that was a regular companion on glider flights. Henniker wished them good fortune one more time, then the ground crew closed the hinged tail of the glider. As their pilots readied for the tow, the sappers were left to look at one another and wonder how the mission would unfold.

After a slight delay, the Halifaxes powered up. Wilkinson steered his onto the runway first, pulling Glider A along behind by a taut, 350-foot-long hemp rope. At 6:45 p.m., with a wave from the crews, the Halifax roared down the runway. The glider followed behind, and at roughly seventy miles per hour, the two aircraft ascended into the sky. Fifteen minutes later, Halifax B and its glider took off. Including air crews and sappers, there were forty-eight men on the mission. Henniker watched from the ground. Early in the operation's planning, he had volunteered to personally lead his men, but his superiors had forbidden it.

Skitten's radio operators messaged Combined Operations and SOE HQs that the two planes had departed. In his diary that night, Leif Tronstad jotted down, "Two small birds following two large ones, off toward an uncertain fate tonight."

"Girl" — the coded message came in to Haugland at the Lake Sand cabin. The sappers were on their way. After Haugland acknowledged reception of the code word, Poulsson led his three men on skis to the seven-hundred-dred-yard-long landing zone they had picked out on the Skoland marshes. When they arrived, it was after dusk and the weather was changing. A moderate wind blew in from the west, and scattered clouds hung in the sky. Visibility was still good, but knowing the Vidda, this might alter at a moment's notice.

Leaving Haugland and Kjelstrup on a hillside to set up the aerials and batteries for the Eureka beacon, Poulsson and Helberg moved down into the snow-covered marsh to mark out the landing zone. Using the length of their strides to measure the distance, they placed six red-beamed flashlights, each 150 yards from the last, in an L shape in the snow. They would be switched on as the planes approached, and Poulsson would stand at the corner of the L, flashing a white-beamed flashlight, to bring in the gliders. Haugland would be the first to know the Halifaxes were coming: When a plane approached, its Rebecca device would send a short-range radio signal to his Eureka. A tone would sound in his headset, and his Eureka would retransmit this signal back to the Rebecca at a different frequency, giving the plane's navigator a bead on the distance and direction to the landing zone.

Everything ready, the Grouse team gathered around the Eureka in the dark and in a cold wind, certain that they could lead the sappers undetected to the target and that the defenses at Vemork would be overcome. The gliders needed only to land safely. Even just one needed to land. But with each passing minute, the scattered clouds lowered, obscuring the moon, and the northwest wind rose into a scream.

10

The Lost

HAUGLAND WAS KNEELING in the snow beside the Eureka when a distinct tone sounded in his headset. It was 9:40 p.m. Through the rising winds, he shouted to Poulsson, "I hear the Rebecca. They're coming now." Poulsson skied down toward the landing site. As he went, he waved to Helberg and Kjelstrup, who were already in position, awaiting the arrival of the Halifaxes. "Up with the lights," Poulsson called out. "Up with the lights." Quickly, they lit the red L in the snow.

Poulsson stood at the corner of the L, covering and uncovering the white beam of his flashlight. The wind whipped around him. He stared skyward, the low clouds breaking occasionally to reveal the moon. Although he worried the flashlight beams were too weak for the pilots to see through the cloud cover, he knew that the Eureka radio beacon would bring them in close nonetheless. A few minutes passed before they heard the low grumble of a Halifax approaching from the southwest.

"I can hear it!" Haugland cried out, though he knew his voice was lost to the others.

The engines grew louder—the Halifax was surely flying right above them. Spirits high, they waited for the glider to appear out of the darkness. Gradually, though, the roar of the engines faded, and Haugland's headset went silent. Poulsson continued to flash his signal, and the red lights continued to shine upward into the empty night. No glider. Nothing. If the Halifax had not released, would it come around for another pass? Would the second Halifax come soon? Had the navigator been unable to zero in on their location with his Rebecca device? Had the glider pilots hesi-

tated to release because they lacked visual confirmation of the landing
site? They waited several more minutes, no answers to their questions. At
last, another tone sounded in Haugland's headset.

"Number two is coming!" Haugland called out.

As before, the drone of engines cut through the night, this time from
the east. But the sound never grew any louder, nor did any gliders appear.
Over the next hour, the Eureka toned a few more times, and they heard
engines from several different directions. The Eureka continued to drain
life from its battery, and they did not have a way to recharge it quickly.
Then there was only silence. Poulsson, Helberg, and Kjelstrup eventually
shut off their flashlights and made their way back to Haugland. Even with
the poor weather, they could not understand how the planes could have
come so close and yet remained unseen.

Flying with the moon behind them, their visibility diminished, the crew
of Halifax A found it impossible to make out where they were on their
map. Every valley, mountain, and lake looked alike. They might have had
as much luck tracking a wave in an endless sea. By the navigator's calculations, they should have been within twenty to thirty miles of the landing zone. But they never saw any red L on the ground, and their Rebecca
was not working. In sum, they were wandering in the dark, and their fuel
gauges were running low. Cooper, who was in the cockpit, decided that
they should turn back to Scotland. It was approaching midnight, and after almost five hours of flying, the plane would just about make it home.

When they had first taken off, Wilkinson had steered the Halifax southeast from Skitten. He threaded through holes in the cloud layer until they
were in open, clear sky at ten thousand feet. The intercom between the
plane and the glider was not working, so they could only communicate
by Morse with lights. Otherwise, the journey across the North Sea was a
calm ride. Because of the high ceiling of clouds on Norway's west coast,
the navigator charted a course around the southern tip. But before they
reached landfall, the power to the Rebecca failed, and no matter what they
did they could not fix it. They would have to find the Skoland marshes by
sight alone.

East of Kristiansand they pinpointed their location, and Wilkinson
turned north, the idea being to follow a path of lakes toward Vemork. For
the first half of the journey, they kept a good bead on their position. There
was only a scattering of clouds, the moon shone brilliantly, and the val-

leys were clear of fog. However, the farther they went, the harder it was to mark where they were on their maps. They maintained course, but at roughly 11:00 p.m., the time they'd estimated they'd be over the target, they were unable to identify any landmarks between the clouds. Wilkinson turned to the east; Cooper hoped they would cross over some lakes from which he could navigate. When this failed, they veered southeast for twenty minutes until they spotted the coast. Then they reversed direction back toward Vemork. It was then that Cooper decided to head back to Scotland.

Having set a new course, they found themselves enveloped in clouds at nine thousand feet. When Wilkinson adjusted the controls to ascend, the glider in tow behind, the aircraft failed to respond. Ice was beginning to form on the wings of both the plane and the glider. At full throttle, engines roaring, the plane finally rose. They reached twelve thousand feet, but the Halifax could not maintain its altitude nor its speed. It dropped back down into the clouds. The four propellers flung off ice in shards that crashed into the fuselage with the same terrifying sound as antiaircraft fire. Cooper knew there was no choice. They needed to get to a lower altitude to clear the ice, or they would not make it. Wilkinson dipped down to seven thousand feet, but turbulence was even worse at the lower altitude. In thickening clouds, the plane shook violently.

The Horsa glider rocked back and forth, surged upwards, and plummeted down, its two pilots at the mercy of the towline. The sappers in the back were tossed about in their seats. The wooden fuselage creaked and groaned, threatening to rip apart at any moment. One minute of this terror followed another, and no prayer to God could make it stop. Then they lurched ahead one final time — and the icebound towline snapped. It was 12:11 a.m., November 20. Halifax A disappeared into the clouds, and the Horsa began a precipitous spin.

The two Scottish pilots never had a chance. They already had very little control of their glider. In the dark, with scarce visibility and an unknown terrain, they were aiming blindly for their landing. The glider came down fast, the wind shrieking through the fuselage. Although strapped to their seats, the sappers in the back might as well have been riding on the back of a bucking bronco. Their equipment was flung through the air. The pilot called out, "Ditching Stations!" and the men hooked arms to brace for the landing. There was little hope that would do any good; they were essentially dropping through the sky in a wooden box. Soon after breaking loose from the Halifax, the glider crashed into the mountains. The pilots died instantly, the glider's glass nose providing their bodies with no protection. Six sappers perished in the crash as well. Of the nine survivors, most were too badly injured to move. A few managed to crawl out into the snow. The glider's wings had been sheared off, the fuselage broken apart. Their gear had spilled out all over the mountainside, and the subzero temperatures bit at their skin. They had absolutely no idea where they were.

Flight Lieutenant Parkinson banked Halifax B to the east as he searched for his own glider, which had also broken loose. Earlier, they too had circled around the landing zone, in their case with their Rebecca device sending out its signal, but they had failed to zero in close enough on the target to release their tow. Running low on fuel, Parkinson aborted the operation and decided to return to Scotland. He experienced the same treacherous, thick layer of clouds, and at 11:40 he lost his glider near Egersund on the southwest coast. To better see where it might have landed, he lowered altitude.

Crisscrossing over the valley, he and his crew tried to locate the glider in the darkness. Suddenly, they found themselves staring straight out at Hæstad Mountain. Pushing into full throttle, Parkinson attempted to maneuver the Halifax away but failed. The plane clipped the top of the mountain with terrible force, throwing the rear gunner from the aircraft. Still traveling at great speed, the Halifax hurtled over the summit, then across a plateau littered with huge stones that tore it apart over the course of eight hundred yards. The bodies of six more crew members, some eviscerated, others with their limbs ripped off, were scattered about the flaming wreckage.

Four miles away, across the valley, Glider B rested on its side in steep mountain forest, its nose sheared off, its two pilots dead. Their success in landing in the dark and fog had saved the lives of all but one of the fifteen sappers. As the weather worsened, the fourteen surviving Royal Engineers tended to one another's injuries as best they could, and wrapped their dead crew members in their sleeping bags.

Lieutenant Alexander Allen sent a pair of his men to find help. They scrambled down the hillside. Through severe gusting winds and showers of ice and snow, they slipped and fell as they traversed the rough terrain. Finally they reached Helleland village. Trond Hovland, a man in his midthirties, answered the knock on his door. His father, Theodor, the local sheriff, joined him soon after. Neither spoke much English. The two sappers tried to explain that their plane had crashed. They asked for help and wanted to know how far they were from Sweden. Very far, they learned.

Sheriff Hovland volunteered to organize a rescue party but said that he would also have to alert the German command in Egersund, ten miles away. It would be impossible to keep their help a secret. The two sappers agreed. They knew there was no way that they and their battered and half-frozen mates would manage an escape to Sweden now. They would have to surrender to the Germans.

An hour later, several Norwegians and a patrol from the German garrison at Slettebø arrived at the house. One of the sappers remained under guard in a sitting room. The other led the patrol and its Norwegian guides into the mountains. At 5:30 a.m. they came upon the crash site of Glider B. Allen and the others had decided to surrender, even though they were heavily armed and could easily have surprised the dozen approaching Germans. The German lieutenant promised Allen, whose men were in uniform, that they would be classified as prisoners of war and that a doctor would tend to the injured. The sappers offered their captors cigarettes. The courtesy would not be returned.

Through the night, in the cigarette-smoke-filled RAF Operations Room at Wick airport, five miles southeast from its satellite field at Skitten, Henniker awaited news. The first sign that the mission was steering toward disaster came when Halifax A radioed that its glider had been released at sea. Henniker tried to scramble planes for a search, but none would be ready until morning. A flurry of messages followed. Confusion reigned as to where exactly the glider had broken off. Another transmission, this one from Halifax B, asked for a bearing on Wick. This plane's location was unclear as well, though some determined it was over the North Sea. No further contact, however, was made with its crew.

At 3:00 a.m. Halifax A landed. Cooper was driven immediately to Wick to give his report on the flight. He explained that he had radioed that his glider had been released off the Norwegian coast so as to throw off potential German patrols. At dawn, when the first search planes set out, the assembly in the Operations Room was convinced that the other Halifax must have crashed.

There was still a chance that Glider B had made it to the landing site. "If

Grouse does not call up, it will probably mean there is a party on," an SOE message read. At noon, however, the Grouse agents reported by wireless. They had been ready at the site at the appointed hour. They had received a signal on their Eureka and had heard the sound of engines overhead, but no gliders had arrived.

Late in the afternoon, Henniker paced the deserted Skitten airfield beside the lone Halifax that had returned. Its four propellers were still and quiet. Most of the search planes that had been sent out had come back. None had reported any sign of the other Halifax or its glider. Henniker had to accept the dark fate of his men and their mission. The only question now was how many of them were still alive. Some were sure to have been captured. Others might have escaped. It was possible.

That same afternoon, on that blackest of days, Tronstad remained with Wilson in his Chiltern Court office. Reading through the Freshman messages, he could not help remembering the meeting with Henniker several days before the sappers had left, when the prospects for success had been so high. Now those plans were shattered. Operation Freshman was a disaster. Those aircrews were lost. Those brave young sappers, many of whom Tronstad had come to know at Brickendonbury Hall, were yet another terrible sacrifice in this awful war.

In silence, Tronstad and Wilson contemplated the giant map of Norway on the wall. It was dotted with symbols of operations in progress by Kompani Linge. Some would go well. Others badly. That was the nature of things. Undaunted by the disaster of Freshman, the two men were determined to learn from it and to try the Vemork operation another way. There was no other choice: they needed to stop its production of heavy water or the threat of greater losses — unimaginable losses should the Nazis obtain a bomb — might become real.

The two men could not know whether the Germans had discovered the sappers' target. If they had, the risks for the next operation would multiply. The Nazis would crack down on any person in the area, putting Skinnarland and the entire Grouse team in jeopardy. Wilson made it clear to Tronstad that it was unlikely Lord Mountbatten would try to send another team of Royal Engineers after this tragedy. Which, they agreed, was for the best. A small group of commandos would have the best chance of slipping inside and destroying the plant, they decided. They should be Norwegian, comfortable with and able to navigate the winter terrain.

They would be dropped into the mountains by parachute, if possible by the next phase of the moon, to hit Vemork and get out.

Their plan came quickly together that afternoon. Over the rest of the day, they set about mustering support for it. Wilson telephoned one of the officers in charge of Freshman at Combined Operations HQ. He expressed how sorry he was for how the mission had unfolded, then asked if the SOE could "take over the job." There was no hesitation. "Thank God for that," said the officer on the other end of the line.

That evening, Wilson met with Major General Gubbins at 64 Baker Street, SOE headquarters, a few hundred yards away. At first his boss was skeptical, but Wilson persuaded him that he and Tronstad had a good plan — and the right men to execute it. Convinced, Gubbins immediately sent a letter to Mountbatten's deputy: "We consider that we might now be able to attempt the operation ourselves on a smaller, but, we hope, effective scale by SOE methods before the end of the year . . . From the point of view of the scientists further delay might be dangerous." Tronstad also met with his superior, General Wilhelm Hansteen, who recommended that he work closely with SOE to deal "with the same problem by other means."

That evening, Tronstad messaged Grouse: "Your work has been done magnificently. Change in weather meant gliders had to be released over 100 km from target. Operation cancelled for this moon period. We are planning to effect it with our own men next moon."

The following day, November 21, the BBC received an official German communiqué boasting about the elimination of the saboteurs. "During the night of 19th–20th November, two British bombers, each towing a glider, flew over southern Norway. One of the bombers and both the gliders were forced to land. The sabotage squads brought by them were engaged in combat and finished off to the very last man." Tronstad was certain that the report was a lie but could not deny that it held more than a grain of truth.

Mountbatten informed the prime minister of the mission's outcome. Churchill, who knew too well the tragedies and setbacks of war, wrote a single word on the report: "Alas."

The German communication was indeed a lie, and it obscured the even uglier facts. After Lieutenant Allen surrendered, the fourteen sappers of Glider B, some too severely injured to walk, were brought down from the mountain and loaded onto two trucks. One of the men flashed a V sign to Sheriff Hovland before being driven off to Slettebø, ten miles away.

Walther Schrottberger, the Wehrmacht captain in charge of the Slettebø garrison, did not know what to do with them. Clearly, his prisoners were British soldiers. Given that their khaki uniforms were without insignia, and given that explosives, radio transmitters, insulated wire shears, Norwegian kroner, and light machine guns were collected at the crash site, they were clearly saboteurs as well. Unwilling to decide the men's fate, he rang his superior in Stavanger, Colonel Probst, who contacted his divisional head. According to the *Kommandobefehl,* "no quarter should be given" to any commandos. Hitler's orders stated that enemy agents falling into the hands of the Wehrmacht were to be delivered "without delay" to the Security Service's intelligence-gathering arm, the Sicherheitsdienst (SD).

While orders on how Schrottberger should handle the situation made their way through the chain of command, the Gestapo in Stavanger got word of the British prisoners. They sent SS Second Lieutenant Otto Petersen to the Slettebø garrison. Petersen, who was known by local Norwegians as the Red Devil, wanted the saboteurs placed in his custody. On orders from Probst, Schrottberger refused. Instead Petersen gave him an hour to interrogate the British. One after the other, the sappers were brought before the Gestapo officer, harangued, beaten, and threatened. They revealed only their names and ages.

Then, in the late afternoon, Schrottberger and his squad of soldiers led them out the garrison gates to follow through on Hitler's commando order. The sappers were brought north along the road from Slettebø. When they reached a sparsely forested valley pocked with boulders, they were spaced out along the road, two soldiers guarding each British prisoner.

The first sapper was brought over to a slight hill beside some granite and concrete sheds. An execution squad came out from behind one of the sheds, fifteen feet away. They raised their rifles. "Feuer frei!" came their order, and a hail of bullets tore into the soldier. After he fell, the squad commander, armed with a pistol, put a final bullet in the soldier's head. As the body was hauled away, the squad disappeared back behind the shed.

The next prisoner was brought forward and this routine played out again. One prisoner pleaded for mercy, showing the Germans a picture of his wife and two children. They shot him anyway. Another, too injured to stand, was seated on a rock. Then killed in the same way. Bill Bray stood before the rifles, knowing he would never see his wife again nor meet his unborn child. Then the bullets came.

Bodies still warm, the fourteen Royal Engineers were stripped to their underwear and brought to a beach where Polish prisoners of war buried them in a shallow trench of sand next to a line of concrete antitank barriers known as Hitler's Teeth. The pilots and the sapper who had died when the glider crashed were tossed in the trench as well.

Heinrich Fehlis was incensed at the executions of the British soldiers. He wanted all Wehrmacht officers involved brought up on charges, and a damning note was sent to his chief in Berlin. "Towing aircraft's crew is military, including one Negro; all dead. There were seventeen men, probably agents . . . Glider's crew was in possession of large sums of Norwegian currency. Unfortunately, the military authorities executed the survivors, so explanation scarcely is possible." Heinrich Himmler, Reichsführer of the SS, was alerted to the lost intelligence opportunity as well.

When it became known, on November 21, that there were nine survivors of a second glider crash, Fehlis wanted to make sure that every bit of intelligence was wrung from them. The thirty-six-year-old lieutenant colonel, with a lantern jaw and thin, bloodless lips, was known for his rigid efficiency and even temper. He had none of the charm of his Gestapo bloodhound, Siegfried Fehmer, and his subordinates only knew he was angry when the saber scar on his left cheek turned livid. For a man of his young age, Fehlis held extraordinary power over life and death in Norway, and he meant to use it.

Born in the industrial city of Wuppertal, he was one of four children. Their father had died from injuries sustained in World War I. Fehlis studied law, and joined Hitler's Brown Shirt paramilitary arm in 1933. From the start, he made sure that his superiors knew how devout he was to the cause. He renounced any ties to the Catholic Church. His wife, who had borne him only a single child despite numerous visits to various doctors, was a disappointment; after apologizing for her to the SS, Fehlis joined their Lebensborn association, where he bred with "racially pure and healthy women" to strengthen the Aryan race. His personnel file read: "Overall race impression: very good, Nordic; appearance: very correct and according to SS-standards. Very quiet and secure, ambitious, reliable, and good at negotiations."

In Norway, Fehlis was head of German security services, including the Gestapo, Kripo (criminal police), and the SD. Although close to Ter-boven, he was also charged with spying on him for their masters in Germany.

Only General Falkenhorst seemed to not be in favor of Fehlis's mete-oric rise. The grizzled Wehrmacht veteran made it clear that he found the young man too immature. Now Fehlis would see that Falkenhorst's army would not interfere in the investigation. Pressured from Berlin, the general notified his troops that the saboteurs were to be delivered directly to the SS for thorough interrogation. Fehlis was told that five saboteurs from the second downed glider were in good enough shape for questioning. He ordered them brought to Oslo, along with everything collected at the crash site. The other four survivors of the crash, he said, could be executed forthwith.

James Cairncross, Paul Farrell, Trevor Masters, and Eric Smith — all in their twenties, all married, apart from the boisterous Cairncross — were unsure of when, or if, they would be brought to a hospital. It was now Monday, November 23, and they were still in a Stavanger prison cell, lying on stretchers and in considerable pain. After the crash landing of Glider A, they had spent the night on the mountain, suffering from their injuries as well as frostbite and exposure. In the morning, several in their party left to find help. Descending a steep hillside, they came to a farm. A group of Norwegians, including a doctor, came up to assist them. The sappers learned that they were on the northern side of Lysefjord, ninety miles southwest of Vemork and hundreds more to the Swedish border. On hearing the news, they were unable to conceal their anguish.

A German patrol, and several Gestapo, arrived at the crash site soon after, weapons drawn. They thoroughly searched the glider, then brought the nine surviving sappers down to a coastal patrol ship. This took them across the fjord to the prison at Stavanger. The injured four were separated from the others. A Luftwaffe doctor, Dr. Fritz Seeling, visited them but left before tending to their cracked skulls, broken ribs, and shattered legs and arms. Then they were left to wait.

On Monday afternoon, Seeling returned to the cell, this time with a Gestapo officer. It was Petersen, the Red Devil. The doctor carried some syringes and bottles labeled Typhus. Petersen told Cairncross and his mates that the doctor was going to inoculate them. Given the extent of their injuries, a typhus shot was the last thing they needed, but they were too weak and in too much pain to resist. The doctor delivered the injections, then he and Petersen left.

Some time later, the door clanged open. Petersen and two prison guards took three of the sappers down to a first-floor office. There, Seeling gave them another shot. The syringes all contained morphine. Fehlis wanted them dead, and Seeling had been commissioned to do it. One of the soldiers succumbed after his third injection. Seeling and a prison guard tried to carry out his body, but the other two prisoners refused to let go of their friend. One of them, sensing that something was very wrong, began shouting. His voice resonated out into the halls. Petersen instructed Seeling to give him another shot. Seeling hesitated, not sure what to do. In a short time, Seeling assured him, the prisoner would be dead anyway. This was not quickly enough for Petersen. He and a prison guard wrapped a leather belt around the man's neck and tied the end to a radiator. Then they strangled him to death. After witnessing this horror, Seeling injected air into the veins of the other prisoner to speed his death. Whether it was this act, or the blow to the neck he received from Petersen's boot, or the morphine that killed him in the end, Seeling did not know. The fourth prisoner, the one who had been left in his cell, was taken by car to Gestapo headquarters, where Petersen pushed him down some cellar steps, then shot him in the back of the head. That night, Petersen and several others took the four bodies out to sea. They tied a heavy stone to each corpse, then tossed them overboard.

The five other sappers who had survived the crash relatively unscathed were dispatched to Oslo for interrogation — and torture, should it be required. Fehlis already knew most of what he wanted to know. Among the gear scattered around the glider, the patrol had found a folded silk map with a planned escape route. Circled in blue: Vemork.

Part III

11

The Instructor

P OULSSON AND HIS MEN packed up their gear in the Lake Sand cabin. Their latest instructions from Home Station were to retreat into the Vidda as soon as possible. "Sabotage troops were engaged and annihilated . . . It is vitally necessary that you should preserve your safety . . . It is almost equally important that we should have earliest possible information in regard to increase of enemy troops in neighborhood of target . . . Advise you to move yourself and your station . . . Keep up your heart. We will do this job yet."

Reports from London of the failure of the glider operation, and the deaths of all those men, struck the Grouse team hard. They wondered how the operation had gone so wrong. They questioned what else they could have done to bring the planes in to the landing zone. Could they have sent more precise weather reports? That there were clear skies for the next two nights made the disaster even more bitter to accept. Their only solace was that the mission to sabotage Vemork had not been canceled. According to another cipher message from Tronstad, the next attempt was to be by men in their own company. It was set for mid-December. Poulsson assured Tronstad that his team would do anything they could to help.

For now, the four commandos needed to get away. On the night of November 22, they left Lake Sand. From Skinnarland, they had a freshly charged battery for the radio and the key to a cabin on the Vidda owned by Olav Skogen. They skirted around the German troops stationed at Lake Møs and trudged a dozen miles northwest to Grass Valley, deep and high enough into the Vidda that few dared venture there in wintertime. The cabin was surrounded by nothing but snow and a few scraggly juniper bushes that struggled to survive in the windswept hills. Inside, they found some salted reindeer in a barrel, but otherwise it was as barren and cold as an ice locker.

The next day, their rucksacks emptied of all but the essentials, they headed west toward the Songa Valley, where they had first parachuted in over a month before. They needed to pick up the food and supplies they had left in their depot. On the way, they spent a night in a dilapidated hay barn. The next evening, they reached their landing site and dug a snow cave for sleeping. The morning of November 26 welcomed them with a misty fog, and they searched for several hours in the deep snow before finding their containers. The stores of food were limited — some sacks of coffee, sugar, and flour — but they were desperate for it. Starvation and the tempestuous Vidda were their enemies now.

After dividing up the supplies, Haugland and Helberg skied east, back to Grass Valley, their packs filled with most of what they had found. Poulsson and Kjelstrup went in the opposite direction. Back in Scotland, Knut Haukelid had given them the names of some locals, including his cousins, who lived around his family's mountain farm. They would help build up underground resistance cells in the area.

That night, a blizzard hit.

Crossing an ice-blue lake that had been blown clear of snow, Helberg and Haugland became separated. Alone in the howling, blinding storm, each man struggled forward in the darkness. Helberg, whose skis had steel edges, managed to make it to the other side without incident. Haugland did not have the same advantage. He was driven across the lake in whatever direction the ferocious gusts of wind happened to be blowing. At one moment, unable to get a grip on the ice, he found himself being hurled toward a patch of open water at the lake's edge. Stabbing the ice with his poles to steer away, he narrowly avoided a plunge that would certainly have killed him.

To the west, Poulsson and Kjelstrup were also at the mercy of the treacherous storm. They skied down into a gully but each time they tried to climb out on the opposite side, the wind threw them back in. Hounded by squalls, they were forced to crawl on their knees to keep moving. At last they came across a shack in the valley. The dirt-floored shelter barely fit the two men, but it was a retreat from the wind. They found a half-rotten reindeer shoulder on the wall. Starving, they cut off strips of the flesh, combined them with a shard of pemmican, and called it dinner. In the morning, they managed to push the door open far enough to discover that the shack had been buried in the storm. Handful by handful, the shack filling up with snow, they cleared a path and eased the door open. Then Poulsson squirreled up and out into a bright, clear day.

On November 24, back in Scotland, Joachim Rønneberg was called into Drumintoul Lodge, Kompani Linge's headquarters at STS 26. Before his British SOE commander, Major C. S. Hampton, the young second lieutenant stood to attention, all six feet three inches of him. With a long dome of a forehead, narrow-set gray-blue eyes, and the edged jawline of a silver screen star, Rønneberg had presence in spades.

"You're to be the leader of an operation," Hampton said.

Rønneberg was unsure what his commander meant. He already had a mission: Operation Fieldfare. "What is it?" he asked, his graveled voice belying his years.

"I haven't the faintest idea, but you're to pick five men for the job."

"Yes, but Lord God," Rønneberg said, all questions, "I have to know if we'll be operating on the coast or in the mountains. Are we going in by boat or plane? Will we be there a long time? Do they need to be good skiers?" He knew every member of Kompani Linge well, and there were many to choose from, depending on the conditions. "Any advice you can give me in my choice of men?"

Hampton shook his head. All he knew was that one of the five had to be Knut Haukelid, who would be Rønneberg's second in command. Why? Hampton gave no answers. A few days later, Rønneberg was summoned to London, where he hoped to find some.

When Martin Linge had sent Rønneberg back to Stodham Park to instruct recruits only weeks after he had finished his own training, new arrivals were shocked that someone so young, with no military experience, who seemed as "gentle as a lark" (as one said) was already an instructor. Then they got to know him. Rønneberg was intelligent, tireless, strategic, thorough in his preparations, and professional in every way. What distinguished him most was his innate ability to lead. "He had a quality that made him stand out alone," said one in his company, "without being envied, without making enemies or rivalries." He did not need to dominate, raise his voice, or entreat. He simply gave his best effort and inspired others to do the same. Wilson and Tronstad were confident that he was the man to lead this most important of jobs.

"Where are we off to?" Rønneberg asked straight out when he arrived by overnight train at Chiltern Court. He tended to speak with his whole body, his shoulders rolling back.

"To Vemork," Tronstad said. "To blow up the plant there."

Rønneberg was to organize, train, and lead a six-man team that would parachute into the Vidda. They would be met by the advance party, Grouse, hit the heavy water facility at Vemork, then escape on skis to Sweden. The operation was codenamed Gunnerside, the name of a hunting cabin owned by a chief SOE officer.

Neither Wilson nor Tronstad told Rønneberg why the Germans needed the heavy water, but apart from that detail they held little back, particularly about the cold-blooded killings of the Operation Freshman sappers. They wanted him and his team ready to leave by December 17 so they could arrive at Vemork on Christmas Eve. Any element of surprise was now unlikely. The Germans would be aware that the plant was a target, and the Christmas holiday might prove their best chance of finding diminished defenses.

It was a tight schedule: three weeks until launch.

Rønneberg already knew which four men he would select for the mission. He had instructed them all at STS 26, and he cabled them now from London, telling them to get down to some hard training in the Highlands.

Birger Strømsheim was his first choice, and the easiest one to make. Like Rønneberg, he was from Ålesund. With his wife, Strømsheim had escaped to Britain on a fishing boat. A building contractor with a broad, honest face and curled sweep of blond hair, he was levelheaded, quiet, and, as reported in his SOE file, "reliable as a rock." Able to fix or make just about anything, he was also a fine skier and tireless worker. The two had spent many hours together in Scotland planning their now-delayed operation to sabotage German supply lines. At thirty-one years old, he would be the grandfather of the group.

Next was Fredrik Kayser, a thin stalk of a man with clean-cut looks. He had trained with Strømsheim at Stodham Park, Meoble, and parachute school. Kayser grew up in a family that prized well-roundedness. And so he fished, played the mandolin, danced, rowed, played soccer, and joined the scouts — often finding that he was second-best at everything. His scouting carried him into the military, where he served in the King's

Guard. Afterward he returned to his hometown, Bergen, to take a job as a clerk at an ironworks factory. When Russia invaded Finland in November 1939 he volunteered to join a Norwegian squad to help push back the aggressors. He was shot at, suffered frostbite, put an ax through his leg while chopping wood, and still kept fighting. The day he returned from that war, Germany invaded his own country. He kept up the fight there too. When Norway surrendered, he took a fishing boat across the North Sea to Britain. Rønneberg knew few men who had proven themselves more adaptable and cheerful under pressure.

Third was Kasper Idland, a former postman from a small town outside Stavanger. In Britain, he had trained with Kayser and Strømsheim. Even as a boy, Idland had been taller and bigger than most. His mother warned him not to get into fights at school out of fear he would hurt somebody. Encouraged by his passivity, the other kids teased and tormented him relentlessly. Loyal to his mother, Idland did not strike back. After one particularly bad afternoon of being bullied, he asked his mother's permission to defend himself. She consented. The next day, Idland challenged the two worst offenders, knocking them about. He was never bothered again. As a soldier, Idland was intelligent, an excellent shot, tough, and what Rønneberg prized most of all, loyal.

His fourth choice was Hans "Chicken" Storhaug, a short, pencil-necked twenty-seven-year-old. Although he was not the deepest of thinkers, there were few better in the woods, hunting or skiing. What was more, he came from Hedemark, a district bordering Sweden that the team would need to cross when making their escape. A man who knew the terrain would give them an advantage.

The fifth member of his team, Haukelid, was cause for concern. His help would be key: he knew the area and was close to the advance team. But he was considered a loner by some within Kompani Linge, and Rønneberg worried about whether he would fit in as second in command on the mission.

The two shared a deep, unbending patriotism that had driven them both to travel to Britain to join the fight against the Germans, a risky move taken only by a small number of Norwegians. Both had been tapped by Linge to join the elite Independent Company. But in many other ways — age, temperament, and experience — a vast gulf separated the two. Unlike Haukelid, who was eight years his senior, Rønneberg was not a rebel. He did not bristle under authority, nor did he seek to break the rules at every turn. Further, Haukelid was a veteran of the resistance fight and the battle for Norway. Given Rønneberg's military inexperience, beyond field exercises in Scotland, he wondered how Haukelid would take to not leading the mission.

Regardless, Rønneberg suffered no hesitation on his return to STS 26. He gathered the five men together and said, "Now, I do not know you all equally well personally, but if there's any disagreement between you, then put it aside until we're done with the job. Or get out." The men stayed where they were.

He then outlined the operation and its purpose. Whatever this heavy water was, it must be important if SOE wanted to make another attempt on a target that had already cost so many lives. They were to understand that, if they were caught, the Germans would show them no mercy. Afterward Rønneberg drew them aside, one by one, to give them the opportunity to back out. No takers.

In a clearing in the woods six miles northwest of Oslo stood the Grini concentration camp. The long, five-story rectangular brick building at its center, formerly a women's prison, was crammed with thousands of Norwegians, most arrested by the Germans for political crimes. Two sets of high, barbed-wire fences surrounded the grounds, and guards with machine guns watched over the prisoners from several towers.

Gestapo officer Wilhelm Esser entered the cell of the five British sappers who had survived the glider crash almost two weeks before. Stripped of their uniforms, they were dressed in blue trousers and wool jerseys. One had his arm in a sling. Another had a bad burn across his forehead. Otherwise, Wallis Jackson, James Blackburn, John Walsh, Frank Bonner, and William White were in good shape. Using an interpreter, Esser began his interrogation with one simple question: What was the object of your mission? The men refused to say. Esser then spread out several of the maps found at the crash site, their target clearly marked. He asked his question again.

Over five days and five nights he interrogated the British soldiers, sometimes together, sometimes separately. He promised them they would be treated as prisoners of war and sent to a camp in Germany if they told him everything they knew. When promises failed, Esser had methods that the Gestapo had perfected to get men to talk. And typically they talked.

When it was done, Esser sent Fehlis his report, and the details of the evidence and information collected on the operation were delivered to Berlin. The men were Royal Engineer paratroopers. The month before, they had been separated out from their company at Bulford and trained to blow up Vemork. Their plan was to land several miles from the target, and after crossing the suspension bridge, a lead team was to silently take out the guards on the bridge, using knives or manual assault. One group was to disable the plant's generators, another to sabotage the electrolysis plant, and a third to destroy any "special liquids." All of this was to take eight minutes. Then, in teams of two to three men, they were to escape to Sweden in civilian clothes. No cooperation with Norwegians had been envisaged.

Fehlis was unconvinced that Norwegians had not been recruited to assist in the operation. He ordered the Gestapo to move into Rjukan and the surrounding area, supported by hundreds of Wehrmacht soldiers.

Every day at the Kummersdorf testing facility, a small army of scientists, engineers, skilled tradesmen, and laborers, all exempt from service at the front, showed their IDs to the sentries at the gate. Then they headed in to do their day's work. There were five groups of buildings staggered about the estate. Each had its own workshops, laboratories, warehouses, and offices, dedicated to designing the future weapons of the German army. These brick structures, their slanted roofs planted with grass to hide them from Allied air raids, were connected by shatterproof corridors. To transport the workforce and material, a narrow-gauge trolley ran around the perimeter of the grounds, past test sites where scientists tried out the latest weaponry. The facility had its own fire brigade, water supply, power plant, and infirmary.

In late fall 1942, in two rows of buildings at Kummersdorf, Kurt Diebner and his handpicked team of young experimental physicists and engineers labored over their first uranium machine, the Gottow I (G-I) experiment, assembled with whatever castoff materials they could procure. Inside a cylindrical aluminum boiler, eight feet in diameter and eight feet high, they stacked honeycombs made of paraffin wax around an empty center. The team then spooned uranium oxide into each cell of the honeycomb, wearing heavy, breath-stealing gear to shield themselves from the uranium's toxicity. The work, which took weeks, was "dreadful drudgery," one participant said.

When complete, the honeycomb was made up of nineteen stacks, with

6,902 uranium oxide cells, weighing almost 30 tons (25 tons of uranium, 4.4 of paraffin). The team, tired of wartime rations, joked, "If only all of it were lard!" Ready to test the design at last, they lowered the boiler into a mantle of water, and Diebner told his men to insert the radium–beryllium neutron source into the center of the honeycomb.

Since the meeting in June with Speer and the generals, Diebner had been largely working in secret to see if his design had more merit than those of the others in the Uranium Club. In early summer, Heisenberg and his research partner, Robert Döpel, had demonstrated an increase in neutrons of 13 percent in their fourth experiment, a machine with two spherical layers of powdered uranium metal submerged in heavy water. Heisenberg claimed their machine generated more neutrons through fission than it absorbed.

Then, on June 23, bubbles rose to the surface of the heavy water tank in which their machine was submerged. When they lifted the machine from the tank and inspected it, flames shot through an opening in the outer sphere. They immediately dropped it back into the water. It was clear to them that hydrogen gas was leaking from the vessel. While Heisenberg and Döpel were trying to figure out what to do, the machine's aluminum shell swelled like a balloon. They ran from the laboratory, escaping just moments before the machine exploded. Streams of flame and red-hot uranium powder shot through the ceiling. A call to the fire brigade followed, then a cascade of barbs about their success in building the first atomic bomb.

The disaster did not negate their experiment's achievement. Still, Diebner thought their design was inferior to his own because the fast-moving neutrons in Heisenberg's layered machine could only escape the uranium mass into a moderator in two dimensions. Diebner's cube design made the neutrons move through heavy water in three dimensions, which increased the probability they would be slowed down to a point where they would split other U-235 atoms rather than be absorbed by U-238 — or lost outside the machine.

Near the year's end, after months of labor, his theories proved correct when his team tested the design. His G-I machine beat the neutron-production rates of machines like those Heisenberg had built. Up until this point Diebner had used cheap, spare materials like uranium oxide and paraffin. Now he asked the Reich Research Council for pure uranium metal and heavy water. With these, he promised he would make fast progress.

On December 2, unbeknownst to Diebner and any other physicist in the German atomic program, the Americans realized their first self-sustaining reactor. In a soot-black squash court underneath the football stands of the University of Chicago's Stagg Field, Enrico Fermi and his team listened to the rapid clickety-clacks of their neutron counters as their twenty-foot-high stacked pile of graphite bricks, many hollowed out and filled with uranium (metal and uranium oxide), went critical at last. After four and a half minutes, the pile was producing half a watt of energy — increasing every second. Fermi calmly called out to his assistants, "Zip in!" to halt the steady multiply of splitting atoms. They returned several cadmium rods completely into the pile, which soaked up the bombarding neutrons and brought the machine under control again. As one physicist involved in the momentous day said, "Nothing very spectacular had happened. Nothing had moved, and the pile itself had given no sound . . . We had known that we were about to unlock a giant; still, we could not escape an eerie feeling when we knew we had actually done it. We felt as, I presume, everyone feels who has done something that he knows will have very far-reaching consequences which he cannot foresee." Fermi and his team celebrated with a paper-cup toast of Chianti. Now that the self-sustaining reactor was no longer the stuff of fiction, the United States and its allies forged ahead with even greater fervor toward the creation of an atomic weapon.

12

Those Louts Won't Catch Us

MINUTES BEFORE DAWN on December 3, air-raid sirens blared throughout Rjukan. Residents awakened to find German soldiers marching through their streets. The Gestapo and several hundred Wehrmacht troops had arrived in the night on motorcycles and transport trucks, sealing off the Vestfjord Valley. Brandishing machine guns, they stomped from house to house, building to building. Soldiers stood on street corners and on bridges. Bundled in heavy coats, their breath frosty in the air, they looked none too happy to be there. The Germans ordered all residents to remain inside their homes — or they would be shot.

Throughout the town, the same scene played out over and over again. A hammering on the door. Shouts to open up. Soldiers, sometimes led by Gestapo officers, storming into the house, asking for the names of everyone who lived there, rummaging through rooms, closets, and chests, looking for illegal material — guns, radios, underground newspapers. If the Nazis discovered any contraband, they arrested the occupants and took them away by truck. Often, the soldiers broke furniture, punched holes through walls, and stole whatever food they could find.

Rolf Sørlie, a twenty-six-year-old construction engineer at Vemork, pleaded innocence when they came to his house, using the fluent German he had learned while studying in Leipzig. A short, slight young man with fair hair, Sørlie had been born with a condition that left the muscles of his hands and feet in constant spasm. Surgery had corrected his twisted feet, but not his hands, which were stuck in a half-clench. Nonetheless, he had never let his disability stop him from wandering the Vidda with his friends Poulsson and Helberg.

As soldiers filed into his house, past his younger brother and the housemaid, Sørlie tried to conceal his fear. Not only was he close to the local Milorg leader, Olav Skogen, but there were two radios hidden in his attic. Fortunately, the soldiers gave the house only a cursory search, either out of laziness or a lack of suspicion about this young Norwegian who spoke such nice German. They left without incident.

Down the street, Sørlie's friend Ditlev Diseth, a sixty-seven-year-old pensioner who had formerly worked for Norsk Hydro and now fixed watches, was not so lucky. Diseth was also a member of Skogen's Milorg cell, and the Gestapo found a radio and weapons in his house. The Germans hauled him away, along with twenty-one other Rjukan residents, including several members of the Milorg cell. All were to be questioned and, if warranted, brought to Grini for further interrogation.

Hans and Elen Skinnarland threw a little party that day to celebrate their son Olav's thirty-second birthday. Einar, who was working at the Kalhovd dam, was the only member of the family not present when the Gestapo roared up on their motorcycles. They arrested Torstein, thinking he was the Skinnarland rumored to be involved in resistance work. Nothing was said about Einar, but the family knew the Gestapo would soon mark him for arrest as well.

On December 8, Knut Haukelid arrived at Kingston House with a leg of venison slung over his shoulder. He was not the first Kompani Linge member to bring Tronstad a portion from a luckless animal that had "strayed into protective fire" at STS 26. Over the past year, the two had met a few times over drinks, to talk about building up Norwegian resistance forces. They shared the view that now was the time for ambition, not timidity. Operation Gunnerside, Tronstad knew, was particularly ambitious. Although Haukelid did feel that given his extensive experience, he should be the one leading the mission, not Rønneberg, an order was an order, and he was man enough to swallow his pride if it meant returning to Norway at last.

Tronstad greeted him warmly and thanked him for the meat. Then he turned in his chair and reached down into his safe, taking out a folder stamped TOP SECRET. He showed Haukelid a few of the diagrams and drawings of the Vemork plant. "Heavy water is very dangerous, you know," he began. "It can be used for one of the dirtiest things man can make, and if the Germans get it, we shall have lost the war and London will be blown to pieces."

Haukelid was not sure what to think of the possibility of such a weapon, but he made it clear that he was committed to the Gunnerside mission. He had come to London to talk about what happened next. He did not intend to make his way to Sweden; rather, he wanted to follow through with Grouse's original mission: establishing a base in western Telemark and recruiting guerrilla groups. If Poulsson chose to stay, he could command eastern Telemark. The other Grouse members could be split between the two regions.

Tronstad wasn't sure. He had helped develop this earlier plan, but things had changed. The Germans had launched an extensive manhunt after Operation Freshman — one could only imagine what they would do if the heavy water plant was actually blown up. "They will do all they can to catch you," Tronstad said. "We can't run such a risk as to have our men operating on the Vidda."

Fearing that his request might be denied, Haukelid grew desperate with emotion. "They won't find us. We're used to the mountains. We can live in the wilderness. I won't come back to England. I shall never come back again however long the war may last." Finally, he vowed, "Those louts won't catch us!"

Tronstad agreed to think about it.

Since Haugland informed Home Station that Swallow — the new codename the SOE had given Grouse, for security purposes — was heading up into the mountains after the glider disaster, there had been no further wireless communication. For more than two weeks, Wilson and Tronstad feared that the four had been caught up in the razzia at Rjukan. From what little was reported, the Germans were everywhere in Telemark, searching villages and setting up roadblocks. Terboven and Falkenhorst had made a great display of checking on Vemork's defenses, and a state of emergency that further limited travel had been declared. It was clear the Germans had learned the target of the glider operation, intelligence no doubt extracted from the captured Royal Engineers.

On December 9, station operators at Grendon Hall received communications from the Swallow team at last. "Our working conditions difficult," the first message began. Then came details of ski patrols searching around

Lake Møs and the setting up of a Gestapo D/F station near the dam to locate any wireless radios operating in the area. A second message reported Torstein Skinnarland's arrest. The Home Station boss instructed his operators that any traffic from Swallow was of the "highest possible priority," to be delivered to Tronstad and Wilson in London immediately.

A few minutes before 10:00 a.m. on December 10, the telephone rang in Olav Skogen's office at a Norsk Hydro factory in Rjukan. It was Gunnar Syverstad, calling from Vemork: "Four Gestapo are on their way to Kalhovd." Skogen knew what this meant: the Germans, who had been unraveling the resistance networks throughout the area over the past week, were finally closing in on Einar Skinnarland. Skogen immediately called the Norsk Hydro office in Kalhovd. He was told that Skinnarland had left the day before to visit his parents at Lake Møs. Skogen asked the switchboard operator to ring the dam-keeper's house, but was told that this was not possible; the Germans had suspended that line. Skogen feared the Gestapo would soon be on their way to the Skinnarland home, if they were not already. In the factory, Skogen tracked down Øystein Jahren, his main courier and a relative of the Skinnarland family, and asked him to take a bus up to Lake Møs to warn Einar. If anyone asked him why he was there, he was to say that he was buying some fish for a special dinner.

Jahren hurried from the factory and, with only seconds to spare, caught a bus at the depot. When he arrived at Lake Møs, there were no Germans in sight. He knocked urgently on the door of the Skinnarland house. Elen answered. Jahren told her that the Gestapo was on its way for Einar. He was not at home, she said, but she would warn him. His duty done, Jahren left, eager to be gone before the Germans came.

The bus taking him back to town had traveled only a few hundred yards when it was stopped by German soldiers. Several Gestapo officers, who had been surveiling the Skinnarland home, boarded the bus and arrested Jahren.

At that same moment, Einar Skinnarland was skiing up into the hills behind his family home. Despite his mother's words to the contrary, he had indeed been at home when Jahren called, and had fled out the back door. Speeding through the woods, he headed for a remote cabin known as High Heaven. It was owned by his brother-in-law. Unless the Germans were skilled cross-country skiers and trackers—and very lucky—they would never find him there.

Once safely at High Heaven, Skinnarland relaxed, shaken by his close escape. For the first time since beginning to live a double life, seven months before, he was a wanted man. Worse was the arrest of Torstein, who had acted as his cover. Skinnarland knew that he needed to inform London of the heightened security around Lake Møs and Vemork. He also had a decision to make: whether to head to Sweden and on to Britain or remain in Norway, living on the run. The next day, Helberg and Kjelstrup arrived at the cabin, having learned of Skinnarland's narrow escape from the Nazis. The two stayed the night and told Skinnarland of the new mission to sabotage Vemork. Skinnarland's decision was made.

In the late afternoon of December 11, the six Gunnerside men arrived at the ivy-clad Brickendonbury Hall. It was a week before their scheduled drop, and there was an uneasy feeling among them; only weeks before, the British sappers who perished in Operation Freshman were trained at this same school, by the same instructors, for the same target. The fact that the school had been cleared of other students because of their mission's secrecy only added to the sense of foreboding. Waiting to welcome them on the manor steps was George Rheam, newly promoted to lieutenant colonel.

Rheam invited them to drinks and dinner that evening; then his adjunct led the six men to their second-floor dormitory room. On each bed was a kit bag with the clothes and gear they would need for their training, including a factory-new Colt .45 with a red belt holster. To a man, the six took out their new guns and tested the action.

Rønneberg cocked his, but when he pulled the trigger the gun fired. As his ears rang and plaster dust fell about him, he realized that he had by mistake fired the loaded gun he had brought with him to Brickendonbury. One of the school's guards rushed into the dormitory, quickly followed by Rheam's adjunct. "What the hell's going on?" he asked. Straight-faced, Rønneberg looked up, pointed at the hole in the wall, and said, "I've tried my new weapon and it works perfectly." The adjunct shook his head and walked out. Crazy Norwegians.

It was a rare moment of carelessness for Rønneberg. In the ten days since being charged with Gunnerside, he had been punctilious in his preparations for the mission. He had gathered several sets of maps of southern Norway from the high command's intelligence office in London. One, at 1:250,000 scale, had been nailed to the wall at the base in Scotland and used to chart out the team's routes. The others, at 1:100,000 scale, were given to the team, so each man could memorize every valley and mountain.

Then there were their weapons. Given the close-quarter fighting they expected, short-range guns were going to be the most important. They tried out the simply designed British Sten submachine guns but found them too heavy and unreliable, often firing in bursts when set on single shot. They selected the Thompson (Tommy) submachine gun, made famous by American gangsters. Within two hundred yards, these could be fired as accurately as a rifle, and, since they took .45 caliber bullets instead of the Sten's 9mm, they could use the same ammunition for their Colt pistols. Rønneberg had the team meticulously clean their Tommy guns, dry-sand them, and paint them white. For the mission, he also asked for hand grenades, a sniper rifle, killing knives, chloroform pads to knock out guards, and lots of spare magazines of ammunition.

While his team trained and practiced shooting, Rønneberg gathered the equipment they would need to endure the harsh Vidda winter. When something didn't meet his exact specifications, he redesigned or altered it. He picked out the best wooden skis, had them sealed with fresh pine tar and then painted white. To improve the design of the steel-frame rucksacks, he added gaiter pockets, drawstring tops, extra-long shoulder straps, and white coverings. He found rabbit-fur-lined underpants, wool trousers, and white camouflage ski suits. He had the quartermasters change the leather linings on their peaked ski caps for khaki, which was warmer. A shoemaker in the East Midlands rushed an order of sturdy, waterproof leather boots.

For rations, Rønneberg went to Professor Leiv Kreyberg. Working with dieticians at Cambridge University, the Norwegian professor had pioneered a method of compressing dehydrated food into blocks to make combat rations. The meals were lightweight, and to prepare them one only had to add water. Using these kinds of rations meant they would only have to carry a cup and a spoon. Given how far they would have to travel in escaping from Vemork to Sweden, every ounce mattered. Rønneberg also had his Gunnerside team build two sleds to minimize what they would have to carry on their backs.

In his quest for mission-compatible sleeping bags, he went to a London bedding manufacturer near Trafalgar Square. The owner sent him to their workshop in the Docklands. There, he sketched out what he needed:

essentially two bags woven into one, the outer a waterproof shell, the inner filled with down and large enough to sleep in while fully dressed, with room for their gear. He also wanted a hood over the top with a drawstring tie that would close almost completely, allowing only a small opening for breathing. The head of the workshop took a look at his design and, recognizing the second lieutenant's urgency, said, "Well, you just come back tomorrow afternoon and see what I've done." The next day, Rønneberg climbed into the first prototype. It was perfect, apart from some seams that were stitched into the down that would allow water to penetrate. He made the design change, then ordered six bags to be picked up as soon as possible.

By the time Rønneberg and his men arrived at Brickendonbury Hall, they knew all the possible routes to and from Vemork, and most of their gear was ready to be packed for the drop.

Now they could focus on the operation itself. Rheam had taught hundreds of SOE operatives everything there was to know about incapacitating the German war machine by targeting their communication lines, railways, and factories. "Seven people properly trained can cripple a city," said one instructor.

Rheam wanted his students to be able to walk into a plant and, within minutes, identify which machinery to disable. A well-placed swing of a sledgehammer might do the trick, or some handfuls of sand in the machinery. Most often, though, explosives were required. Nobel 808, a rust-colored explosive that smelled like almonds, was Rheam's first choice. Soft and malleable, 808, or "stagger juice" (as some at STS 17 referred to it), could be cut, shaped, stretched, thrown against the wall, and even shot at, and it would not explode. But set off a small explosive charge (essentially, a detonator) buried within it — even while underwater — and . . . *boom.*

The Gunnerside team, Rønneberg foremost, was well trained in sabotage techniques and in the use of explosives. Once Rheam had satisfied himself on this, via some test exercises, he began to instruct them on how to blow up the heavy water facility at Vemork. Using the same wooden model on which the Freshman sappers had been trained, Rheam showed where on the base of each high-concentration cell they should place the explosive charges.

The facility had two rows of nine heavy-concentration cells. The team needed to place a daisy-chained series of nine half-pound charges connected by detonator cord on each row. This cord would then be rigged with an initiator of two-minute fuses to allow them time to clear the room before the explosion. The aim was not only to destroy the machinery but also to puncture the cells and drain them of their precious contents. Working with children's modeling clay, the team trained in pairs to rig the explosives as quickly and efficiently as possible. They repeated the maneuvers so many times, they could nearly do them in the dark.

Almost every day, Tronstad came down to the school to answer any questions, bringing drawings and blueprints of the plant as well as aerial photographs of the surrounding area. Whenever a question was asked to which he did not have the answer, whether it was about layout, doors, guards, or patrols, he would leave the room and return soon after with the answers they needed. In hiding at Brickendonbury was Jomar Brun, the source of the information. He remained unknown to the team.

When the latest intelligence from Swallow detailed more guards at Vemork as well as new searchlights and a machine-gun nest atop one building, the team revisited potential approaches to the target. They could cross the suspension bridge, come down from the penstocks, or climb up from the gorge to the cliffside railway line that connected the plant with Rjukan. Plans for each were hashed out, but they decided to postpone the final decision until they could reconnoiter the site.

One realization they did have was that six men were not enough for the operation. They would need a covering party to give the demolition team the time — and security — to carry out the attack. Swallow could fill this role.

When they were not practicing for the sabotage, the men kept fit, exercising around the extensive grounds. One day, a burglar on loan from a local prison showed them how to break through locked gates. Another morning, they arrived at breakfast to find a monk-like figure sitting at one of the tables. He was Major Eric Sykes, a former Shanghai policeman who was now one of the British Army's leading weapons instructors. He asked Rønneberg if his men wanted to show him what they could do. The six brought him out to the street-fighting range, where pop-up dummies appeared in open doors, behind windows, and across alleys. As they took aim through the sights on their pistols and Tommy guns, Sykes stopped them. "That's doomed from the start," he said. He urged them to shoot from the hip, as they had been taught at Stodham.

Rønneberg assured him they knew what best to do. Sykes waved them onto the range. Firing over two hundred times, Haukelid hit almost 100 percent of his targets, as did the others who followed him. Sykes was dumbfounded, and it was clear that there was nothing he could teach them.

Claus Helberg rummaged in a dark cupboard, desperate for food. He had been at Ditlev Diseth's cabin beside Lake Langesjå a few days before and found some rakfisk, but he was hoping he might have overlooked some flour or oatmeal. There was nothing. Then he heard voices. His skis and boots were outside the door; there was no hiding his presence. He drew his pistol and waited. There was a knock. "Who's there?" he asked sharply.

"Skogen."

"Olav? Is that you?" Helberg asked.

"Yes, it's me."

Helberg yanked open the door, pistol still in hand. There were three men outside the cabin, each of them carrying a hunting rifle. With the sky darkening, it was hard to see who they were.

"It's me, Claus," Skogen said. "You can take it easy."

Helberg stepped outside and started to lower his gun when another of the men approached. Skogen held him back. "Wait until the gun's away, you'll get yourself shot." Then Helberg recognized his childhood friend. "Well, if it isn't Rolf Sørlie!" The two embraced, then Skogen introduced the third man: Finn Paus, a member of Milorg. Because Diseth had been arrested during the Rjukan razzia, Skogen, Sørlie, and Paus had come to his cabin to hide contraband that was being kept there. They also had plans to hunt reindeer.

The four men got in out of the cold, and the new arrivals shared their bread and butter with Helberg. He asked after his family in Rjukan, and Sørlie told him they were safe. They informed him about the German raid and the string of arrests. Clearly, the Gestapo was not targeting people at random. Helberg said very little about why he was in the mountains. Already Skogen knew too much, and Swallow could not risk an expanded circle.

The next morning, they buried the weapons Diseth had been hiding in his cabin. Sørlie gave Helberg his Krag-Jørgensen bolt-action rifle and some ammunition. Given his team had only brought pistols and machine guns from Britain, they were desperate for such a hunting rifle. Helberg said his goodbyes, telling Sørlie, "You will hear from me soon." Then he skied back to Grass Valley to rejoin his team.

Each of the four members of Swallow had a lucky escape during the razzia. Haugland had been out in the woods, searching for supplies, when a patrol missed him by only a few yards. Poulsson and Kjelstrup narrowly avoided some Germans when they were returning to Grass Valley from the mountains to the west. And Helberg, traveling to meet Einar Skinnarland by his house at the dam, had almost arrived at the same time as the Gestapo arrest party the day they had taken Øystein Jahren.

But when it came to food, their luck ran out. They had long since finished their treasured supplies of pemmican. The little bit extra they had recovered from the Songa Valley was already gone. Haugland had managed to find a shotgun and had killed several grouse, but they had picked those bones clean within a day. Helberg had gone up to Lake Langesjå in search of reindeer, but had spotted no herds. He had been searching cabins in the area for stores of rakfisk or meat when he'd encountered the Milorg team.

The four suffered constant hunger and had resorted to digging in the snow for the rust-colored moss the reindeer ate. "It's full of vitamins and minerals," Poulsson promised the others, who were more accustomed to using the moss as a bed when camping. Nonetheless, they boiled it in water with a handful of oatmeal. It made a bitter soup.

When Helberg arrived back at the Grass Valley cabin, Poulsson delivered the good news from London: They were to take an "active part" in the new operation against Vemork. There was much to be done before the standby period began, on December 18, in six days' time. They had to secure some food, get the latest intelligence about security at Vemork, retrieve the Eureka device that had been left at Lake Sand, and recharge their wireless radio batteries. Then they needed to leave for a new hideout, twenty miles northwest of Vemork, far up in the Vidda, where Gunnerside was to be dropped.

Misfortune hounded them at every turn. German patrols delayed their fetching the Eureka device. Helberg was caught in a storm while returning with a freshly charged battery. Struggling to continue with the thirty-pound battery in his rucksack, he finally had to leave it behind in the woods for retrieval the next day. They failed to track down any reindeer, and some salted meat they found in a cabin by Lake Møs turned out to be rotten.

All of them became too sick to hold down the little food they had left. Helberg and Kjelstrup were in particularly bad shape, their malnutrition having caused edema. They grew so bloated they were unable to button their shirt collars and had to urinate six times a night. Still they went out every day to prepare for the new operation. Skinnarland provided them with what little food he could spare, as well as recharged batteries and intelligence on the new defenses at Vemork.

On December 17 Haugland received a signal from Home Station that Gunnerside was set to leave on the next clear night. It was time to head north, even though the charge on the Eureka battery was dwindling. Starving, sick, and with the clock against them, they soon moved deeper into the Vidda.

At Brickendonbury Hall, the Gunnerside team readied to leave for Gaynes Hall outside Cambridge to await their drop into Norway. They had come through the school, earning Rheam's high, albeit understated, praise: "If the conditions are at all possible, they have every chance of carrying out operation successfully." Tronstad visited one last time to go over the mission and to bid the men farewell. It was a solemn moment. Each of the saboteurs had been given a cyanide capsule; each knew that their chances of hitting the target and escaping with their lives were, at best, even. Tronstad reminded them of the executions of the Freshman sappers and warned them that they would likely be treated the same — or worse — if caught alive.

Then he concluded, "For the sake of those who have gone before and fallen, I urge you to do your best to make the operation a success. You do not know now exactly why it's so important, but trust that your actions will live in history for a hundred years to come. Be an example for those that will later participate in the recapture of our country, and thus in the defeat of Germany. What you do, you do for the Allies and for Norway." There followed an awkward pause. Some on the team felt Tronstad was looking at them like they would never come back.

"You won't get rid of us so easily," Rønneberg said.

13

Rules of the Hunter

THE FOUR BEARDED, haggard men struggled with every lift of their skis in the sticky, wet snow on Lake Store Saure. Over the surrounding peaks, a mist hung like cotton wool, and the cloud-ridden sky hid the relief of the sun. Trudging forward, they cursed their hunger, their waterlogged skis, and the cut of the wind. Finally, they spotted Fetter, the hunting cabin Poulsson had built with his cousins before the war from precut wooden planks they had brought across the lake. Positioned high on the plateau, near a copse of birch trees, unmarked on the maps and miles from any other cabins, there was no better place to hide and await Gunnerside. There might also be a good chance of finding reindeer in the area.

Eager to rest and hopeful there might be some food in the cabin, Poulsson pulled ahead of the others. After taking off his skis, he took a hacksaw to the padlock on the door. Sweat dripping from his temples, he cut away at the steel, his hands stiff from the cold. When at last the padlock gave way, Poulsson pulled the bolt to open the door but found it stuck. He took a small ax from his rucksack and hammered at the lock. Not a couple of strokes later, the ax handle broke. He kicked at the door. It didn't budge.

This was *his* cabin. He only wanted to rest. Frustration took hold, then rage. Poulsson pulled the Krag rifle from his back and fired two bullets into the lock, which broke free at last.

On entering, he found Fetter looking much the same as it had when he'd left it in the summer of 1940. He had written in the logbook, "En route home from Lake Langesjå. Have come for a bite of food and to leave the canoe." The canoe was now riddled with bullet holes, but otherwise everything was in order. There were four beds, bought from an Oslo hospital, a square, rough-hewn table, some three-legged stools, and a stone-lined stove that heated the ten-by-twenty-foot cabin. Poulsson scoured the cupboards for food but found only a bottle of cod-liver oil and a scoop of leftover oatmeal.

The others clambered inside soon after, more exhausted from the journey than they should have been. If they continued to starve themselves, they would have no strength for the operation. While Haugland readied the wireless, the others cut up some firewood for the stove and got a pot of snow stirred into a boil. The sun set early, and the four lit a single candle and ate reindeer moss for dinner. Again. Gusts of wind howled around them as they awaited radio contact with London. Finally they learned that Gunnerside would not drop that night, December 19. Stomachs rumbling, they settled into their beds.

"Just wait until there's game in the area," Poulsson vowed. "Then we'll have plenty to eat." The others were less sure.

Days passed in the same pattern, and still Gunnerside did not come. Each morning, soon after sunrise, Poulsson left the cabin to hunt. He knew well, from the memoirs of Helge Ingstad, a Norwegian explorer who had survived the northwest wilds of Canada in the 1920s, that one could live on reindeer meat for a long time. Stopping occasionally to search the horizon for any signs of a herd, Poulsson kept this fact firmly in his mind. He knew that the winds blowing in from the north and west were against him. Herds traveled into the wind so they could smell any predators, and unless the wind changed direction, the reindeer would continue to drift far out of range of the cabin. Even so, Poulsson trudged from mountain to valley to mountain, often through heavy fog, hoping for a shift in the wind direction that would give him an advantage. But he continued to return to Fetter empty-handed. His team was becoming dangerously weak, their eyes listless, their skin yellow. Unless their luck turned, they would not be able to hold out much longer.

At Gaynes Hall, Rønneberg and his team got word yet again that they would not be leaving that night. In the five days since they'd arrived at STS 61, bad weather had shuttered any chance of a North Sea flight, and the men waiting at Gaynes Hall had grown increasingly anxious.

The Gunnerside team members continued to train, taking long runs through the softly rolling meadows of the estate. On occasion, they sneaked through the barbed-wire perimeter fence to hunt pheasant in the neighboring fields. At night, they played cards. One evening, they took a car into Cambridge, drank champagne and dined at a good restaurant. They knew they might not have another chance. "Let's go home," Rønneberg said, cutting off the evening too early for Haukelid. But he went along with the group, on the possibility that they'd finally be given the go-ahead to depart the following night.

The men ruminated over when the weather would improve, how the approach to Vemork would go, whether they would be engaged in a firefight, if it was better to be shot or captured, and whether they should have another go at the letters to their wives and families in case they did not come back.

One way or the other, Haukelid would not be returning to Britain. Tronstad and Wilson had given him the go-ahead to build a guerrilla organization in western Telemark. All he wanted now was one clear night for the drop. Then he would at last be reunited with his original team members.

After a week of snowstorms and cloudy, fog-swept days, Poulsson stepped outside to a beautiful morning, December 23. It was still freezing, but the sun felt good on his face. Skis fastened, he was just lifting his Krag rifle over his shoulder when Kjelstrup came out of the cabin to get some firewood. Of the four, Kjelstrup was in the worst state, roiled by cramps and edema that left him too worn out to leave his bed for more than a few hours a day. "Crisp and clear," Kjelstrup said, his eyes ringed with red.

"If I only knew where the reindeer would be tomorrow," Poulsson said. "I could build myself a snow cave and lie in wait for them."

Kjelstrup managed a smile. "Maybe today will be your day."

Poulsson headed away from Fetter. A frost that had blanketed the Vidda in the night made for fine skiing, and he settled into a steady tempo. The crust over the snow broke with each placement of his poles. As he crossed the Vidda in his white anorak, one mile after the next, the air sharp in his lungs, the rules of hunting his grandfather had drilled into him at a young age governed his mind. "Your rifle is a weapon, not a toy. When hunting, you carry the rifle over your right shoulder with your arm over the barrel. If you are having a rest in the woods [or] crossing over a fence, unload your weapon. When you are in position, you don't touch the safety catch before you see your target. Never shoot unless you can clearly see what you are shooting at. Never get so excited or eager that you forget that you are holding a weapon in your hands. If you do, you will forget yourself." Over the many years since his grandfather bought him his first gun, Poulsson had not neglected these lessons. He had tracked and killed reindeer many times on the Vidda. But now, when he most needed a kill, he could not even find a herd. His team had three jobs: to maintain radio contact, to collect intelligence on Vemork, and to stay alive. They could not do any of these without food.

Then, five miles out from the cabin, his legs beginning to weaken, he came across a band of fresh reindeer tracks in the snow. He nearly leaped in excitement. On inspection of the tracks, he surmised that the herd was a good size, led by an old cow and followed by some young bulls and griz-zled ones. Grouped behind would be other cows, yearlings, and calves. They would be moving slowly, maybe a couple of miles every hour, graz-ing on moss and resting as they liked.

His men would eat this night, Poulsson thought. They would feast for Christmas. He set off for the nearest hilltop but saw nothing in his bin-oculars apart from the endless stretch of white. The reindeer were likely down in a valley or on top of a plateau. He followed the tracks for several miles, almost reaching the northern end of Grass Valley. Still nothing. If they had caught wind of a wolf or some other predator, they might already be far away. Bolting reindeer could travel twenty-five miles in less than an hour. "They are like ghosts," he remembered reading in one of Ingstad's books. "They come from nowhere, fill up all the land, then disappear."

After zigzagging to the top of the next valley, Poulsson stopped. He wiped his binoculars clear with a square of flannel he kept under his watch and examined the horizon again. A noon sun shone down, and the snow-bound landscape played tricks on his eyes. A rock stretched as high as a tree. The roll of a hill flattened into smooth terrain. But far to the north, in a small valley, he spied some dark spots. At first he thought they were only stones, but they were assuredly moving together.

The herd.

Slowly, deliberately, Poulsson skied closer. At times he lost his vantage, but he knew the reindeer would remain in the valley. There was almost no wind. The few wisps of cloud overhead remained almost still. He made his way to the eastern side of the valley, to keep downwind of the herd and prevent them from picking up his scent. He ascended a ridge and, after unfastening his skis, eased himself up onto a boulder. The herd stood in the valley over a half mile away. They were roughly seventy in number. Several grazed on moss beside a small patch of lake. Others rested on the ice or stood like statues. Their winter pelts had faded to a long, shaggy whitish-gray, camouflaging them within the landscape. Farther to the north, up on a raised plateau, was a second herd, smaller in size.

For a good shot with the Krag, Poulsson needed to be within two hundred yards of his target. He did not see how he could get within range without the herd spotting him across the flat approach or catching his scent. He would have to hope they came toward him or moved off to steeper, uneven ground where he would have more cover to stalk them. Perched on the boulder, he waited ten minutes, then thirty, then an hour. The cold settled deep into his bones, and he grimaced and contorted his face to ward off frostbite. If he felt any numbness, he removed his glove and pressed a finger against his flesh until it thawed. He curled his toes frequently and rubbed his beard free of ice.

More time passed. The herd looked to be staying put for an eternity, munching on moss, milling about. Poulsson had to act. After almost two months of living off the sparest of meals since their arrival in Norway, his men needed food. With the dwindling light of day, he must make a kill soon or risk losing the herd in the dark — and being caught out on the Vidda. If a storm hit, like the many that had plagued them over the past month, without shelter and alone in the open, he would surely die.

Resolved, he climbed down from the boulder, then crept into the valley, hiding behind narrow ridges and slight knolls as he came to them. As he made his approach, two bulls strayed from the herd. Poulsson stood motionless, still leeward and too far away for their eyes to distinguish his white anorak from the terrain. When they turned back to the herd, he eased his way down toward a small mound that would bring him within range. Nearly there, he slipped and almost tumbled down the hill before catching his fall.

To the reindeer, which survived through caution, the slight movement on the hillside was enough. The two bulls stamped the ground, then took off toward their herd. Their flight spurred the others. With the pounding of hooves, the lot of them vanished over a hillock toward Grass Valley.

Poulsson rested back in the snow, whatever strength he had left seeming to drain from his body. He swore into the empty sky. When his anger settled, his desperation almost brought him to tears. Chasing after the beasts was fruitless. They might have stampeded too far away to reach before dark. He turned his eye to the high plateau, half a mile away to the north. The smaller herd was still there, grazing.

He clambered back up the slope to gather his gear and then skied fast. When the rise to the plateau grew steep, he climbed slowly on foot. There was no way to know the direction of the wind once he reached the top, but he prayed that he would be downwind of the herd. Heart twisting in his chest, he crept up the last few yards on his belly. Raising himself up on his arms, he spotted some thirty reindeer, their breath a cloud of mist hovering over them.

The wind on the plateau shifted every few seconds, leaving the herd ill at ease. They were still slightly out of range, and they might catch his scent at any moment. His rifle in his right hand, Poulsson crawled forward inch by inch in the snow. In his mind he played out the kill. He would hit the first reindeer in the diaphragm. Past experience told him that if the shot hits its mark, the animal would collapse gently to the snow, as if suddenly taken by the need to sleep. The herd might mistake the crack of his rifle for the breaking apart of frozen mud. Then he would have another shot or two before they fled.

He was within range. Carefully he rose to his knees, took aim, eased his finger onto the trigger, and fired. His target did not drop. The panicked herd fled toward the peak of the plateau. He took aim at another reindeer and fired, then at another. None fell. The herd kept moving, leaving a trail of swirling snow behind them. And then they were gone.

Rising to his feet, Poulsson felt confused. It was impossible to think he had missed all three. He trailed the herd in the direction where they had fled. The snow was speckled with blood along three separate tracks. He had indeed struck his targets, but the military-issue, steel-jacket bullets in the Krag that Sørlie had provided had passed right through instead of expanding on impact like the soft lead bullets he typically used for hunting. He had no idea how far they had run. The blood trail would tell.

He followed one track a hundred yards over the crest of the plateau and found a wounded cow, its hooves scrabbling at the snow in an attempt to rise. Poulsson fired another shot. The cow stilled. He chased after the next track of blood for a short distance but then turned back. He despised leaving wounded animals to a slow death. However, this was a matter of his and his men's survival. He returned to the cow. It was a fair size. Relief, then joy, stirred within him. He marveled at his hard-fought good fortune. For the first time in two months, his team would eat their fill, and would be revived.

Poulsson took his tin cup from his rucksack and filled it with the blood spilling from the reindeer's wound. He drank deeply, the warmth spreading through his body. Then he drained more blood into a small pail before skinning and quartering the carcass with a knife and ax. Its head and tongue, rich in nutrients and flavor, went into his rucksack, followed by its stomach — with four chambers filled with half-digested moss. Then the heart, liver, kidneys, ribs, and legs. He cut off slices of fat while he worked and drank milky white marrow from the small bones near the hooves. He left the leanest cuts of meat in a heap in the snow to be retrieved the next day. First and foremost his men needed fat and nutrients.

With almost fifty pounds of reindeer in his rucksack, Poulsson made his way back toward Fetter. He was exhausted from the day's exertions, but euphoria over the kill made his burden seem light. At one point he came across the herd that had eluded him; catching his scent, they once again thundered away.

Night fell before he reached the cabin. After wiping his hands clean in the snow, he left the rucksack by the door and entered. He said nothing to the others, who assumed that once again his efforts had been for naught. They gave him a pitiful look, sorry for him, sorry for themselves. Haugland reported that Gunnerside would not be coming that night. With the number of days in the moon phase dwindling, the chances the mission would launch that month were slim.

A few moments passed. Something in Poulsson's look gave the men pause. Then Kjelstrup spotted a smear of blood on his white anorak and erupted in a shout. They rushed out of the cabin, and Kjelstrup lifted up the heavy, blood-soaked rucksack. Cheers rang out.

The next night, Christmas Eve, the four gathered at the table under the light of a kerosene lamp. For a centerpiece, they had decorated a twig of juniper with little paper stars. They listened to Christmas carols on the radio for a short while, the wireless headphones set on a tin plate so they could all hear. Then the radio went off — the power in the battery needed to be conserved. They dined on fried reindeer tongue and liver, blood soup with the half-digested moss from the stomach, boiled meat, and marrow. Helberg had found some salted trout in a nearby cabin, a further treat.

Sated, they sat in silence. Poulsson puffed on his pipe, and they all listened to the wind swirling around the cabin. It shook the corrugated-iron roof and sent a dusting of snow underneath the door. One of them began to hum a tune. Soon they were all singing a song they had learned while in Scotland, "She'll Be Coming 'Round the Mountain When She Comes." Not daring to test their emotions on a Norwegian song, they still felt like kings of Norway.

In his office on the third floor of Oslo's Victoria Terrasse, the building emblazoned with swastika flags, Fehlis reviewed the December 14 petition from Bjarne Eriksen, Norsk Hydro's director general. Eriksen wanted two of his men, Torstein Skinnarland and Øystein Jahren, released from custody immediately. Fehlis knew that Terboven wanted to keep Eriksen happy. He had recently hosted a holiday dinner in his home at Skaugum, previously the royal residence outside Oslo, where he promised the assembly of industrial chiefs, including Eriksen, that once victory was complete, Norway — and their companies — would find a bright economic future waiting for them. Until then, the German war machine needed their help. The aluminum and saltpeter provided by Norsk Hydro made keeping Eriksen an ally worthwhile, but there was another key Norsk Hydro resource — designated SH-200 — that Fehlis had to worry about. The British saboteurs had clearly targeted this substance, and Skinnarland and Jahren were key to uncovering the resistance network in the area.

Every week, 99.5 percent pure heavy water was tapped from the high-concentration plant into five-liter aluminum bottles. These were packed into wooden crates (four bottles per crate) and shipped by train from Vemork to Mæl, by ferry across Lake Tinnsjø, then by rail again to the Oslo office of Norsk Hydro. From there, the Wehrwirtschaftsstab Norwegen took control of the crates. They relabeled them for delivery to Hardenbergstraße 10, the Berlin headquarters of the Army Ordnance Research Department.

Once the crates left Norway, they were no longer the responsibility of Fehlis. Up until that point, his job was to protect Vemork and its 150 kilograms' (and rising) monthly SH-200 output from sabotage. He had allies in his efforts. The Wehrmacht was responsible for supplying men to defend industrial plants of significant importance to the war effort, and Vemork was high on this list. After the failed glider operation, General Falkenhorst had visited the plant and had sent out a stream of directives on how to prevent future attempts. Access points must be identified, minefields laid, searchlights and sound alarms set up, and guards trained and armed with submachine guns, hand grenades, and even brass knuckles.

"Our security teams must be mobile and capable of fighting within the plant and its rooms," Falkenhorst ordered. "They must be able to quickly pursue the enemy, overtake him in the course of his flight, and swiftly overpower him in hand-to-hand combat." He also warned that "the gangsters will choose precisely the most arduous approach route to infiltrate a facility, because that is where they expect to encounter the least protection and flimsiest barriers . . . The enemy spends weeks, even months, meticulously planning sabotage operations and spares no effort to assure their success. Consequently we too must resort to every conceivable means in order to thwart their plans and render their execution impossible."

But it was not Fehlis's style to depend on Falkenhorst. It was his own interrogators who had wrung the plans for the Vemork attack from the five British saboteurs brought to Grini. Although they had denied any Norwegian involvement, he was committed to crushing any resistance network that might be aware of or on hand to aid such an operation. Informers had already supplied the names of the Skinnarland brothers, Einar and Torstein. Jahren, who had been tortured by the Gestapo, led them to the Rjukan Milorg leader, Olav Skogen. If anybody knew anything, it was Skogen. That all four were Norsk Hydro employees was not lost on him. On the morning of December 27, Fehlis had his men arrest Skogen. Let that be an answer to Eriksen and his petition.

A thirty-minute, steep hillside trek from the shore of Lake Møs found Einar Skinnarland bunkered down in a small, dark cabin set beside a dribble of a stream. The cabin, nestled between boulders, was called Nilsbu, and was indistinguishable from the surrounding terrain. In winter it was almost entirely buried in snow. In summer it was well hidden amid the heavy pine forest. If he lay flat on the floor, Skinnarland could almost touch every wall with his outstretched arms and legs. There was no place he felt safer. Two local families, the Hamarens and the Skindalens, were helping him, both in obtaining supplies and keeping an eye out for Germans.

Since he had left High Heaven, Skinnarland had largely been on his own at Nilsbu. He celebrated Christmas Eve with a "great steak and pancakes," as he wrote in his diary. Otherwise, the days passed one after the other with only a jotted note about the change in weather: lots of fog, snow, and winds from the east and west. Tronstad had given him the option to flee to Sweden, but Skinnarland would not leave, not with his brother held hostage by the Germans, not with Jahren in prison, and all because of him. He would stay and do what he could to help the sabotage of Vemork. He understood that if the wrong person caught sight of him (and he was known well in the area), the Gestapo would surely be on to him.

On December 27, a telephone call to the Hamaren farm brought word from Rjukan that the Gestapo had seized Olav Skogen. Disheartened over the arrest of his friend, Skinnarland headed toward Lake Store Saure to inform the Swallow team. After several hours of skiing through a biting southwest wind, he finally arrived at Fetter. Inside, he found Jens-Anton Poulsson stirring a pot of stew: reindeer flank with cuts of intestine, windpipe, and fat; bits of fur floated on top. Skinnarland welcomed the meal nonetheless, and they invited him to stay as long as he wanted. With the moon phase at its end and the next chance of a drop a month away, the men could all use some company.

14

The Lonely, Dark War

ON JANUARY 6, 1943, Tronstad was working late at the Norwegian high command, Kingston House. Every once in a while, he glanced down the corridor toward the empty office of Lieutenant Commander Ernst Marstrander. A few days before, after landing Odd Starheim and forty other men on the Norwegian coast for the launch of Operation Carhampton, Marstrander's ship hit a floating mine and sank. All hands perished. "We grow harder and harder," Tronstad wrote in his diary, "and every adversity just steels us to make new efforts."

Over the holidays, Tronstad had missed his family terribly. Bassa had written before Christmas, saying how difficult it would be to celebrate without him. A courier brought some recent photographs of her and the children, and Tronstad sent some coffee and hot chocolate. But he could not manage a letter expressing the grief caused by his absence from them. On December 29 Bassa sent another letter, pleading for him to write soon. "Do you think we'll soon have peace?" she asked. "The best of life is gone . . . What we talked about in the past, I do not know, but now it is only of war and food." Sidsel, now ten years old, followed with a letter of her own: "We've been mostly healthy, both Brother and me. We're often skiing . . . We had a very pleasant Christmas, but certainly it would have been much better with you . . . You don't know how tall I've become . . . We have a dog, with a foxlike head, a sheep's body, and a monkey's tail, and it's very, very strange . . . Hearty greetings from Snoopi and Brother and Mother. We'll meet again."

At the start of the new year, Tronstad wrote back, telling his daughter that the photographs showed how "big and pretty" she had become and then, "I have it very good and hope to come home soon when this ugly war finally will end. You must not be afraid of anything. Keep your head high." He promised her that he would keep fighting until Norway was free.

To cripple the Germans, Tronstad wanted to launch operations daily against them if he could. Carhampton promised to be a sizable blow against their ability to drain his country of its critical resources. Gunnerside would be an even greater strike. Although the bad weather across northern Europe continued to prevent a drop, forcing Swallow to endure the harsh Vidda for longer and allowing security at Vemork to tighten further, the additional time gave Rønneberg and his men weeks more to train. There was comfort in this, and one had to find comfort wherever and whenever the opportunity arose in this dark fight.

In a small, bare cell, numbered D2, at Møllergata 19, Olav Skogen waited for his torturers to come. Formerly an Oslo police station, the stout Romanesque structure, which looked like a half-flattened wedding cake, now served as the Gestapo prison. Again and again Skogen promised himself that he would not break. He was in the right, and his torturers would one day be judged for their crimes. He must never reveal what he knew and whom, nothing about Milorg or Rjukan or Swallow or Skinnarland. Not a single name would be wrung from him. Keeping his silence was the one fight left to him in this war. He knew it now. He accepted it. He would not break.

Two weeks before, he had gone to Bergen to review some contracts for Norsk Hydro. When he entered the office there, he was met with a pistol in the face. Taken away to a waiting car, he chastised himself for not having gone into hiding after the raids on Rjukan. He had been foolish and too bold by half.

At Bergen prison, a junior Gestapo officer tried to get him to confess his involvement in the underground resistance. "I'm not involved in any such illegal activities," Skogen said.

The officer shrugged. "We have people who will get the information they are looking for." With that, he sent him to stew in a cell.

Skogen knew what was ahead for him. In resistance circles, accounts of Gestapo torture were widespread. On his second night, staring out the window in his cell, he questioned whether he should kill himself with the pills Skinnarland had given him. It would be the best way to ensure the Germans got nothing from him, and he would save himself a lot of pain. Making his decision, he took out the pills sewn into his breast pocket, cupped them in his palm, then dropped them to the floor and crushed them underfoot. The Germans would have to kill him all by themselves.

The next day, New Year's Eve, they brought him to Oslo, then into Møllergata 19. He was fingerprinted, photographed, and stripped of his belt and shoelaces. There was a brief first interrogation — a lot of bellowed threats, but no fists. "The arrest is a misunderstanding," Skogen said. He was thrown into a cell and given chunks of stale bread to eat. Day after day he waited, but nobody came to his dark, lonely corner of the prison.

Then, on the night of January 11, the iron door of his cell swung open. He was handcuffed and driven through the deserted streets of Oslo to Victoria Terrasse. Guards led him up four flights of steps and into a room where several Gestapo were waiting for him. On the wall opposite the door hung a portrait of Hitler. Skogen was pushed down onto a stool near a desk. The lead interrogator, a bullnecked man with bushy eyebrows and a jaw that stuck out like it wanted to be punched, sat down behind the desk and asked in German, "Name?"

Skogen spoke German well enough, but they did not need to know this fact. "Interpreter?" he asked.

There were sighs. Then one of the Gestapo, who spoke Norwegian, began to translate for Bullneck. He asked if Skogen knew two names: Øystein Jahren and Einar Skinnarland.

"Yes," Skogen said. They probably thought they had him already.

"From where?" Bullneck asked.

"Rjukan Sports Association," Skogen said.

"That was the first lie!" Bullneck shouted, red in the face. He looked to the interpreter who translated.

Skogen said, "I don't lie."

Questions followed about Einar Skinnarland. Skogen pleaded ignorance. Soon after, a tall, athletically built German in a tracksuit came into the room, and the torture began. The men pushed up Skogen's left pant leg and clamped a vise onto his lower leg. Its jaws were sharp half-moons of steel.

"Do you confess now?" Bullneck asked.

"I've nothing more to say," Skogen said.

Tracksuit rotated the wingnut on the vise, squeezing Skogen's leg between the two jaws.

"Do you now?" Bullneck asked.

Skogen grimaced and shook his head. Tracksuit tightened the vise. The half-moons dug into his shin and calf muscle, distorting the shape of his leg. He groaned.

"We won't stop until we've squeezed the truth from you," Bullneck said.

Again and again, Tracksuit rotated the vise's screw. Skogen saw bright lights of pain, and his leg turned a violent shade of purple. Eventually he fell off the stool. The men then started in on him with rubber batons, striking his bare feet, legs, and back. Then he was lifted back onto the stool.

"Speak! Last chance!" Bullneck screamed.

Skogen shook his head. He would not speak. Not a word. This was his lonely war now. He would win it.

Tracksuit freed Skogen's battered leg from the vise, and a shock of agony followed. Then Bullneck took a bamboo stick from the wall and struck him across the chest, back, and shoulders until he was out of breath. "Tell us what you know!" They punched him in the face. When he fell off the stool again, they dragged him across the floor by his hair. They beat him over and over with truncheons and sticks. He kept silent. After a time, the world backed away from him and he fainted. Giving up for the night, his torturers reported to their superiors on the floor below of the failure of their "enhanced interrogation."

For three weeks now, the five surviving Royal Engineers from Operation Freshman had been held in solitary confinement at Grini. Having extracted the information he could from them, Fehlis ordered them shot. On January 19, they were told that they were to appear in front of a military commission before being sent to a prison camp in Germany. Instead, soldiers drove them a couple of hours north, through heavy snow and ice. They arrived at Trandum, a forested area the Gestapo used for executing political prisoners and dumping their bodies. The men were led blindfolded into the woods, each flanked by two Germans, then told to stop. They did not know it, but their feet were at the edge of a grave, dug earlier by their executioners.

Among them, in the same blue rollneck sweater and trousers that had been issued to him in Scotland, was Wallis Jackson. In the twenty-one-year-old's pocket was a fine red-threaded handkerchief, obviously a gift from a loved one back home. The stomp of boots sounded around them. They heard a clank of rifles. On the call to fire, the execution squad shot the British soldiers, then buried them in the unmarked grave.

Huddled in his sleeping bag on yet another icy January morning in Fetter, Poulsson was watching the steam from his breath. The reindeer pelts hanging over the bare wooden planks failed to keep the cold at bay, and a thick layer of hoarfrost covered the ceiling and walls. At 7:30 a.m., while the others still slept, he climbed out of his sleeping bag. It was his turn to make breakfast. He lit the kerosene lamp and, sitting on a stool, fed slivers of birch into the stove. When the kindling was crackling, he added a couple of logs. The stove heated up. He placed on top a pot of porridge mixed with a paste made from reindeer bones ground and boiled for two days. While this cooked, he dressed. Then he grabbed the hatchet by the door and stepped outside.

A few feet from the cabin was an area that looked altogether like a slaughterhouse. Blood covered the snow, and there was a pile of frozen reindeer stomachs against the cabin wall. The half-digested moss in the guts was the men's only source of vitamin C and carbohydrates. Poulsson cut off a chunk, then returned to the cabin. Already the frost on the walls and ceiling was beginning to melt. Water dripped on his face and collected in pools on the floor. He stirred the porridge, heated up the organ meat in another pot, and started on coffee. As this brewed, he called the others to the table.

They emerged from the half darkness, all grunts and groans. None of them had bathed in months, and only Haugland had kept up a shaving regime. With his rangy red beard and grimy face, Kjelstrup looked more beast than man.

They pulled on their clothes, and a couple of them made a quick escape outside in untied boots to relieve themselves. On their return they took their usual seats at the table: Poulsson near the kitchen counter, Haugland opposite him, Helberg and Kjelstrup squeezed together on one side, and Skinnarland across from them.

They spoke little as they ate their meal: porridge, loin with strips of suet, and offal. In the weeks since Poulsson had made his first kill, they had become connoisseurs of reindeer. By taste alone, they could distinguish an old bull from a calf from a yearling. They liked the first the best, for its rich taste. Eyelid fat and bone marrow were the finest of delicacies, and *gørr* — a soup made of the contents of a deer's stomach, rich in moss, mixed with meat, blood, and water — was a favorite as well. Truthfully though, they were indiscriminate. They ate the heart, kidneys, liver, larynx, brain, tongue, tooth nerves, eyes, nose, every sliver of meat on the bones, and then the bones as well. Other than the hooves, horns, and pelts, nothing escaped their plates.

When dawn crept up through the east-facing window, they extinguished the lamp. After breakfast, Poulsson and Skinnarland headed outside to survey the weather. Northwest of Fetter, heavy gray clouds hung over the mountain peaks. "Murky weather," Skinnarland said.

"No operation today," Poulsson replied, knowing from experience that the clouds foretold there would be no flights from Britain.

When the December moon phase ended, they had spent two weeks skiing back and forth across the Vidda through heavy wet snow and snapping winds. They tracked reindeer, secured recharged batteries for the wireless set and the Eureka device, gathered more gear from their Songa Valley depot, and collected intelligence about Vemork. Sometimes they spent the night away from Fetter, breaking into cabins or, if caught out, sleeping in a hastily built snow cave. Through sources in Oslo, Skinnarland knew that Olav Skogen was being tortured and also that he had bravely remained silent. If it were otherwise, the Germans would have been scouring the Vidda by plane and ski patrol, hunting for them.

On January 16, when the new standby period came into effect, Tronstad sent a message: "Weather still bad but boys eager to join you."

As they waited for the weather to change, one day blended into the next. They cut wood, hunted, cooked, and sat by the wireless, eager for news. By 4:00 p.m. each day, the sky darkening, they sat around the table for dinner. Poulsson had run out of pipe tobacco, so at least the air was clear of smoke. After eating, they retreated into their sleeping bags to endure another night of the raging storms that threatened to tear the cabin apart.

There were petty flares of temper. One man was accused of not keeping the cabin in order. Another didn't get up early enough to cook breakfast. Yet another couldn't get the damp wood to burn in the stove. Crowded into a small space, isolated from civilization, and always either cold, wet, tired, or hungry, the men could easily have let these conflicts escalate.

Skinnarland's presence helped. During the standby period, he alone could venture away from Fetter, and after miles of skiing he always seemed to come back in good cheer, often with a pat of butter or some dried apricots that he added to their reindeer stew for a change of flavor.

Without Poulsson, the team would have fallen apart. During the short days, he kept the men busy with chores, but it was the nights, which stretched for sixteen hours, that were the real danger. They could not sleep the whole time, and they had neither the kerosene nor the candles to keep the cabin lit all night to whittle, play cards, or the like.

One evening, to occupy them, Poulsson gave an interminably long and intricate discourse on the art of hunting. This inspired Kjelstrup to instruct everyone in the science of plumbing. Haugland followed with lectures on radios, Skinnarland with a lesson on building dams. Having explained how the different places on the Vidda got their names, Helberg then proved to be at least a second-class lyricist. His description of Kjelstrup had the cabin in stitches of laughter. "As number three in the gang / We have Arne, oh the devil / Expert in heating and sanitation / Hates frost and winter weather / But he knows what to do at night / Sleeps with his balaclava on all right." When they ran out of lecture subjects and poems, they talked about their lives at home, and their families. Instead of tearing them apart, the winter nights thus drew them closer together. Still, they could not stay there forever.

Through the window of a Halifax bomber, Haukelid spied at last the surf breaking against the Norwegian coastline. Jostled about in the fuselage, he kept his eyes trained on the view. He spotted a fishing boat and wondered if its captain heard the roar of the plane overhead — maybe the swinging of his lantern was to welcome them. To avoid German radar, the pilot flew low over the valleys toward Lake Store Saure. The forests and mountains were as clear under the bright moon as they were during the day. The navigator should have no trouble finding the drop site on this perfect night, January 23.

The Gunnerside team had stayed so long at Gaynes Hall that the staff there joked they might as well be part of the furniture. But for men who woke every morning not knowing whether today was the day, the joke had long ceased to be funny.

Haukelid had been on edge almost all the time. Every delay meant further suffering for Poulsson and his men. They would be out of rations and surviving on whatever they could hunt. He knew well the trials that the Vidda imposed on those who dared spend even a day there in the winter. His friends had been out there for months.

Rønneberg had used the time well. As a team, they had endlessly gone over the drawings and photographs of Vemork, and by the January moon phase could have drafted the architectural plans of the plant themselves. They knew the location of every window and access tunnel. They knew in which direction the doors opened and whether they were made of wood or steel, and they knew where they could climb in the plant without the supports or equipment giving way. They knew which fence wires were electrified, where the covering party should position themselves for the best field of fire while the demolition crew set the charges. Whenever a new question arose, a letter was dispatched to Tronstad.

Now, sitting in silence inside the Halifax, the six men looked to their dispatcher for the sign they were nearing the drop. They drank tea to stay warm. Two more hours passed.

The plane zigzagged across the Vidda. Its pilot relayed to Rønneberg that he could not yet see any signal from their reception party. A scattering of low clouds had come across the sky, and fog hugged the valleys. The Halifax veered back toward the coast to allow its navigator to get a new bearing; then they returned to the drop zone. Out his starboard window, Haukelid saw Lake Langesjå. The dispatcher gave the call for them to prepare to jump, and the hatch in the fuselage floor opened. A six-minute warning was called.

But no signal to jump followed. The navigator was not able to spot the drop site. Haukelid wanted to go regardless. Rønneberg did too. The Vidda passed below them. They urged the pilot that they would jump blindly and "sniff our way to the dance floor," but he would not allow it, fearing that they were too far from Lake Store Saure. Wind gusted through the hole, freezing the men stiff. Still the Halifax continued to circle.

After midnight, they were told that the plane was running low on fuel and needed to go back to Britain. It felt like a crater was opening in each of their hearts. To return, on the brink of action in their homeland . . . they simply could not resolve themselves to it.

Suddenly, there was a burst of antiaircraft fire. The sky around the plane lit up with what looked like sparks. The Halifax banked sharply from left to right, trying to avoid the barrage, throwing the Gunnerside team about in the fuselage. The plane was hit in a wing and rocked violently. One of the engines burst into flames and then cut out. As the cascade of shells continued, the Halifax corkscrewed. Another engine died. Then all was silent but for the din of the remaining two engines. They had escaped. The Halifax continued over the North Sea, limping home.

That night, on a hilltop overlooking the northern end of Lake Store Saure, Einar Skinnarland cranked hard on the hand generator he had rigged to the Eureka in place of their dead battery. His arms ached, and he was bone-weary after skiing seventy miles in three days to retrieve the generator — returning just in time. An hour before, the yellow moon still visible in the sky, they had heard an airplane cross overhead. Poulsson and Kjelstrup marked out an L shape with flashlights on the lake, but there was no drop. Soon after, a freezing fog descended over the Vidda. If the plane returned, the radio beacon was now the only chance the navigator would find them, and the hand generator was the only way to power it.

While Helberg listened at the headset, Skinnarland rotated the handle. At 3:00 a.m., after hours of waiting, the team finally gave up hope that Gunnerside was coming. Frozen and exhausted, they returned to Fetter and collapsed.

Over the next several days, storms blotted out the sun and blew across the Vidda. On January 28, Haugland deciphered the message from Tronstad they knew was coming: "Deeply regret weather conditions have made it impossible to land party. Do hope you can manage to keep going until next standby period February 11. Revert to ordinary sked [wireless schedule]. Take care."

Poulsson and his men knew what to do: survive, maintain radio contact, and collect the latest intelligence from Vemork. Their brief diary entries spoke to the constant, monotonous toil and the isolation of a winter that was far worse than any they had experienced before. From Poulsson: "Jan. 29 — Skinnarland and Kjelstrup went off to Lie. Jan. 31 — Helberg's birthday. He celebrated with a trip to [recharge] the flat battery. Feb. 1 — God-awful weather last night. Feb. 3 — Helberg returned and brought some food. Feb. 6 — Awful weather." Skinnarland mostly jotted down the daily weather and the distances traveled. He recorded journeying from Fetter to Lake Møs to High Heaven and back again, over and over, often ten miles a day, every day, the wind always blowing — the only variation being in which direction it blew.

At the approach of the next standby period, the men found themselves at breaking point — or beyond it. They had plenty of reindeer now, but their bodies were suffering from a lack of diverse nutrients. The cold, the constant effort, the tension — these all took their toll as well. One or another of them was always sick, whether from stomach trouble, fever, muscle strain, edema, or sheer exhaustion. One morning, away from Fetter, Skinnarland found himself almost unable to move his legs. He holed up where he was for two days until he got his strength back. Haugland, who roused himself from sleep in the middle of the night, every night, to radio Home Station, came down with a terrible flu. Still he kept up his schedule, tapping coded messages with dots and dashes by the dim light of a candle while his body shook from fever.

On February 11, an icy, southwest wind cut across Fetter, followed by another blizzard that left them stuck in the cabin for days. At times their thoughts darkened. Unlike soldiers on the front, whose courage was tested in the immediacy of battle, the five men in Fetter were fighting the relentlessness of time and the faceless, remorseless Vidda. There were moments when they wanted to surrender. Enough waiting. Enough cold. Enough hunger and strain and wind and snow. The comfort of their families was only a day's ski away. They could so easily go home, sleep in their old beds, eat at their parents' tables.

"What was the point of it all?" Poulsson wondered to himself as he huddled with the others in the cold, dark cabin. "What was the point of our suffering here in the mountains? What was the point of carrying out the job we had been ordered to carry out? Was there any chance that it might succeed and that we might escape alive?" He dared not give voice to these doubts. This made them no less real.

15

The Storm

AFTER THE FAILED January drop, Rønneberg insisted the Gunnerside men take a break from Gaynes Hall. While waiting for the new moon phase, they spent two weeks in a lonely stone cottage beside Loch Fyne, in western Scotland. Owned by a former intelligence officer and friend of Colonel Wilson, the place had no electricity and was accessible only by boat or on foot across the moors. Surrounded by mountains, it was the perfect training location — far rougher than the easy meadows around Cambridgeshire. The six took daily hikes with full packs to accustom their muscles to the terrain they would face on the Vidda. When not trekking, they fished for salmon and hunted seals and stags. Their time in Scotland both readied them physically and brought them closer together as a team.

On February 12 they returned to Gaynes Hall. The window to depart was open, but if Norway was being ravaged by the same rain and strong winds as Britain, they would not be leaving any time soon. On their first day back, Tronstad met them to review the latest information from Swallow. They had feared that the long delay in launching the operation would give their enemy time to reinforce security at Vemork and that had now happened. "German troops have joined Austrians. Strength of guard at Vemork increased to thirty men," reported Swallow. "Double post on the bridge. During air raid alarm complete state of readiness." Since the Freshman disaster, the Germans had increased their garrisons at Lake Møs from ten to forty men; at Vemork, from ten to thirty; and at Rjukan, from twenty-four to two hundred. These reinforcements were mostly crack German soldiers. An antiaircraft battery had been set up at Lake Møs. A pair of D/F stations searched constantly for radio transmissions, and several Gestapo investigators were permanently on hand in Rjukan, sniffing out any trouble.

The Germans had also laid additional minefields around Vemork, positioned searchlights throughout the grounds, and posted reinforcements at the top of the pipeline and on the suspension bridge. Patrols ran around the clock. The winter fortress was prepared for an all-out assault. All these heightened defenses signaled the importance of the atomic program to the Nazi war effort, yet recent intelligence Tronstad and Welsh had collected from their spies painted no more than a murky picture of the intensity with which they were pursuing a bomb.

Paul Rosbaud, aka the Griffin, sent news that a member of Hitler's inner circle, Albert Speer, had commandeered the German program, but Rosbaud was uncertain what this meant for its status. He also highlighted Heisenberg's intention to build a self-sustaining reactor despite the explosive fire that had ruined his most recent experiment.

Harald Wergeland, the University of Oslo professor Tronstad had recruited as a spy, recounted a meeting with a German physicist who stated that their research was focused on building a power-generating machine, not a bomb. Any such weapon was a distant dream, Wergeland was told. Nicolai Stephansen, a Norsk Hydro executive who had recently escaped to Stockholm, backed this up. He delivered a report that chronicled the continued push for heavy water at Vemork. Yet from the conversations he had with German scientists at Norsk Hydro headquarters, the rise in production was "not to be utilized for bombs or other sorts of devilry connected with the war."

However, couriers also brought secret messages from Njål Hole, a twenty-nine-year-old Norwegian physicist. Tronstad had encouraged Hole to join the Physics Department at the Nobel Institute of the Royal Swedish Academy of Sciences to spy on any Germans who visited or corresponded with its staff. Among the institute's noted scientists was Lise Meitner, whose close collaboration with Otto Hahn helped lead to his discovery of fission. In January, Hole sent a message, detailing recent attempts in Berlin to separate uranium isotopes with centrifuges. He also reported an outright statement by a German physicist involved in the program that his countrymen "intended to make uranium bombs."

Tronstad shared none of these conflicting reports with Rønneberg and his team. Vemork was the single identifiable target open to them in their effort to stop the Germans from obtaining a bomb that could annihilate a city in a single strike.

Rain battered the roof of the truck as it came to a stop at the edge of Tempsford airport. The rear doors opened, and the Gunnerside team emerged in white camouflage suits and ski caps, their weapons at their sides. They crossed the tarmac to the Halifax, where Tronstad stood waiting. "Whatever you do, you must do the job," he told them. "Whatever problems you hit against, think of the job. That is your main responsibility." He wished them luck, and they clambered inside the fuselage. They took their seats, squeezed in beside all their gear, which included an arsenal of half-pound charges of Nobel 808, detonator cords, primers, delays, and pencil time-fuses. At 7:10 p.m. February 16, the plane rumbled down the runway and lifted off. Over the North Sea, the clouds disappeared, revealing the moonlit water below.

After four days at Gaynes Hall, each bringing notice of "No operation today," they had finally got the call to go, despite the weather. During the briefing with the aircrew, both Rønneberg and Haukelid made it clear that if the navigator was unable to spot their new drop site (Bjørnesfjord, one of the largest bodies of water on the Vidda and a short day's journey from Swallow's location at Fetter), they should be called into the cockpit to help. They also told the crew that, whether they spotted Swallow's reception lights or not, sure of the position or not, they would be jumping that night. "We'll find our way ourselves on the ground," Rønneberg said.

"Ten minutes," the pilot called out just before midnight as the Halifax crossed over the Vidda. The team readied themselves to drop. At last, two minutes into the new day, the warning light switched to green. Rønneberg led the way, vanishing into the darkness and rushing wind. In quick and sure order, four of the remaining five men and several containers followed.

Then Knut Haukelid edged toward the hole, his heart thumping in his chest. No matter how much he had practiced at STS 51, parachute school, his nervousness about jumping never abated. Twelve hundred feet of empty air was a lot to fall through, and it was impossible to know what dangers existed on the landing. One Kompani Linge team had parachuted down onto a lake, gone straight through the thin ice, and drowned. Every second he delayed put him farther from the others.

Haukelid then saw that the cord that released his parachute was wrapped around the dispatcher's leg. If he jumped, the dispatcher would come with him. Swiftly, Haukelid rose, shoved the man out of the way to free the line, and then, without further hesitation, leapt through the hole. A moment later his parachute opened, and with a sharp tug, he was momentarily lifted up by his shoulder straps. Sixteen other parachutes, attached to containers and packages of gear, floated down with him.

He landed in a bank of snow, drawing his parachute together before it could sweep him away across the ground. One of their supply packages suffering this fate was carried through the snow for over a mile. They discovered it lodged in a crack of ice. A few feet to the left and the winds would have continued to carry its essential contents — three rucksacks and sleeping bags — too far away to find.

They assembled quickly. Rønneberg asked Haukelid if he knew where they were. He had spent the most time on the Vidda. "We may be in China for all I know," Haukelid joked, but given the expanse of flat terrain surrounded by hills, he suspected they had landed exactly on target. For a brief moment, he sat, cupping a ball of white snow in his hand, savoring his return to Norway at last.

Then the team got to work. First they buried their parachutes. Then Storhaug, who was the strongest skier, scouted the area, while the five others set out to locate their containers. A short while later, Storhaug returned with the news that he had found a cabin a mile away. Over the next several hours, they collected their containers and placed them in a long trench they dug in the snow. They set rods around the depot to mark its location and used a map and compass to take a navigational bearing. By the time they finished, the steady drifts of snow had obscured almost all signs of their arrival.

At dawn, they arrived at the empty cabin. To enter, they removed the doorframe with an ax. It was an expansive space, with a sleeping loft, a well-equipped kitchen, a fireplace, a sitting area, and a cord of cut birch. It would have been a nice place to hole up for a few days, but they didn't have that kind of time: they would need to start soon for Fetter. After a short sleep, they returned to the depot and sorted out the weapons, equipment, explosives, and food they would need for the sabotage. When the deed was done, they would retrieve additional supplies for their retreat to Sweden.

At 6:00 p.m. Rønneberg led them eastward, compass in hand. According to their maps, fifteen miles separated Bjørnesfjord and Lake Store Saure. They were carrying almost 65 pounds each on their backs and towing two toboggans of equipment, each weighing 110 pounds. Four miles into their journey, the winds picked up, blowing against their backs. Soon after, they were caught in a storm that surged across the high plateau with runaway force. With each slide forward of their skis, the westerly winds cut harder, and it became very difficult to see.

Forging ahead, Rønneberg came across a twig sticking up from the snow. He thought it curious but continued, only to come across some underbrush a couple of hundred yards later. Had they been cutting across the Bjørnesfjord — a channel of water — there would be no such vegetation. Then the realization hit: they had not landed at the intended drop site.

He stopped, and the others came alongside him. Through the sweeping winds, he yelled, "We have to turn back to the cabin so —" the rest was lost in the storm.

He started back in the direction from which they had come. The others followed. Now they were headed straight into the gale, and ice and snow bit at their faces. The gusting wind made it almost impossible to breathe without the men shielding their mouths with their hands. Visibility cut to zero, and their incoming tracks erased, Rønneberg led them only by the needle of his compass.

They pushed on, hauling their gear through the snow. The darkness was impenetrable, and the cold overwhelming. If they missed the cabin by even a few feet to either side, they would continue endlessly into the Vidda, into the arms of a tempest.

The huge storm enveloped Fetter. Inside the cabin, the Swallow team and Skinnarland were worried. On the morning of the sixteenth, clear weather over the surrounding mountains had given them hope that Gunnerside would launch that night. Soon after, Haugland received the crack signal "211" on the wireless set — the prearranged code from Home Station that the drop was moving forward.

Poulsson led his team to Bjørnesfjord. They set out the Eureka and prepared the lights, but besides hearing the distant drone of engines, there was no sign of the plane or the drop team.

Throughout the night of the seventeenth, the blizzard continued, almost burying the cabin in snow. They talked of sending out a search party to Bjørnesfjord, but Poulsson thought better of it. It would be near impossible to find one's own hands out in that weather. Further, he doubted that Gunnerside had dropped anywhere near the target. They would have to wait until the storm subsided. But with each passing hour, the blizzard seemed only to grow more angry and murderous. Huddled in their sleeping bags, the walls of Fetter thick with hoarfrost, they shivered and feared the worst.

Rønneberg took the framed map down from the wall. A few hours earlier, he and his team had run blindly — seemingly miraculously — into the very cabin they were looking for. Stamping their frozen feet and trying to thaw their frostbitten faces, they were well aware that they had barely escaped the Vidda with their lives. As the others slept, Rønneberg took first watch, though it was hard to think there was any threat greater than the terrible cold and wind.

Now, staring at the map by flashlight, Rønneberg tried to determine exactly where on the plateau they had landed. Starting at Bjørnesfjord, he ran his finger in a broadening circle, spying for terrain that matched their surroundings. Someplace flat, with a sizable lake, bordered by hills. On the third encirclement, his finger settled on Lake Skrykken, twenty miles northeast of their targeted drop and forty from Vemork.

In the morning, they broke into the small locked side room in the cabin. In it, they found a fishing logbook and learned the cabin was named Jansbu. It was owned by a Norwegian shipping magnate and was indeed beside the lake Rønneberg had identified.

With no wireless set and the storm outside continuing to rage, the team could only sit and listen to the wind, which sounded like muddled screams. The gale blew with such force and duration, they began calculating how much they and their equipment weighed and whether the sum total was enough to keep the cabin fixed to the ground. The structure held, but the rattle and shaking of the walls made them feel as if they were aboard a ramshackle ship tossed on a roiling sea. None of them had ever experienced a storm of such ferocity.

Three feet of snow fell over the course of the blizzard. When the team dared to crack open the door, the storm still howling with undiminished force, they glimpsed a landscape transformed into high drifts and flat, indistinct planes of snow. Their mission receded in their minds; there was nothing but the storm.

All six became ill from the swift change in weather. Only two days before, they had been near sea level in the moist, relatively warm English climate. Now they were suffering cold beyond measure on a plateau almost a mile above sea level. Swollen glands made it hard for them to swallow; their eyes became rheumy, their temples hot with fever.

On February 19 Rønneberg wrote, "Same weather. Storm and driving snow. We made an attempt to reach the depot to fetch more food to save the rations. This had to be given up because of the danger of losing our way."

The storm elevated in intensity that night, a beast set loose in the world. In the middle of it, the smoke from the fireplace began choking the cabin. Rønneberg braved the conditions outside to check the chimney. When he shoved the door closed behind him, he found himself lost in a desolate wilderness. Snow fell in fist-sized clumps. Eyes wide behind his goggles, he could see nothing, and the wind stole his breath. He made his way onto the roof of the cabin, keeping his body low.

The howl of the wind made it impossible for him to think straight. The landscape seemed to be moving and reshaping itself, as if nothing was real or firm. Finally he determined that one of the braces supporting the chimney pot had come loose. He struggled to straighten the pot and secure the brace, acting by touch alone.

While engaged in this task, Rønneberg was suddenly lifted up and back, as if a giant had grabbed the back of his jacket. Then he was flying, head over heels off the roof, tossed away by a gust of wind. He landed in a snowbank. When he staggered to his feet, the cabin had vanished. All was white, swirling white, around him. Heading into the wind that had knocked him away, he eventually found the cabin. Climbing onto the roof a second time, he managed to fix the chimney pot, only to be hit by another gust of wind that sent him flying into the snow.

A slight respite in the storm the next day allowed the team to venture from the cabin to attempt to locate their depot. All landmarks had sunk into the drifts of snow, including the rods they had put up as markers. A three-hour search ended in vain. Another, in the late afternoon, turned up one container, but then the blizzard returned with a fury.

On the fifth day after their arrival, a fraught Rønneberg wrote, "The storm raged with renewed power. Visibility was zero. The general lassitude of all members of the party was still very much in evidence." All the world was snow and wind, and there appeared no escape from its hold.

As quickly as the storm swept into the Vidda, it left. On February 22, the six Gunnerside men woke up to silence. They stepped from the cabin into a clear, windless day. The blizzard had transformed the landscape. Jansbu was now an igloo. Drifts had gathered to make new hillsides. Stalagmites of ice and snow stood like a collection of quiet sentries on the watch. Jutting precipices of white hung over cliff sides. They might as well have emerged onto a planet made wholly of snow.

Rønneberg gave the order that they must depart for Fetter by early afternoon. For six days they had been out of contact, and as far as Tronstad or the Swallow team knew, they might be dead, and the operation off. They must hurry.

They returned to the general area of their depot, and over the next several hours, rummaged about in the drifts of snow until they found one of the rods marking it and were able to retrieve some extra rations. Given the distance and steep terrain they had to travel to reach Lake Store Saure, Rønneberg decided to minimize their loads. They would carry only enough explosives to blow up the high-concentration plant (not the surrounding machinery), uniforms for Swallow, Kreyberg rations for ten men for five days, and their operational equipment — weapons, hand grenades, shears, axes, field glasses, detonators, time fuses, and first-aid equipment.

At 1:00 p.m. the team was finished packing up back at the cabin, ready to head out, when Haukelid spotted a figure in the distance. Towing a sled, he was headed straight toward them. They retreated inside, shut the door, and hoped he would pass without incident. There was no doubt, however, that their ski tracks would attract attention. In the dead of winter, particularly after such a storm, signs of human habitation deep in the Vidda would surely be cause for investigation.

They readied their guns. As the man approached within a few steps of the cabin's door, they sprang out. With six gun barrels pointed at him, the man's weathered face paled. He was fitted out like a typical Norwegian in winter.

"What are you doing in the mountains?" Rønneberg asked.

"I'm a hunter," he said, innocently enough. They searched him and his equipment. His identity card stated he was Kristian Kristiansen, forty-eight, from Uvdal, a valley due east on the edge of the Vidda. On his sled was over fifty pounds of reindeer meat. He was carrying rifles, a bundle of cash, and his pocketbook contained a list of names and addresses in Oslo. He was evidently who he said he was; the list of names, clients for his meat. That did not translate, however, into him not being a threat.

Rønneberg brought Kristiansen inside the cabin. He asked him if he was a member of Nasjonal Samling. "Well," Kristiansen said, still frightened, "I'm not exactly a member, but that's the party I support."

"Are you sure?" Haukelid asked. The man was all but stating that he was their enemy, in cahoots with the Nazis.

"Yes," he said hesitatingly.

Glancing at the others, Haukelid knew they were all thinking the same thing: that they might need to kill the man. Given their mission, there was no way they could simply detain him. If they let him go, he might reveal their presence to the Germans or to the police.

Haukelid tried a different tack. If he were to go to Uvdal and ask around, he asked Kristiansen, would his neighbors say he had Nazi sympathies? "I've so many enemies down there," he said, "they're sure to say I'm not a Nazi, just to make things difficult for me."

It seemed as though the man was saying whatever he thought would win him the sympathy of his captors. He was in all likelihood harmless, but they could not be sure. It was for Rønneberg to decide what to do. "This is a bit of a shame," he finally said. "You think we're Germans, but we've nothing to do with them. We're Norwegian soldiers, and we assume you look forward to the day when the king and government can come home."

"They've never done me any good," Kristiansen said. "They can just stay where they are."

His statement shocked the Gunnerside men. Kasper Idland asked to speak with Rønneberg in private. "I'll shoot him for you," Idland said, once outside the cabin. Rønneberg knew he was trying to unburden him of the responsibility. It was a kindness. But he put himself into Kristiansen's shoes: six heavily armed bearded men seize him in the middle of the Vidda — he is scared and trying to wriggle out of an impossible situation. It would have been one thing if Kristiansen was carrying an NS card, but he wasn't. He might not pose a threat. Even so, Rønneberg's instructions were that if the unforeseen occurred, he must act with the aims of the mission foremost in mind. He was not ready to decide. He told Idland, "We'd better take him with us for now."

Kristiansen immediately proved to be of use to the Gunnerside team. They took some of his store of reindeer and cooked a big lunch, saving their own rations. Then Rønneberg asked if he could guide them on a route toward Lake Store Saure. Kristiansen said he could. They decided to leave that night to avoid any further chance encounters.

At 11:00 p.m. they departed. Kristiansen was in the lead, a sled tied around his waist loaded with rations and equipment. Rønneberg stayed close behind, compass in hand, to ensure they were following a proper course.

Kristiansen proved a better guide than they could have imagined. Not only was his path sure, but he skied a course that used the natural contours of the land, economizing effort. It was, Rønneberg thought, beautiful to watch.

At dawn, February 23, as the sun rose, first bronze, then gold over the mountains, Kristiansen led them to a small, flat-roofed hut owned by his family, where they rested. He chatted easily with them now and even attempted to buy one of their Tommy guns. When they left the hut and came across a herd of reindeer, Kristiansen begged to be allowed to shoot three or four of them — to be collected later. Rønneberg refused, but he had come to the conclusion that their captive was a simple mountain man, without guile — and no threat to the Gunnerside team.

At the entrance to a long valley, seven miles by their map from Fetter, they spied a man crossing the lake below in their direction. They dropped quickly behind some boulders, none more quickly than Kristiansen.

Rønneberg waved Haukelid over and handed him a pair of field glasses. Given that the skier was headed in the direction of Bjørnesfjord, he might well be a member of Swallow out searching for them. Haukelid would know better than anyone if this was the case. Although the skier was only a few hundred yards away, Haukelid could not make out who it was. He had a heavy beard and wore a thick layer of Norwegian clothes. Then Haukelid sighted another skier coming around a bend, a hundred yards behind the first. Rønneberg ordered Haukelid to move forward for a closer look. If he was discovered, and the skiers were not Swallow, Rønneberg told him he should simply say he was a reindeer hunter, much like Kristiansen.

Haukelid crept through the soft snow as the two skiers came up from the valley toward him. Near the crest, they stopped. One, then the other, scanned the surrounding area. They were looking for somebody, for something.

As they started ahead again, Haukelid recognized a weatherbeaten Helberg when he turned his face toward him. Beside him, bearded and unkempt, was Kjelstrup. For a second, Haukelid remained hidden, overjoyed at the sight of his friends. He wanted to say something funny — "Dr. Livingstone, I presume"— but seeing how thin and wan they looked, he thought better of it. Instead, he simply coughed, and the two swung their heads around, startled, hands on guns. When they recognized Haukelid, a shout, a whoop, then a holler rose over the valley.

A message from Home Station had confirmed that Gunnerside had been dropped on February 16, but still stated, erroneously, that they had been released over Bjørnesfjord. Poulsson had sent Helberg and Kjelstrup out to search for them there. By pure luck, Gunnerside came along that same route from Lake Skrykken. The three men embraced and slapped each other's backs, and Haukelid waved for the others to join them.

Crowded into Fetter that night, the men had a feast. The Gunnerside team provided crackers, chocolate, powdered milk, raisins, and, as was welcomed most of all by Poulsson, "tobacco directly imported" from England. The Swallow men offered reindeer of every cut, including marrow, eyeballs, stomach, and brain. Their guests were happy enough with the lean meat.

Earlier, Helberg had gone ahead of the others to warn Einar Skinnarland that the Gunnerside men were on their way. Skinnarland's identity needed to be kept a secret even from Gunnerside in case anyone was captured during the operation. After Skinnarland left Fetter, Rønneberg, Haukelid, and two more of the six arrived while two others kept Kristiansen under guard in the valley.

After discussion with Poulsson, Rønneberg decided to release the hunter with a warning that if he spoke to anybody about them, they would make it known that he had helped guide their party. "Stay on the Vidda and say nothing," Rønneberg told Kristiansen before letting him go.

Nobody felt completely at ease with the situation, but Rønneberg measured the risk against the taking of an innocent life.

The ten men talked and laughed like old friends into the night.

Haukelid thought of the months Swallow had spent surviving on the Vidda, and considered that, despite their rough beards and sallow skin, they were in remarkably good shape. He asked whether they had experienced any trouble, and when Kjelstrup answered, "Nothing," they all understood it was to be left at that.

Poulsson and his team were cheered to see new faces and to enjoy new conversation. Their long wait and struggle had proven worthwhile. Altogether, they were almost overwhelmed by how uplifted and inspired they felt just at being united with Gunnerside.

Of the mission ahead, there was little talk. There would be time for business in the morning. For now, they celebrated.

16

Best-Laid Plans

A FTER A RESTLESS NIGHT — every bed and every inch of floor taken over by curled-up sleeping figures — the ten men had some coffee and a breakfast of reindeer. Their choice of meat: "boiled or roasted?" Then they gathered around the table to hash out their operational plan.

First, Rønneberg assigned each man his task. He himself would lead the demolition party, accompanied by Kayser, Strømsheim, and Idland. They would split into pairs to double their chances of reaching the target. Haukelid would command the covering party of Poulsson, Helberg, Kjelstrup, and Storhaug. They would see to it that no one interfered with the setting of the explosives.

Haugland would go to Jansbu, the cabin beside Lake Skrykken, with the radio equipment to establish and maintain contact with London. Skinnarland, referred to in front of the Gunnerside members only as the teams' "local contact," would join him.

Now they needed to figure out the best way in and out of Vemork. With a pencil and paper, Rønneberg sketched out the plant and surrounding area. He had never actually been there, but he knew every detail from their prep work and drew it true to scale. Rjukan to the right. Vemork in the middle. Lake Møs to the left. The Vestfjord Valley split the rudimentary map from left to right, following the course of the Måna River.

Vemork was perched on a ledge of rock above the river gorge, on the south side of the valley. The eleven pipelines that fed its turbine generators rose along the valley wall above the power station at a sharp angle. A single-track railway line ran east to Rjukan along this same wall. A seventy-five-foot-long single-lane suspension bridge connected Vemork to the valley's north side. Located by the bridge was Våer hamlet, a scattering of residences for the plant's staff. Through it ran the Møsvann Road, connecting Rjukan to the Lake Møs dam. A long, high trek up the valley's northern wall led to the endless Vidda.

While in Britain, Tronstad and his Gunnerside team had pored over maps and photographs to decide on the best course to reach Vemork. The Swallow team had done the same during their time in Fetter. Now, together, they needed to finalize a plan.

They debated three main routes to Vemork. They could approach from across the top of the southern side of the valley and descend to the plant alongside its penstocks. This idea was quickly ruled out because a guard had recently been placed at the top of the pipelines, and numerous minefields lined the approach.

They could make a straightforward attack: ski down to Våer, neutralize the guards on the bridge, and head across to the plant. This approach had the benefit of easy terrain, but if the guards saw them coming and were able to raise the alarm, the team would have to make their attempt on the heavy water cells in the middle of a firefight. Any escape would be doubtful.

Last, they could cross down through the gorge, climb up to the railway track, and enter the plant through a locked gate that, although patrolled, was not under permanent guard. As Tronstad had suggested in London, this was their best chance of reaching the compound unseen, but there were dangers: certain places along the railway line presented the saboteurs with a six-hundred-foot drop straight into the river below.

Rønneberg decided not to make a decision until they had secured the latest intelligence about the plant's patrols and defenses. Helberg was due to collect such from a contact in Rjukan the following night. Once Helberg had the information, he would rejoin the team at a cabin in Fjøsbudalen, a narrow valley a few miles northwest of Vemork. This cabin would serve as their launch point for the operation.

They took a break from their planning to stretch their legs, and all of the Gunnerside team members, with the exception of Haukelid, went off skiing. Before they returned, Haukelid sat down with his original team members. From a thin piece of paper hidden in his belongings, he read out the operational orders he had developed with Tronstad. These detailed setting up guerrilla groups in Telemark after the Vemork mission was complete. Skinnarland, although absent, was included in this work.

After he read the message, Haukelid balled up the paper and moved to throw it into the stove. "Is that edible?" Haugland asked. Haukelid nodded. "Well, we don't throw away food here." Haugland took the ball of rice paper and popped it into his mouth. He had to work a bit hard to chew it, but it was a change to his diet and welcome for that.

On Thursday, February 25, Rolf Sørlie and his family were celebrating his brother's birthday with chocolate pastries. Sørlie checked his watch frequently — he did not want to be late. Finally, as the time approached 7:30 p.m., he told the others that he was going out for a walk. His mother did not question him, understanding that he was likely up to some resistance work about which it was best she knew nothing.

A few days before, Sørlie had been approached by one of his colleagues at Vemork, Harald Selås, who asked him to find the answers to a list of questions about guard rotations and the like at the plant. Clearly, there was sabotage planned, and Sørlie suspected his friends up on the Vidda were most likely involved.

The answers to the list of questions in his pocket, he headed over to the base station of the Krossobanen, the cable car that brought Rjukan residents up to the Vidda. He arrived early and waited, searching the darkness, unsure who was coming to meet him. Then his old friend Claus Helberg approached from the Ryes Road, a series of switchbacks beneath the cable car line that had been cut through the woods during its construction.

"I'm glad it's you," Helberg said.

Sørlie brought Helberg back to his house and asked him to wait in the garage until the birthday party was over. Then they went up to his room. The family's Gordon setter, Tarzan, barked madly at the stranger's presence, but nobody came out of their rooms to investigate.

The two men went through the list, notably the number of soldiers on guard at Vemork at any one time (fifteen), their rotation schedule (every two hours), the security on the suspension bridge (reinforced in the past month), and the patrols around the grounds (the route went past the railway-line gate). The only viable path into the plant, they agreed, was across the bridge, even though it was defended. As for the approach from the gorge, the two locals recalled the time when a car had plummeted off the Møsvann Road into the gorge, and the rescuers had needed ropes to reach the driver and passengers. It was simply too steep to climb.

After they had gone through the information, Sørlie gave Helberg some leftover "beef" (spiced and fried yeast dough) in the kitchen, which he devoured. He stayed overnight and, in the still and dark of morning, slipped out of the house and headed up to Våer along the main road.

Sørlie watched him go, fearful that he might never see him again.

Throughout Friday, Helberg waited in the Fjøsbudalen cabin for the others to arrive. There was nothing more in the cupboards than some dried syrup riddled with ants. High up in the isolated side valley that ran down to the road to Lake Møs, the cabin offered a fine view of Rjukan in the far distance. Vemork, two miles away on the opposite side of Vestfjord, was not visible.

At 6:00 p.m. the others arrived, led by Poulsson. They had spent the night at Poulsson's brother-in-law's place on Lake Langesjå, where Poulsson had found some new skis and woolen stockings. He also found a bottle of Upper Ten whiskey, but they decided to leave it there for when they returned — if they returned.

The nine men barely fit inside the snug cabin; if they were to sleep, they would have to do so in turns. When a rotating one-and-a-half-hour watch was in place and the windows blacked out, they sat down to go over the new intelligence. Helberg began. Machine guns and floodlights were mounted on top of the main building. Two guards patrolled the suspension bridge. The guardhouse was on the Vemork side, where a third soldier had an automatic weapon and access to an alarm, easily activated if there was trouble on the bridge. If the alarm were raised, the entire area would be lit up — including the penstocks above the building, the suspension bridge, and the road through Våer — and the soldiers in the barracks at Vemork would be alerted, as would the garrison in Rjukan.

The gorge was unguarded, and the entrance to the railway line was only lightly patrolled, but as Helberg explained, this did not make the approach a better option. The climb from the gorge to the railway was all but impossible in summertime. In the dark, in the ice and cold, there was surely no way it could be done.

Idland argued for the bridge approach. It was swift and sure. They would kill the guards, then storm the plant. Poulsson and Helberg doubted it would be so easy, but agreed that the bridge was the better of the two options. Rønneberg sided with them. Only Haukelid was convinced that they should attempt the gorge approach, as Tronstad had recommended. Otherwise, he said, they would likely face a pitched battle to reach the high-concentration plant.

From his rucksack, Rønneberg produced a set of aerial photographs of Vemork, taken during the previous summer. The team had studied these with Tronstad, and the professor had suggested places to cross the gorge and points along the railway line where they could hide before the attack. Rønneberg spread the photographs out on the table.

Haukelid pointed to some scrub and trees growing along the sides of the gorge. "If trees are growing," he said, "you can always find a way." The others nodded in agreement. Rønneberg instructed Helberg to return the next morning to scout a potential route that would keep them from plunging to their deaths. He would need to go in daylight if he was to have a chance of determining one. They all understood that the operation would be in jeopardy if he was spotted.

In Kummersdorf, Kurt Diebner and his team were refining their new reactor design. The central concept was to suspend a lattice of uranium cubes in a sphere of frozen heavy water. The setup was far simpler than their previous experiment, and the architecture of the cubes and moderator were better suited for bombardment.

Abraham Esau, the new head of the Uranium Club and now Diebner's boss, had made it clear that he needed results. "If you make a reactor, and you put a thermometer in it, and the temperature increases only one-tenth of a degree," he said, "then I can give you all the money in the world — whatever you need." On the other hand, he warned, if this could not be achieved, then they would not see another pfennig.

As it was, money was not exactly pouring in to the program. Paul Harteck had developed an ultracentrifuge design to enrich U-235 that showed great promise, but he had been unable to get the funds he needed for an expansion. Diebner had also seen requests for additional heavy water supplies hobbled by limited finances. Two hydrogen electrolysis plants were found in Italy that might be able to provide limited amounts of heavy water of 1 percent purity. This could then be enriched in Germany to almost 100 percent, but only if the pilot plant built by IG Farben outside Leipzig was constructed at full scale. This project was on hold while Norsk Hydro was continuing to provide heavy water at a limited cost to the war effort.

For his next test, Diebner secured, with Esau's approval, a special low-temperature laboratory in Berlin and some of Heisenberg's heavy water supplies. Given that Vemork was producing close to five kilograms of the precious fluid a day — and that the nearby Såheim and Notodden plants were shortly to add significantly to this amount — Diebner could scale his experiments up quickly if his design proved its merit, and resources and manpower would pour in to the program once he produced a self-sustaining uranium machine.

Interest in the potential fruits of atomic fission certainly seemed renewed in many quarters. Hermann Göring, who now oversaw the Reich Research Council, described its atomic research as of "burning interest," as it was indeed for high officials in the Kriegsmarine, the Luftwaffe, and the SS. A uranium machine would bring many new champions.

An hour after sunrise on Saturday, February 27, Helberg headed down toward Vemork. The temperature was mild, but the wind warned of an approaching storm. Because it was the weekend, and because he was in civilian clothes, his hiking and skiing in the valley would not attract undue attention. Passing Våer, he continued east through the woods above the road that ran along the northern wall of the valley between Rjukan and Lake Møs. There he came upon a power-line track that he knew ran parallel to the road.

He followed this track until he sighted what might serve as a path down into the gorge. Stashing his skis and poles, he made his way through the trees and crossed the road. He slipped and slid down the slope, using juniper bushes and pine branches to control his descent. Eventually he reached the Måna River. Its surface was frozen, but the ice was very thin in places. In a warm spell, the river would be impassable.

He hiked back and forth along the river's edge, trying to spy a manageable route up to the railway line. Finally, he spotted a groove in the cliff that was somewhat less steep than the surrounding wall. Some bushes and small trees rose out of the splintered crevices of rock, and they could provide hand- and footholds. Weather and luck permitting, he figured it was worth a try.

Smiling broadly, he returned to the Fjøsbudalen cabin after lunchtime to give his report. "It's possible," he said. Overwhelmingly, the team agreed with the proposal to climb up the gorge on the night of the operation.

Now they could turn their minds to how they were going to get away.

They were confident that, whether they arrived quietly or not, one of them was sure to reach the target and set the charges. Nine heavily armed and well-trained commandos, comfortable in the terrain and eager to serve their country, were a good bet against thirty German guards.

However, even though they did not voice it, most of them, including Rønneberg, believed that the odds of their getting away afterward were thin at best. Once the plant blew, they were not likely to get far. Either they would be trapped at Vemork or hunted down by the hundreds of soldiers garrisoned in Rjukan. Even so, not one of them had a death wish, nor was it good for morale to think of their mission as a one-way ticket.

Rønneberg wanted his men to feel like they each had a say in the final plan. As with the approach, there were a few options. They could climb up the penstocks and escape that way, but the steep ascent and the presence of guards at the top, not to mention minefields, made this an easy option to cross off the list. They could retreat across the bridge and head straight back into the Vidda. This was the fastest, simplest escape. However, they would have to kill the German guards, guaranteeing even harsher reprisals on the local population. And their pursuers would know which way they had gone.

Their third option was to go back down the way they had come. They could return to the Fjøsbudalen cabin or, as Helberg suggested, go in the opposite direction down the power-line track, to Rjukan, and then hike from the Krossobanen base station to the plateau along the zigzagging Ryes Road. The Germans would have no idea where to look for them. This route demanded another descent and ascent of the gorge, then a punishing climb. Further, if they were seen, they would be heading straight into the enemy embrace of the Rjukan barracks.

Once all the possibilities had been sketched out, Rønneberg gave each man a vote. It was a close call. The majority favored leaving the way they came, then ascending to the Vidda underneath the Krossobanen.

The plan was set. Rønneberg told them that the operation would start at eight o'clock sharp that night and that they should rest as best they could.

Soon after, Idland drew Rønneberg to one side. Since he was first chosen for Gunnerside, the heavily built former postman had known that the retreat to Sweden was going to be a problem. He was sufficiently skilled to reach Vemork, and he was as fit as anyone else on the team, but he was convinced that the 280-mile trek to the border would prove too much for him. He had always figured it was going to be a one-way journey, he said, and now he wanted his leader to know that he did not intend to slow down the team if they had the chance to escape. He would find his own way. "Nonsense," Rønneberg said flatly. "You've kept up with us until now, and you can keep up with us to Sweden."

For this decision the young leader did not offer a vote.

That same afternoon, at 4:45 p.m., Leif Tronstad caught a train to Oxford at London's Paddington station. He was on his way to St. Edmund Hall, the oldest academic gathering place at any university anywhere in the world. For a brief spell, he would try to avoid all war news and think only of science and history and the pursuit of knowledge for its own sake. It would not be easy. Two days before, he had finally received word that Gunnerside had survived their first days on the Vidda and had met up with Swallow. "Everything in order," Haugland had sent. "Spirits are excellent. Heartiest greetings from all."

So while he wasn't expecting another message from Gunnerside until the operation was complete, Tronstad was hoping for news from Odd Starheim and his forty commandos. From the start, the Carhampton operation had been plagued with bad luck. Their first attempt to commandeer a convoy of ships had been thwarted by a blizzard. Their second had ended in a furious gun battle, followed by a manhunt. Starheim and his men had hid in a remote valley farm. Instead of abandoning the operation and trying to find a way back to Britain, Starheim insisted they attempt another target. He would not leave Norway without having achieved something — Marstrander and a host of others had lost their lives bringing them there.

Tronstad and Wilson had assigned them a raid on some titanium mines, but the terrible winter weather delayed them — and for too long. The Gestapo closed in, and Starheim and his men fled again. On February 25 the Royal Navy dispatched a boat to pick them up, but a North Sea storm had forced it to turn back. Now the men were without food, equipment, or shelter, and the German noose was tightening. According to the latest cipher message, Starheim was going to attempt to seize a coastal steamer to escape to Scotland.

When Tronstad arrived in Oxford and entered into the cloistered and peaceful world that had once been his own, he had trouble settling, not least because of reports he had received of a recent speech by German propaganda minister Josef Goebbels. "Do you want total war?" Goebbels had demanded of his audience, packed into a Berlin stadium in mid-February. The crowd responded with raised salutes and a resounding "Yes!" Goebbels then shook both fists and shouted: "From now on, let our slogan be 'Rise up, people, and unleash the storm!'"

Tronstad knew the Nazis would follow through on that promise as soon as they had the means to do so.

Sitting outside the Fjøsbudalen cabin, wearing his British uniform, Haukelid quietly smoked a cigarette, ready for the mission that was still a few hours away. Beside him, Helberg and a couple of the others were oiling their pistols and Tommy guns.

Their silence was broken by the approach through the woods of a young man. There was a scattering of cabins in the high valley — they should have expected one or two of them to be occupied over the weekend. The men slipped unseen into the cabin and alerted the others. The stranger knocked once on the door of the cabin. Before he could knock again, the door was thrown open, and Poulsson grabbed him by the throat and dragged him inside.

Haukelid pressed a pistol into his belly. "Who are you?" Poulsson demanded.

The man recognized his captor: "We were in the same class at school, Jens."

"Kåre Tangstad," Poulsson said, releasing him. "Yes, I remember you."

Tangstad explained that he had only come looking for the loan of a snow shovel. He, his fiancée, and another couple were staying up in the valley for a couple of days. Poulsson told him in no uncertain terms that he was to stay in his cabin, with the others, for the whole of the weekend and that they were not to leave for any reason. Tangstad agreed.

Haukelid watched him leave the way he had come, disappearing into the forest. He imagined being in the woods for a simple weekend with a girl and some friends — and felt at an enormous remove from such a normal life.

The saboteurs had a hurried conversation about whether or not their neighbors could be trusted. Rønneberg decided to go over to their cabin to talk with them. Their assurances that they were good Norwegians and opposed to the occupation convinced him that they posed no threat. Parting, Rønneberg called out, "God save the King and Fatherland!" He might well have added the hope that God would look after his team that night.

17

The Climb

A T 8:00 P.M., white camouflage suits covering their British Army uniforms, the nine men skied away from the cabin in silence: Rønneberg, Strømsheim, Idland, Storhaug, Kayser, Poulsson, Helberg, Kjelstrup, and Haukelid. They were armed with five Tommy guns as well as pistols, knives, hand grenades, and chloroform pads. In their rucksacks they carried the explosives for the attack and everything they needed for their retreat into the Vidda: sleeping bags, rations, maps, and other survival gear. Cyanide pills were hidden in their uniforms, to be swallowed in the event of their capture. The men knew all too well what became of those who were brought in for interrogation by the Gestapo.

Helberg led the way down Fjøsbudalen Valley. The moon, hidden by low clouds, shone dimly, and Helberg navigated mostly by memory and a natural feel for the terrain. He kept a steady pace, sweeping around boulders and twisting through the scattering of pine and mountain birch. The others followed closely behind, the cut of their skis barely a whisper through the snow.

Rønneberg had made clear that, no matter what unfolded, no matter whether he or anyone else on the team was killed or wounded, those who were able were to "act on their own initiative to carry out the operation." Destroying the heavy water plant was paramount. Each man knew what to do once they arrived at the target. They had practiced their SOE training many times on schemes against mock targets. Stop and listen frequently. Take short steps, lifting feet high. Move silently. The covering party—the most heavily armed—went in first. To prevent accidental shots, no guns were to be loaded until necessary. Two demolition teams, two sets of charges. Rendezvous after the sabotage, passwords to be called out. Now their mission was underway. They were finally getting their opportunity to strike a blow against their country's occupiers, and from everything they had been told, Vemork would be a significant blow.

Roughly a mile from the cabin, the valley became steep and thick with boulders and shrubs. The men unfastened their skis and hoisted them onto their shoulders. They continued on foot. When they were not sinking up to their waists in the snow, they slipped and scrambled to remain upright, their heavy rucksacks and weapons throwing them off balance.

An hour into their descent they reached the Møsvann Road. Free of the woods, they saw Vemork, a mere fifteen hundred feet across the gorge— were they able to fly. Even at that distance, they could hear the hum of the power station's generators. After months of thinking about this leviathan, months of examining every facet and corner of it in blueprints and in photos and in their minds' eyes, there it was. To a man they stood mesmerized by the winter fortress. It was no wonder, Haukelid thought, that the Germans felt they needed only thirty guards on hand to defend it.

Then they were off, back on their skis and heading east along the road to Våer. The mild temperature and the warm wind coming down the valley had turned the surface of the road into a treacherous blend of snow, ice, and slush, and they had to fight to control their skis' edges. They also needed to keep their eyes peeled for headlights coming up in front or from behind. Rolf Sørlie had told Helberg about recent German troop movement in the area, and there was a chance that the Wehrmacht was still transporting soldiers along this road at night. In spite of the risk of discovery, using the road was considerably easier and faster than attempting the whole distance through the uneven hillside terrain.

They made it to the first sharp turn in the Z-shaped segment of the road without incident. To bypass Våer and avoid spying eyes, Helberg steered them back into the forest. For a spell, they followed the narrow path made for the line of telephone poles that advanced like sentries through the thick woods. Still, they often had to trudge through drifts, sometimes sinking nearly to their armpits in soft, wet snow.

The slope through the trees became an almost sheer drop. Backs flat against the snow, feet acting as brakes, they edged themselves downward. It was so steep that if they bent forward even slightly at the waist, gravity would send them pitching into a head-over-heels tumble certain to take down anyone in their path.

Helberg finally dropped onto the road east of Våer, followed by a few of the others. While they were waiting for the rest of the team, headlights suddenly cut through the darkness. They hurried to hide behind a roadside snowbank as two buses rumbled toward them. Those still sliding down the slope frantically tried to anchor themselves to keep from falling into the road. Two of them narrowly missed landing on the roof of the first bus. But the vehicles, carrying nightshift workers, passed, oblivious to their presence.

Once all the men were on the road, they put on their skis again and traveled east, away from Vemork and toward Rjukan, for about half a mile. When they came alongside an open field, Helberg signaled them to follow him, and they worked their way up seventy-five yards to the power-line track that ran parallel to the road. A short distance down the track, they stopped and unloaded anything they would not need within Vemork into a hastily dug snow depot, including their skis and ski poles. They also stripped off their white camouflage suits — army uniforms were better suited for hiding in the shadows. It was also essential that the sabotage be seen as a British-only military operation to prevent retaliation against the local Norwegian population.

At 10:00 p.m. they made their final checks. Rønneberg and Strømsheim each had a rucksack with a set of explosives, detonators, and fuses, either one capable of destroying the high-concentration plant. The covering party carried Tommy guns, pistols, spare magazines, and hand grenades. Kjelstrup had the added burden of a pair of heavy shears to cut through any locks that stood in their way. "All right, let's go," Rønneberg said.

Helberg guided them down from the power line, across the road, and into the gorge. They hung on to shrubs and branches as they descended to the Måna River. Time and time again they lost their footing, creating small slides of snow that advanced ahead of them. Then they were at the bottom of the valley. The wind continued to blow, and melting snow dripped down the rocks on either side of the gorge. There was a danger that the thaw had caused the waters of the river to rise, sweeping away any ice bridges they were planning to use to cross the river. They trekked along the riverbank, seeking a still-frozen section. The cliffs of the gorge soared upward on either side of them.

After a few minutes, they found an ice bridge that looked like it might hold their weight. Helberg went first, quickly stepping across. In single file, the others followed. The bridge did not break, but they knew that it might well be gone when they made it back — if they made it back.

Now Helberg searched for the slot in the gorge he had seen on his scouting mission earlier that day, for the place he suspected they could climb. He felt no relief when he found it: this mountainside, over six hundred feet to the railway line, was even steeper than the slope they had just descended, and even though a few brave trees clung to clefts of the rock, the ascent looked all but unassailable in the dark, without ropes and pitons. This was the approach they had chosen, and there was no turning back. Rønneberg gave the signal with his hand. *Up.*

Each man took his own silent path up the rock wall, guiding his hands and feet into holds, feeling his way along. Water trickled down the cliff, and they often slipped on patches of ice and encrusted snow. On some stretches, the ascent was more like a scramble, where they grabbed tree trunks and rock outcrops, just to gain a fast few feet. On other stretches, they dug their fingers and toes into crevices and inched their bodies sideways, pressing tightly to the gorge wall to avoid the wind that gusted around them, always reaching higher. Sweat soaked their clothes as they forced themselves up from ledge to ledge. Now and again they rested, flexing numb fingers, rubbing cramped muscles, waiting for pulses to calm, before venturing up again.

A quarter of the way up, Idland was trying to pull himself higher when the fingers of his left hand slipped off a rock handhold. His breath caught in his chest, and he searched frantically for some crevice or spur of rock to grab hold of but found nothing but slick, wet stone on which he was unable to gain purchase. He pressed himself close to the wall, made sure his feet were secure, then switched hands on the piece of rock. His rucksack and Tommy gun suddenly felt very heavy. With his right hand, he stretched out, running his fingers across the wall in every direction, hoping, needing to find something to grab. There was nothing.

To extend his reach, he moved his body in a slow but widening arc, like a pendulum, from side to side. At last the tips of the fingers of his right hand brushed up against what felt like a clump of roots. With the grip on his left hand weakening, he had to act quickly — and boldly — or he was lost. After a few short breaths, he started the pendulum swing again.

Then, when he had the momentum, he forced himself to do the most unnatural thing in the world: he let go. In the same instant, he stretched
out to snag the roots with his right hand. There was a moment when his
hands were empty, when he was sure he was going to pitch backward into
the gorge. Then his fingers tightened around the roots. They held — long
enough for his left hand to grab another hold. He hugged the wall, a swell
of wind coursing around him. Then he continued up.

A half hour into their climb, more than three hundred feet up and the
railway line still out of sight, they were all starting to weary. Their fingers hurt. Their toes were numb. Their limbs ached. They had skied and
trekked for miles through rough, snowbound terrain before even reaching the base of the gorge. Now they were climbing a steep mountain gorge
in the pitch dark with heavy, awkwardly balanced rucksacks on their
backs. Any missed hold or slipped foot could mean a fatal fall.

When training in the mountains of Scotland, they had been taught
never to look down when they were climbing for fear of losing their nerve.
But one or two of them did look down the way they had come. The gorge
seemed like terrible jaws, ready to devour them. The sight froze them until the sounds of the efforts of their mates broke through their momentary
terror. Shaking it off, they continued. No matter what individual battles
they had to fight on the wall, they were not alone. If one of the men found
an easy path, the others followed as best as they could. When one of them
sank into a pocket of snow, needing a good shove from behind to get moving again, he didn't need to wait long. Or if someone was searching fruitlessly for a secure hold, help came quickly, either in the form of advice
about a possible foothold or a hand stretched out from above.

At last, a few minutes past 11:00 p.m., the first man scrambled up the
final bit of scree to the railway line. The others followed — dazed, exhausted, relieved to be at the top. For a spell, nobody spoke. They rested
on the tracks and looked to the fortress at the end of the line.

In their cabin retreat, thirty miles from Rjukan, Skinnarland and Haugland sat on their beds as a snow squall blew outside. Neither had wanted
to remain behind just to report on the operation. They wanted to be in on
the job; they wanted to help. But that was the curse of the radioman: he
was the eyes and ears of the mission but rarely had a hand in its execution.

Well-trained radio operators were limited in number within the SOE and were considered too valuable to risk in sabotage operations. Their transmissions — and the fact that they were burdened with such heavy equipment — made them traceable targets for the Germans. In fact, radio operators suffered the most casualties among SOE agents. At that moment, though, Haugland and Skinnarland were safer than many of their countrymen, high up in the Vidda, in the middle of winter.

Skinnarland could not help but think of his brother Torstein and his best friend Olav Skogen, both being held at Grini. They had suffered so much, while he was still free. Both Skinnarland and Haugland knew their families might face reprisals if the Vemork sabotage succeeded — or even if it did not. The Germans would suspect any commandos would have had local help. There was nothing to do but wait and wonder and hope that all the intelligence they had gathered and all the preparations they had made before the Gunnerside team arrived were contributing to the operation's success.

Careful with each step, Haukelid led the others down the railway line toward Vemork. The wind had blown the snow clear from the track closest to the gorge's edge, and they kept to the frozen gravel so as not to leave footsteps. There was very little moonlight, and Haukelid was sure that they would not be seen coming along the track. He was even more sure that they would not be heard — he could hardly hear his own voice over the wind, the din of rushing water, and the plant's massive turbine generators.

Across the valley, the lights of a vehicle snaked along the road that they had crossed an hour before. As Haukelid came around the bend, the team following silently behind him, he saw the suspension bridge down below to his right and the silhouettes of two soldiers guarding it. Straight ahead, five hundred yards away, was the railway gate, and beyond that stood the hulking shapes of the power station and hydrogen plant. Squeezed between these behemoths was the small German guard barracks.

At 11:40 p.m. Haukelid stopped beside a snow-covered transformer shed and waited for the others to catch up with him. This was the perfect vantage point from which to watch the bridge and wait for the midnight changing of the guard. Rønneberg wanted them to move on the target half an hour after that. This would allow enough time for the new sentries to ease into their duty and get lulled into the regular routine. Using the shed as a windbreak, the nine men settled down on the tracks. They ate chocolate and crackers produced from their pockets. A few of them ambled out of sight to relieve themselves.

Rønneberg then gathered them close together so he would be heard over the loud drone of the power station and asked if they understood their orders. Despite their nods, he went over the key points one more time: The demolition party would enter through the basement door and lay the charges. The covering party would hold their positions until the sabotage was complete. No matter what, the target must be destroyed. If they were captured, suicide orders were in effect. They must not be taken into interrogation.

A few minutes before midnight, a new set of guards headed down to the bridge. Soon after, the men they had relieved from duty plodded up to the barracks. Bundled in coats, caps drawn tightly over their heads, they looked bored, complacent. They carried their weapons in the listless manner of men who felt they had nothing to fear.

The saboteurs returned to their waiting. There were nerves, to be sure, but beside some gallows humor about what would happen if they were caught, they each kept their fears to themselves. Instead, they gazed out at the surrounding valley and made small talk about anything and everything apart from the operation ahead. Kjelstrup cleaned his teeth with a matchstick. Helberg joked about the ant-infested syrup he had found at Fjøsbudalen. Others talked about old pranks and romances gone sour. From the conversation, they might have been mates shooting the breeze after a night on the town.

The minutes ticked by, and eventually Rønneberg's watch told him it was almost half past midnight. With barely a nod from him, the men rose to their feet. They checked their guns and explosives one last time. "In a few minutes we'll be at our target," Rønneberg said. He echoed the words Tronstad had told them all before they left for the mission: "Remember: What we do in the next hour will be a chapter of history for a hundred years to come . . . Together we will make it a worthy one." Then Rønneberg gestured to Haukelid, who left first with the covering party.

Now within range of the land mines, Haukelid was even more deliberate with each step. A few yards down the track, he found some footprints in the snow, probably from one of the plant's workers. These were his safe passage. Kjelstrup followed close behind him, then the seven others in single file. One hundred yards from the gate, they stopped behind a short row of storage sheds. At this proximity, the hum of the generators was now a roar.

At 12:30 a.m. sharp, Rønneberg signaled to Haukelid and Kjelstrup to cut the gate lock. "Good luck," he whispered before they headed over the final yards in a crouch. With the shears, Kjelstrup snapped off the sturdy padlock like it was a dry twig, and Haukelid pulled the chain out and pushed open the twelve-foot-wide gate. The five-man covering party entered Vemork quickly, Poulsson the last man in. The demolition party— Rønneberg, Kayser, Strømsheim, and Idland—followed soon after.

Within moments, Haukelid and his team had scattered to take up their positions, ready to lock down the guards if the alarm sounded or if there was any approach on the demolition party. Tommy gun at the ready, Kjelstrup covered the steps down from the penstocks. Helberg stood at the railway gate, protecting their way out. Storhaug was posted on a slope with a clear bead on the road leading down to the suspension bridge.

Haukelid and Poulsson slid into position behind two large steel storage tanks, fifteen yards from the barracks. While Haukelid lined up a row of hand grenades, Poulsson trained the barrel of his Tommy gun on the barracks door. They also kept their eyes peeled for any roving patrols. "Good spot," Poulsson said.

In the meantime, the demolition party cut a hole in the fence fifty yards down from the railway gate. Then, farther down again, they snapped the lock off another gate that led to some warehouses. These would both serve as alternative paths of escape. When they were done, Rønneberg stopped still for a long moment. He cast his eyes about for any sign of movement in the darkness; he also listened intently. The plant's machinery drummed on. So far, they were undetected. With Kayser close by him, Rønneberg crossed the open yard to the eight-story hydrogen plant. Strømsheim followed, Idland covering him with a Tommy gun. The two men with the explosives were to be protected above all the others.

The four saboteurs went around the side of the plant. Little flecks of light glittered in the windows where the blackout paint was chipped. In places it did not completely cover the glass. Rønneberg peered through a sliver of a gap in a window at the northeastern corner of the building. A lone individual watched over the room.

The demolition team edged around the eastern wall of the plant at basement level until they reached a steel door near the corner. His gun at the ready, Rønneberg pulled at the handle. The door didn't budge. "Locked," he said. He sent Kayser up an adjacent concrete stairwell to see if the first-floor door was open. Returning down as fast as he went up, Kayser reported, "No."

Rønneberg tensed. They needed a way into the plant, one that did not involve blasting the thick steel doors or breaking the windows, either of which would alert the Germans.

18

Sabotage

RØNNEBERG RECHECKED THE plant's basement and first-floor doors himself. Kayser stayed with him, searching the shadows for any sign of an approaching patrol. Strømsheim and Idland looked for another way in. They knew they had only so much time before a guard crossed their path or ran into the covering party. Growing desperate, Rønneberg was struck by the thought of the cable tunnel on the northern wall. Racing down the steps, he waved for Kayser to follow him.

At Brickendonbury Hall, they had discussed alternative entry points into the plant, and Tronstad had told them about a narrow tunnel filled with pipes and cables, which ran between the basement ceiling and first floor and out a small access hole in the exterior wall facing the gorge. Rønneberg thought that if the tunnel had not been blocked during the recent security upgrade, it might provide an access point.

He hurried around the building and searched through the snowbank along the outside wall for the ladder he had been told led up to the tunnel. After a couple of minutes, his hands came across a rung. "Here it is," he said to Kayser. The two climbed the slippery steel, Rønneberg first. Fifteen feet up, he found the tunnel entrance, half-filled with snow. It was unbarred. He swept the snow out of the opening, then crawled inside. There was barely enough room for his body, and he had to drag the rucksack of explosives behind him. Kayser squirreled into the tunnel after him.

They made their way on their bellies over the cables and pipes for several yards. Rønneberg tried to turn his head to see if the others had come in after them, but the space was too cramped. Kayser confirmed that Strømsheim and Idland were not in the rear. The only direction in which they could go was forward. They would have to execute the sabotage alone.

Rønneberg kept crawling. After a few minutes, he saw some water pipes that bent through a hole to his left into the ceiling. Through the hole, he could just make out some of the plant's high-concentration cells. They were close.

They continued to worm their way forward. Suddenly Rønneberg heard the sharp ping of metal. He went still. Behind him, Kayser had slipped, and his Colt .45 had fallen out of its shoulder holster. Although attached by a string to his body, the pistol had dropped far enough to hit a pipe. For a long spell, the two remained completely motionless, worried that the sound might have given them away. But the longer they waited, the more certain they were that the reverberating drum of Vemork's machines had masked the noise. Kayser returned the pistol to its holster, and they continued ahead.

Twenty yards into the maze of pipes, Rønneberg arrived at a larger opening in the floor. He looked through it into a cavernous hall. After making sure there were no guards in the room, he slipped through the opening and dropped the fifteen feet to the floor. Remembering his parachute training, he collapsed into a roll to blunt the fall. Kayser came down after him.

They reached the room with the eighteen high-concentration cells. A sign on the double doors read: NO ADMITTANCE EXCEPT ON BUSINESS. Colts drawn, Rønneberg and Kayser flung open the doors. The night-shift worker overseeing the plant, a portly, gray-haired Norwegian, swung around from his seat at the desk. Kayser was beside him in an instant. "Put your hands up," he barked in Norwegian, his pistol barrel aimed straight at the worker's chest.

The man did as ordered. Clearly scared, his eyes flicked back and forth between the two saboteurs. "Nothing will happen to you if you do as you're told," Kayser said as Rønneberg locked the doors of the room behind them. "We're British soldiers." The worker looked at the insignia on their uniforms. "What's your name?" Kayser asked.

"Gustav Johansen."

As Rønneberg got to work, Kayser kept guard over Johansen, dropping a few remarks about life in Britain to reinforce their cover. Rønneberg unpacked the explosives and fuses from his rucksack. The two rows of nine high-concentration cells on their wooden stands looked exactly like the replicas Tronstad and Rheam had assembled at Brickendonbury Hall. Each cell tank was fifty inches tall and ten inches in diameter and was made of stainless steel. A twisting snake of rubber tubes, electrical wires, and iron pipes ran out of the top.

Rønneberg did not need to know exactly how the cells worked — only how to blow them up. The eighteen sausages of Nobel 808, each twelve inches long, set out in front of him, would do the trick. They might only weigh ten pounds in total but they would bring about an almighty bang.

After putting on rubber gloves to avoid electrical shock, Rønneberg moved to the first cell and pressed the plastic explosive to its base. Once this was secure, he moved to the second, then the third, his movements almost automatic after so many hours practicing at Brickendonbury.

Hands still high, Johansen was becoming increasingly nervous as he watched Rønneberg's work. Finally he blurted out, "Watch out. Otherwise it might explode!"

"That's pretty much our intention," Kayser replied drily.

Rønneberg had just finished fastening the ninth band of explosive to its cell when glass shattered behind him. He made a grab for his gun as he and Kayser turned swiftly in the direction of the threat.

Out in the plant yard monitoring the guard barracks, Haukelid glanced at his watch. Twenty minutes had passed since they broke through the railway gate. It felt like hours. He wondered if the demolition team had made it into the plant and if the operation was going according to plan. At any moment, a sentry could raise the alarm, bringing searchlights, sirens, and machine guns. All Haukelid could be certain of was what was around him. Darkness. The relentless drone of the generators. The barracks' closed door. His eyes searched constantly for any sign of the guards who were patrolling the grounds. He had ampoules of chloroform at the ready to take them down. So far they had seen no one.

He was reminded of the first weeks in the fight for Norway. One hundred miles north of Oslo, he had been part of a group of Norwegian soldiers who surrounded a wooden house occupied by German soldiers. When they refused to surrender, Haukelid and the others in his party had opened fire on the structure. Its thin walls gave scarce protection, and the soldiers were soon dead to a man. Some dangled out of the broken windows. Others lay in pools of blood on the floor. War was ugly, and Haukelid knew that the guards in the barracks in front of him would not survive the barrage of his Tommy gun and grenades.

With each slowly passing minute, the gnawing fear in him grew that something might have gone wrong. Every man in the covering party had his gun primed to fire or his grenade pin set to be pulled, as they waited for the demolition team to finish the job.

With the butt of his gun, Strømsheim knocked a few more pieces of glass out of the broken window. Still recovering from the surprise, Rønneberg ran to help. Strømsheim and Idland had decided to force their way into the high-concentration room — and only narrowly missed being shot by their own compatriots in the process.

Rønneberg helped clear the glass so Strømsheim could enter. In his rush, he cut the fingers on his right hand on a shard of glass. He told Idland to stay outside and block the light shining from the broken window. The guards would surely come running if they saw it.

Then Rønneberg and Strømsheim resumed setting the explosives to the cells. Working together, they finished quickly, then double-checked that everything was in place before securing the 120-centimeter-long fuses to the charges. When these were lit, they would burn at a rate of one centimeter a second, which would give them exactly two minutes to get clear of the room.

Strømsheim suggested they use a pair of thirty-second fuses to ignite the explosives, to ensure that nobody snuffed out the bomb after they left. "We can light the two-minute ones first and see that everything is okay," Rønneberg said. "Then we each strike the short ones and get out." Strømsheim agreed.

As they set about fixing the thirty-centimeter fuses, Johansen broke their concentration: "Where are my glasses? I must have them." Rønneberg glanced at the man. Johansen continued: replacement glasses would be very hard to come by because of the war; he must be allowed to look for his. There was a brief moment when the ridiculousness of the request struck the saboteurs. So many had suffered, had risked their lives — or indeed, lost them — in order for them to reach this critical moment, and now they were being asked about spectacles.

Rønneberg rose and searched the desk. He found the glasses case and passed it to Johansen, then returned to his task, attaching the fuses with insulation tape. His right glove was soaked with blood from where the window glass had cut him.

"But the glasses aren't in the case," Johansen whined.

Rønneberg turned to the man, clearly irritated. "Where the hell are they then?"

"They were there," Johansen pointed to the desk, "when you came in."

Rønneberg once again crossed to the desk and found the glasses wedged between the pages of Johansen's logbook. Johansen thanked him meekly.

Close to finishing their work, Rønneberg told Strømsheim to unlock and open the exterior basement door so they could make a quick exit. Johansen pointed out the key on his lanyard, and Strømsheim took it and disappeared from the room. At the same time, Kayser hauled Johansen out of the room to get clear from the blast. As the three crossed the large hall, they heard footfalls coming down the interior stairwell. A German guard?

Kayser and Strømsheim leveled their pistols toward the stairwell. A moment later, they found themselves aiming their guns not at a German guard but at the night foreman, a stunned Olav Ingebretsen, who flung his hands up and whelped in surprise. While Kayser guarded the two hostages, Strømsheim opened the basement door slightly. A rush of cold air blew inside.

In the high-concentration room, Rønneberg made one final check of the chain of explosives. Almost forty-five minutes had passed since they first entered by the railway gate. They were pressing their luck. Confident that the explosives were properly set, he tore off his bloodied gloves and dropped a British parachute badge on the floor. Then he took out a box of matches and, with a quick flick of his wrist, struck a light.

He brought the flame first to the two-minute fuses, then to the thirty-second fuses. Then he barked at Idland, who was still standing outside blocking the window, to get clear. Rønneberg dashed into the hall, counting down the seconds in his head. To the two prisoners, he said, "Up the stairs. Then lie down and keep your mouths open until you hear the bang. Or you'll blow out your eardrums."

As the Vemork workers raced up the stairs, the three saboteurs pushed through the basement-level steel door. Kayser flung it closed behind him, and the men sprinted away from the plant. Idland joined them on their run.

They were twenty yards away when they heard a muffled boom and saw flames burst through the shattered windows of the high-concentration room. Alive with the thrill that they had done the job, they made their escape toward the railway line.

On hearing a faint thud in the distance, Haukelid and Poulsson looked at each other. "Is that what we came here for?" Poulsson asked. From the light now streaming from the basement of the plant, it was clear that the windows had been blown out, but that was no guarantee the sabotage was complete. Then again, they whispered to each other, their eyes now trained back on the barracks, the high-concentration room was housed inside thick concrete walls, and the explosion would have been muted by the wind and the drone of the power station. But they were expecting something more forceful. Had something gone wrong?

In that instant of doubt, the barracks door opened, casting an arc of light onto the snow. A guard stood in the doorway for a few seconds, looking left and right, before stepping out into the cold. He wore a heavy coat but was unarmed and had no helmet. Even so, Poulsson cocked his Tommy gun for the first time that night, and Haukelid placed his index finger through the ring of a grenade's safety pin.

Both waited to see what the guard would do next.

At a slow walk, seemingly not in the least alarmed, the guard crossed the fifty yards between the barracks and the hydrogen plant. He gazed up at the plant building, then at the surrounding area. If he saw the light coming from the broken basement windows, he did not react. Seconds later, he returned to the barracks and closed the door behind him. Perhaps he had taken the noise to be a land mine set off by a wild animal or by a fall of thawing snow.

Haukelid knew that the demolition team could have already retreated to the railway line through the hole they had cut in the fence. There was no sign of them now, but enough time had passed for them to get away. He was about to tell Poulsson that they should fall back, when the barracks door swung open again.

This time the guard came out wearing a steel helmet and carrying a rifle. Advancing from the barracks, he shone the flashlight beam close to where Haukelid and Poulsson were positioned. Poulsson brought his finger to the trigger and took aim. The guard was only fifteen yards away. A single shot would take him down and might even go unnoticed. The guard swept his light in an arc and slowly approached their hiding place behind the storage tanks.

Poulsson looked back at Haukelid, his expression clear: *Shall I fire?* Haukelid shook his head sharply and whispered, "No." They were not to kill unless absolutely necessary. Until that beam of light exposed them, they would wait. The guard swung around again, the beam creeping across the snow, almost touching their feet. Poulsson drew a tight bead on the man with his submachine gun. Then the guard turned on his heel. Glancing once again about the grounds, he returned to the barracks.

Haukelid and Poulsson waited another minute, then bounded toward the railway gate to rendezvous with the others.

On reaching it, they heard a voice call out from the darkness: "Piccadilly!" They would have answered with the matching code phrase, "Leicester Square," but before they could speak they were already upon Kjelstrup and Helberg.

"Piccadilly," Kjelstrup urged, his training ingrained.

"For God's sake, shut up," Haukelid and Poulsson said together with a laugh, jubilant that they had made it this far. Helberg told them that Rønneberg, Kayser, Strømsheim, Idland, and Storhaug were already moving back down the railway track. Haukelid closed the gate and looped the chain in place. Hiding the direction of their retreat from the Germans — even for a few seconds — might make all the difference.

The four were a couple of hundred yards down the track when the first sirens sounded. The alarm caused their steps to quicken, and soon after they caught up with Rønneberg and the others. They all shook hands and pounded one another on the back. The mission was a success, without a single bullet fired or grenade thrown. They could hardly believe it themselves.

But there was no time to celebrate. Sirens now sounding throughout the valley, they flung themselves into the gorge, hurtling down the southern wall with little thought to avoiding injury. They simply wanted to get away. Helberg found a slope that was slightly less steep than the one they had ascended, and they hopped and scrambled from ledge to ledge through the banks of heavy, wet snow. There were still some stretches they needed to climb, but mostly they exercised a controlled fall down into the gorge.

As he moved, Rønneberg thought through their chances of escape. Since the men had neither crossed the bridge nor retreated up the penstocks, the guards might believe they were still inside the plant. Once they discovered their footprints or found the broken padlock, they would know differently. How much of a head start would they have? Would the Germans have dogs? When would troops arrive from Rjukan? Where would they be stationed? On the northern side of the valley? At the foot of the Krossobanen? Why had the plant's searchlights not yet been turned on? Without answers to these questions, all they could do was move as fast as possible, and as far away from Vemork as possible, before the manhunt started.

When they reached the valley floor, they encountered deep pools of water on the surface of the ice bridge across the Måna River. Helberg crossed first. The others followed. By the time they were over, their boots were soaked.

They reached the other side of the gorge and began their ascent, clutching at whatever they could find — a root, a boulder, a tree — to pull themselves up. Once again their clothes stuck to the sweat on their bodies. Their throats ached with thirst, but still they climbed.

As they reached the road, they turned to see flashlights moving along the railway line, roughly 150 yards from Vemork. The direction of their escape was known. They would have to move, and faster still.

Back at Vemork, Alf Larsen, the chief engineer, stepped over the steel door that had been blasted off its hinges and shone his flashlight around the high-concentration room. Everything lay in ruins. The two rows of heavy water cells — what was left of them — stood at an awkward angle on the floor, their wooden stands in splinters. The pumps were broken, the walls were scorched, the windows were shattered, and the network of tubes overhead was a twisted wreck. Shrapnel had sliced through the copper pipes of the cooling system, and water was spraying the whole room.

Half an hour before, the thirty-two-year-old engineer, who had replaced Jomar Brun after his mysterious disappearance, had just finished a long game of cards in one of the worker houses between the plant and the suspension bridge when he heard the explosion. It was exactly 1:15 a.m.

He rang the hydrogen plant, and a few moments later he was on the line with Olav Ingebretsen. Still catching his breath, the night foreman had explained that three men broke into the plant and took him and Johansen prisoner. They spoke Norwegian — "normal like we do" — but were wearing British uniforms. "They blasted the plant into the air," Ingebretsen told him.

As the sirens blared in the background, Larsen rang Norsk Hydro's plant director in Rjukan, Bjarne Nilssen, and informed him of the events. Nilssen said that he would drive up to the plant straightaway after alerting the local German army commander and SS officer. The minute he hung up, Larsen headed to the plant himself.

Now Larsen, drenched through from the shower of water, stepped across the debris-strewn floor to a row of high-concentration cells and bent down to inspect the damage. All nine of the steel-jacketed cells were in shreds. The other row was the same. All the precious heavy water inside the eighteen cells had poured out and swirled down the room's drains. Whoever these saboteurs were, they knew exactly what to destroy, and they had done their work well.

German police troops march into Oslo in May 1940.
NATIONAL ARCHIVES OF NORWAY

Kurt Diebner, first leader of the German atomic bomb program.
NATIONAL ARCHIVES AND RECORDS ADMINISTRATION, COURTESY AIP EMILIO SEGRÈ VISUAL ARCHIVES

Werner Heisenberg, Nobel Prize–winning German physicist.
NORGES HJEMMEFRONTMUSEUM

The dam at Lake Møs. NORGES HJEMMEFRONTMUSEUM

Professor Leif Tronstad.
NORSK INDUSTRIARBEIDERMUSEUM

Jomar Brun, head of the Vemork
heavy water plant.
NORSK INDUSTRIARBEIDERMUSEUM

Before the war, Knut Haukelid was a bit of a lost soul.
PRIVATE COLLECTION, HAUKELID FAMILY

SS Lieutenant Colonel Heinrich Fehlis (left) and Reichs-kommissar Josef Terboven (center).
NORGES HJEMMEFRONTMUSEUM

Martin Linge, founder of the Norwegian Independent Company No. 1.
NORGES HJEMMEFRONTMUSEUM

The members of what became known as the Kompani Linge. NORGES HJEMMEFRONTMUSEUM

The Norwegian king, Haakon VII, and Tronstad, in exile in Britain. NORGES HJEMMEFRONTMUSEUM

Lieutenant Colonel John Wilson, SOE Norwegian branch leader.
NORGES HJEMMEFRONTMUSEUM

Einar Skinnarland.

Odd Starheim.

Galtesund, captured ship.

Norwegian commandos
complete a parachute jump.

Jens-Anton Poulsson,
leader of the Grouse mission.
NORGES HJEMMEFRONTMUSEUM

Arne Kjelstrup.
NORGES HJEMMEFRONTMUSEUM

Knut Haugland, Grouse radio operator.
NORGES HJEMMEFRONTMUSEUM

Claus Helberg.
NORGES HJEMMEFRONTMUSEUM

Parachute landing on the Vidda.

The Vidda.

A wireless radio set used by the
Norwegian resistance.
FREIA BEER/ORKLA
INDUSTRIMUSEUM

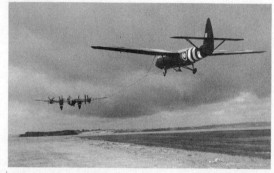

Halifax airplane
towing a Horsa glider.
HULTON ARCHIVE/
ROYAL AIR FORCE MUSEUM/
GETTY IMAGES

Members of the 261st Field Park Company Royal Engineers from Operation Freshman.
Seated: Lieutenant Colonel Mark Henniker (bottom row, fourth from left). DENIS BRAY

General Nikolaus von Falkenhorst and Reichskommissar Josef Terboven visit Vemork.

Olav Skogen,
Rjukan resistance leader.

PRIVATE COLLECTION / NORSK
INDUSTRIARBEIDERMUSEUM

Møllergata 19, a Gestapo prison.

NORGES HJEMMEFRONTMUSEUM

Reindeer herd in Norway.
ERIC CHRETIEN/GAMMA-RAPHO/GETTY IMAGES

Clockwise from top
left: Fredrik Kayser,
Kasper Idland, Birger
Strømsheim, Joachim
Rønneberg (leader),
and Hans Storhaug.
NORSK INDUSTRI-
ARBEIDERMUSEUM

Knut Haukelid,
second in command of
Gunnerside.
PRIVATE COLLECTION,
HAUKELID FAMILY

Vemork.

The suspension
bridge and gorge at
Vemork.

Dramatization of
the Gunnerside
saboteurs.

Ruins of the heavy water cells.
NORGES HJEMMEFRONTMUSEUM

Tronstad (seated, center)
with his team after
the Gunnerside mission.
NORGES
HJEMMEFRONTMUSEUM

The building at Gottow,
near Berlin, where Diebner's
team built their G-I and G-III
piles using heavy water.
PHOTOGRAPH BY SAMUEL
GOUDSMIT, COURTESY AIP
EMILIO SEGRÈ VISUAL ARCHIVES,
GOUDSMIT COLLECTION

Kurt Diebner's G-III machine.
AIP EMILIO SEGRÈ VISUAL ARCHIVES,
GOUDSMIT COLLECTION

Haukelid and Skinnarland
outside Bamsebu.
NORGES HJEMMEFRONTMUSEUM

Owen D. "Cowboy" Roane,
flight officer.
COURTESY OF THE 100TH BOMB GROUP,
WWW.100THBG.COM

The American planes ready
an attack on Vemork.
NORGES HJEMMEFRONTMUSEUM

The *Bigassbird II* and its crew in October 1943.
COURTESY OF THE 100TH BOMB GROUP,
WWW.100THBG.COM

Rjukan after the American raid. NORGES HJEMMEFRONTMUSEUM

Rolf Sørlie, Rjukan resistance member.
THE LONGUM COLLECTION / NORSK
INDUSTRIARBEIDERMUSEUM

Knut Lier-Hansen,
Rjukan resistance member.
PRIVATE COLLECTION / NORSK
INDUSTRIARBEIDERMUSEUM

The D/F *Hydro*.
NORGES HJEMMEFRONTMUSEUM

Left to right: Haukelid, Poulsson, Rønneberg, Kayser, Kjelstrup, Haugland, Strømsheim, and Storhaug receive official honors for their service.
NORGES HJEMMEFRONTMUSEUM

Celebration after the war at the hotel owned by the Skinnarlands.
O. H. SKINNARLAND FAMILY PHOTO

The saboteurs' memorial at Vemork.
JÜRGEN SORGES / AKG-IMAGES

Part IV

19

The Most Splendid Coup

THE NINE SABOTEURS ducked behind a bank of plowed snow as a car rushed past from the direction of Rjukan. The car disappeared around the bend, and they set across the road, which had become little more than an icy stream in the thaw. Just as the last of them made it to the far side, another car came barreling down the road. They jumped into the ditch to escape its headlights. Trucks full of soldiers were sure to follow.

After locating their supply depot, they once again donned their white camouflage suits and collected their gear, then skied along the icy power-line track toward Rjukan. Poulsson and Helberg led the way. Both local boys, they were thinking about their families in town. What would the Nazis do in retaliation for the operation? How easy it would be to slip into Rjukan, sit down for a meal with their parents and siblings, protect them if needed. Nobody knew they had participated in the sabotage. They had not been seen. Putting these dreams away, they skied on, toward the Krossobanen.

The sirens continued to sound, and a truck sped past on the road below. Down in Rjukan, the beehive of Germans was stirring. If they so much as suspected that the saboteurs were making their escape under the funicular, they would be caught. They could not risk coming onto the Ryes Road beside the base station. Roughly a mile and a half down the power-line track, they took off their skis, hoisted them onto their shoulders, and headed into the woods. After a short hike, they came to the steep, zigzagging road. Weary from the long operation and weighed down by their gear, they still had a half-mile vertical climb to reach the Vidda. It was already past 2:00 a.m., and they wanted to be at the top of Vestfjord Valley and away before dawn in five hours. Each switchback on the path was a fifteen-minute slog. The men trudged in single file, each trying to follow in the footprints of the man in front. On some stretches, the path was slick. On others, they sank into the snow. At each bend, they took a short break and then forged ahead.

Three-quarters of the way up the northern wall of Vestfjord, the men were long past exhaustion. Pure will and the fear of capture drove their bodies now. When one of their team fell back, another slowed to urge him on. At times, a gap in the trees gave them a view either of Vemork or of the Krossobanen base station. Mysteriously, floodlights had yet to illuminate the area around the plant, and the funicular station was still dark. The Germans might easily have sent a band of soldiers up in the Krossobanen to cut them off at the top. There would be a bitter fight if it came to that.

After four hours of climbing, they reached one of the last switchbacks. Avoiding the station at the top, they returned to the woods, an even tougher hike. At last, they were at the top. They celebrated only briefly, then refastened their skis. The wind was beginning to gust in their faces, and the temperature was falling fast. A storm was on the way. They struck out into the open hills of the Vidda. Then, as dawn was breaking, Rønneberg called for a break. The men sat down on the hillside and rested. They ate some chocolate, raisins, and crackers, and gazed quietly across toward Vestfjord Valley. Silvery blue clouds hung overhead, and to the southeast the towering peak of Mount Gausta was framed in red by the rising sun. Unseen, a bird chirped.

Lingering there, the men thought back on their mission. With the operation behind them, there was a general sense of amazement that they had all come through it alive. And that they now had their lives ahead to consider.

Helberg prepared to ski back to the cabin in Fjøsbudalen to retrieve his passport and the civilian clothes he had used during his reconnaissance around Rjukan. The others had kept theirs at Fetter, where Helberg planned to join them after they stopped at Lake Langesjå to rest. If there was any trouble, he would meet Poulsson at a café in Oslo in a week's time. Before there were any farewells, he'd skied away.

The other men set a course northwest through the Vidda toward Lake Langesjå. At around 7:00 a.m., the storm finally hit, and winds struck with such force that the men could barely stand. Bent over their poles, they fought for every step. At times they were flung backward — almost off their feet. The snow turned brittle and slick in the sudden freeze, making their advance harder still. If there was any advantage to the storm, it was that it would remove all trace of their movements. Hour after hour, the men struggled on, until at last they sighted the cabin on the edge of the lake.

Inside, they stripped off their rucksacks and collapsed. They stayed awake just long enough to raise a toast to their success with the Upper Ten whiskey they had left behind. There was no need to post a watch. The Nazis would never venture into the Vidda during such a storm. Lying in his bag, waiting for sleep to come, Poulsson felt a curious mix of emotions. He was proud the mission had come off so well. However, for five months he had endured almost nothing but hardship. Now, within a few short hours, it was all over and without an actual fight. He couldn't help feeling a tinge of disappointment that he had yet to test his mettle against his country's invaders.

The men slept for almost eighteen hours straight, the raging wind clamoring through their dreams. In the morning, the storm still blowing, they set off for Fetter, wanting to get deeper into the Vidda. As they made their way, the relentless wind broke hunks of ice and snow from rocks and ridgelines, sending them hurling through the air, pummeling their faces and bodies. So far, the Nazis had been nothing compared to the wrath of the Vidda.

When his car swept along the road to Vemork, thirty minutes after the reported explosion, SS Second Lieutenant Muggenthaler had failed to see the nine saboteurs hiding behind the bank of snow at the side of the road. Formerly a Munich policeman, Muggenthaler was Fehlis's right hand in Rjukan, responsible for collecting intelligence and breaking up resistance cells. The sabotage of a key industrial site was a major problem for him.

As soon as Muggenthaler arrived at the plant, Bjarne Nilssen and Alf Larsen brought him down to the basement of the hydrogen plant. The pumps for the cooling system had been shut off, but the floor was a swimming pool. Larsen informed him that they had lost all the heavy water in the cells and that rebuilding the high-concentration facility would take months.

Muggenthaler then proceeded out toward the railway line with Nilssen

and Sergeant Major Glaase, the head of the German guard, to investigate
how the saboteurs had come in and gone out. A severed lock and shears
had been found in the snow by the gate, and there were traces of blood
along the tracks. Guards were looking for where the saboteurs had gone,
but the truth was that they might be moving in any direction now: up the
southern wall of the Vestfjord, across toward Rjukan, down the gorge and
up the northern wall. It would be dark for hours, and waking soldiers in
Rjukan to hunt through the snowbound, wooded hills was sure to be an
exercise in futility.

When they returned to the plant, Muggenthaler began his interrogations. He started with Johansen and the foreman who had encountered
the saboteurs. Then he questioned Larsen at length. He kept his gun on
the table, toying with it, making clear to the men what they faced if they
lied to him. On the surface, the events were clear: three men with heavy
weapons, wearing British uniforms but speaking flawless Norwegian, infiltrated the high-concentration plant. They had clear knowledge of the
facility. They set the charges with efficiency, and then made their escape
without firing a shot. This was not the work of amateurs, a fact made all
the more clear when a rucksack of explosives and detonators was found
by a patrol.

Who had helped them? How? And where were the saboteurs now?
These were open questions that Muggenthaler needed answered. By the
end of the day, he had arrested several Vemork workers, including Johansen and Ingebretsen, who were present during the attack. He had scores
of locals brought in to his headquarters in Rjukan's Grand Hotel for interrogation. The best description he could glean — "three strongly built men
who spoke Norwegian" — was essentially useless.

Nevertheless, house-to-house searches began. Identity cards were
checked. Roads and railway stations into and out of the valley were closed.
The telephone lines were shut down. Notices were posted on walls and
signposts throughout Rjukan, instructing residents to notify officials immediately if they had any knowledge that could lead to the arrest of the
perpetrators. There would be "sharp coercive measures" taken if they did
not obey.

Soldiers tracking the saboteurs had discovered a trail down into the
gorge, almost a mile from the plant, and another up to the power-line
track on the other side, but the storm that continued to blow through
the dark, half-frozen valley had erased any further traces. The saboteurs

were clearly skilled skiers. If they had retreated up to the Vidda, there was nothing the Germans could do while the current weather conditions prevailed.

Muggenthaler notified his superiors in Oslo of the sabotage and his investigations. A report was sent to SS headquarters in Berlin, stating, "an installation of importance to the war economy was destroyed" by attackers likely from British intelligence and the Norwegian resistance. The report continued that they had targeted the "most important part of the plant." While Muggenthaler waited for reinforcements and orders, a list of five key townspeople, including Nilssen, was drawn up and posted throughout Rjukan. Those whose names appeared on it would be shot if there were any further attacks.

Early the next day, March 1, General von Falkenhorst himself arrived, in spite of the blizzard. It was his men who had been on guard at the plant, and he wanted answers too. He had alerted his superiors in the Wehrmacht about the events, and notified his officers throughout Norway to expect further attacks. Falkenhorst inspected the wreckage of the high-concentration room and remarked that it was "the most splendid coup." The attackers were clearly military.

His admiration for their work did not dull his wrath toward those guarding the plant. Even though Larsen and several other Norwegians were present, he berated his men, chiefly Sergeant Major Glaase and Captain Nestler. "When you have a chest of jewels, you don't walk around it. You plant yourself on the lid with a weapon in your hand!" Then he walked along the icy path to where the saboteurs had entered through the railway gate. Turning to Glaase and his men, all of them wrapped in heavy fur coats, caps drawn tightly over their heads to ward off the cold, he barked, "You lot look like Father Christmas. How can you see or hear saboteurs with all those clothes on?"

Glaase tried to defend himself, saying that they had completed laying land mines around the penstocks but had not yet finished doing so around the railway line.

"Why have you not finished it?" Falkenhorst demanded.

"We haven't enough men to do these jobs."

"Men? You know in Russia they use women for that sort of work?"

Glaase tried to list the plant's many defenses: The barbed-wire fences. The double guard on the bridge. The machine-gun nest. The floodlights.

"Then put them on," Falkenhorst ordered.

Glaase sent a guard to switch on the lights. Several minutes passed. The lights stayed off. Falkenhorst stamped his feet impatiently. Glaase sent off one of the plant's workers to see what was wrong. The guard had been unable to find the switch.

Falkenhorst left Vemork in a rage. He ordered Glaase and some of his guards be removed to the Russian front and instigated a range of new security measures. In Rjukan, he ordered the release of most of those who had been arrested and warned that there must be no reprisals against the local population. The sabotage of Vemork had been a military operation, and revenge killings would not change that fact. A manhunt for the saboteurs and anyone who had aided them would be launched, and SS Lieutenant Colonel Fehlis was in charge of its execution. As for the plant, whether it would be razed to the ground or reconstructed was for the German scientists who needed the materials produced in Vemork to decide.

At Kingston House that same day, having returned from Oxford on the early-morning train, Leif Tronstad was passed a note from the BBC Monitoring Service. They had missed the first part of the Swedish radio broadcast, but what they did transcribe provided enough information to spark celebration throughout the office: ". . . perpetrated against Norsk Hydro installations. The damages are said to have been extensive, but at one point where the attack was made the destruction is said to be complete. The attack was performed by three Norwegian-speaking soldiers in British uniforms who are now being searched for."

The Nazi atomic program had been delivered an assured blow, and Tronstad was eager to know more.

It was turning out to be a banner day. A wireless message transmitted to London reported that Odd Starheim and his team had captured the seven-hundred-ton coastal steamer *Tromøsund* and were making for Aberdeen. The Royal Navy had dispatched destroyers to escort *Tromøsund,* and a pair of long-range RAF fighters was also en route to intercept any German planes that might attempt to thwart their escape. Now all that Tronstad wanted was for the Vemork saboteurs and the Carhampton commandos to make it back to safety.

Through the blur of snow and biting winds, Poulsson led his team across the Vidda to Fetter. They had to turn back earlier in the day during the storm, but as soon as it eased, out they went again. When they arrived at the cabin, the winds strengthened again. There they stayed for thirty-six hours. Helberg did not arrive, and even though Poulsson always said there was no better man in getting out of trouble, worry crept in that he had either got caught out in the blizzard or been captured.

When the weather cleared, the eight men departed for Lake Skrykken in the north. As previously arranged, Rønneberg left a message for Haugland hidden in a hut along the way, to pick up and send to London: "High-concentration plant totally destroyed. All present. No fighting."

At Jansbu, the cabin where Gunnerside had spent its first nights in Norway, the team got ready to split up, portioning out food, weapons, ammunition, clothes, and other supplies. All of Gunnerside, apart from Haukelid, were setting off on the ten-day trek to the Swedish border. Haukelid and Kjelstrup were off to the southwest to organize resistance cells. Poulsson was heading east, to Oslo, in the hope that Helberg would meet him there.

The next day, March 4, the Gunnerside five departed first. They continued to wear their British uniforms under their white camouflage suits, in case they were caught. In parting, the men shook hands and wished one another luck. Before Rønneberg turned to go, Haukelid said, "Give our best regards to Colonel Wilson and Tronstad. Tell them we shall manage whatever happens." Any doubts Rønneberg had about Haukelid's suitability for second in command had long since been dispelled. Together, the two had orchestrated a flawless operation.

Wearing civilian clothes and carrying only a few supplies in his rucksack, Poulsson was next to leave. He grabbed Haukelid's hand and wished him well. Then he moved to Kjelstrup. They had spent months together through the worst hardship and danger. "Remember the day you carried those heavy batteries through miles of fog and snow?" Poulsson said, choked with emotion. "You looked just like a snowman when you got back."

"I felt like one," Kjelstrup said.

"Well, Arne," Poulsson said finally. "Goodbye and good luck. If we don't meet sooner, we'll meet after the war." Skis fastened, he started away.

Haukelid and Kjelstrup watched him disappear into the distant white of the countryside. They felt very much alone themselves and worried that Poulsson was even more so.

• • •

After thirty miles of skiing, a night in a cold hut, and spare rations, all Poulsson wanted was a good sleep in a soft bed. The welcoming lights outside a small roadside inn were too much to resist. Nobody had seen him during the Vemork operation who might provide a description, and he had a fake passport — it was in the name of Jens Dale. A hot meal alone was worth the risk, he decided.

He went in. The fireplace crackled with flames, and the smell of fish and potatoes wafted out from the direction of the kitchen, comforts that softened the reticence he felt when he saw the Quisling newspaper, *Fritt Folk*, open on the reception counter.

"On holiday?" the innkeeper asked.

"That's right. Good skiing," Poulsson replied. There appeared to be few guests. After dinner, he retired to a room on the second floor. He took a bath, washed out some of his clothes, and relaxed on the bed, half-dressed. Months on the Vidda made the simple room a luxurious experience. Then he heard voices downstairs. Easing open the door, he made out two men asking who was staying there. Then footsteps, fast and heavy, coming up the stairs. Poulsson shut the door, threw on pants and a shirt, and considered leaping from the window. There was a sharp rapping on his door. Sticking his pistol in his right pocket, he opened it. A police sergeant with big teeth and a face that seemed to retreat into itself entered the room, followed by a burly, younger man — his deputy. The sergeant asked Poulsson for his identification, while the deputy took a hard look around the room. Poulsson followed his eyes, from his clothes hanging out to dry to his rucksack half-open on the floor. Unwisely, it contained Kreyberg rations, printed maps, chocolate, and English cigarettes. While the sergeant checked his passport, Poulsson sat on the edge of the bed, his hand gripping the pistol in his pocket. If either officer moved to inspect his bag, he would shoot them both.

The deputy stepped toward the worn sleeping bag on the floor, and commented on its high quality. Stitched on the inside was an English label, but he did not examine it that closely. "What are you looking for?" Poulsson asked, his palms beginning to sweat.

"Something happened in Vemork. Saboteurs attacked the hydro plant," the sergeant said, passing back Poulsson's passport. He explained that the Germans wanted any strangers in the area inspected.

"Hope you have luck finding those men," Poulsson said.

"No," the sergeant said. "I think they're armed . . . I'd rather not meet them." The two policemen apologized for the disturbance and closed the door behind them. Poulsson let go of his gun and eased his hand from his pocket.

Leaving the inn now would only raise suspicions. Instead, he got dressed, packed his bag, and slept under the feather-filled comforter, his pistol close at hand.

The wireless radio set and the Eureka device on his back, Einar Skinnarland was headed for Skårbu, a cabin some fifteen miles northwest of Fetter. He needed to rest. On each of the past three days, he had skied roughly a marathon back and forth across the Vidda, bringing food, equipment, and weapons over to the cabin, which would serve as a temporary base for him and Haugland. There was still no word from Gunnerside about the operation. Haugland had gone to check the prearranged dead drop, but if this day, March 6, was like the previous ones, there would be no note.

Approaching Skårbu across a frozen lake, Skinnarland spotted a pair of tracks in the snow. As he neared the cabin, he saw skis leaning against the wall. Then, out through the door came Kjelstrup and another man, a stranger. He was introduced to Knut Haukelid — and given the good news about Vemork. Almost a year had passed since Skinnarland started his double life, spying on the plant. Almost a year since he had brought so much hardship on his family and the friends who had helped him. Finally, the sabotage was done. He quietly celebrated the moment; the sacrifices had not been in vain.

Inside the cabin, Skinnarland brewed some coffee. Haukelid and Kjelstrup recounted the night of the operation, after which the conversation turned to what lay ahead. All three were dedicated to remaining in Norway to foster the resistance.

Near midnight, Haugland returned to the cabin in a terrible temper. For several hours, he had dug through the snow at the dead drop, looking for the tin box with the message that Gunnerside was supposed to leave indicating how the mission went. And to think that all the while Haukelid and Kjelstrup had been here, in Skårbu. He was glad to see his friends, but he wanted to know what happened, why the delay, where was everybody else? "Don't worry, Knut, keep calm," Haukelid said, easing his feet onto the table. He paused. "It all went according to plan." The news sent Haugland into a stomping, cheering dance joined by everyone in the cabin.

The men then began drafting a message to be sent to Home Station. Tronstad and Wilson would be desperate for news. Haugland tried to get a connection on the wireless transmitter but something must have broken when Skinnarland was transporting it. It would need to be fixed — and if one thing was certain, it was that a manhunt would soon be underway, if it had not begun already. "You can bet the Germans are in a fury," Haukelid said. "They'll search every corner of the mountains."

"Those backcountry peasants and factory hands aren't worth much up here in the wilds," Kjelstrup said with contempt. Still, they would have to hide themselves — and well.

At noon on March 7, the fourth day of their march to Sweden, Rønneberg and his men were crouched in the corner room of a farmhouse, waiting for a skier to leave the area. The minutes ticked by slowly. They had watched the skier enter a cabin only thirty feet away from the farmhouse; he had remained inside for over an hour. Only when he'd left to go on his way did they relax. At 6:00 p.m. they moved on themselves — they had been waiting for night to fall before they crossed the long Hallingdal Valley.

The first stretch through the woods was easy enough. One man skied ahead of the group as a scout, making sure there was nobody else on their path. Then the pitch of the slopes became difficult to navigate. Rønneberg in particular struggled, his injured hand swelling badly. He said nothing to the others.

They came across a lumber road and followed it down to the valley floor, where they attempted to cross the Hallingdal River over patches of frozen ice. But the ice broke apart, and they were forced back to the bank. They found a boat, but to take it might have attracted attention, so they kept on going. Farther north, they spotted an ice bridge that ran all the way across the river and used it to cross.

They climbed the eastern wall of the valley, becoming lost in a labyrinth of lumber tracks, unable to get their bearings in the dense woods and darkness. They continued to zigzag upward until they reached the top of the valley's eastern side. Then they skied for a couple of miles before calling a halt beside a lake. Their pants and boots soaked, exhausted after the night's trek, they crawled into their sleeping bags.

Rønneberg had always known that the escape to Sweden would be a trial over many days. Five men in British uniform, heavily armed in case of a fight. Two hundred and eighty miles through an enemy-occupied country on alert for their presence. A punishing terrain of steep valleys and half-frozen bodies of water that none of them had traveled before except for short stretches. They were exposed to freezing temperatures and to storms, and when no empty cabins or farmhouses could be found, they had to sleep outdoors without an open fire. Their course steered clear of country inns, towns, and bridges, and far from anywhere that had a German garrison.

Back in Britain, Rønneberg had prepared exhaustively for what he foresaw as a ten-day, circuitous journey: they would travel northwest from the Vidda, across the Hallingdal Valley, then northeast until they circled the town of Lillehammer (a Nazi stronghold). From there, they had to cut southeast through three long valleys, after which they would finally reach the Swedish border. They brought Silva compasses and twenty-five topographical maps. But their maps and compasses could not predict thawed ice bridges, random patrols, Norwegian hunters, and blinding storms.

The next day, the winds quiet and the sky clear, they made good progress through low hills and gentle valleys. After sunset, the temperatures fell, and a stiff wind blew, clearing their path of snow. The bare ice beneath them ground at their skis. They found another unoccupied farmhouse, this one with stores of flour and bread. In the fireplace, they set alight the map that covered the area they had crossed through during the day — a ritual celebration.

On the sixth day, they set a quick pace, never venturing more than a few hundred yards from the planned route. They crossed paths, unavoidably, with two skiers on the bare mountainside and hoped that they were mistaken for German ski-troops, in their camouflage whites, weapons visible. Then they came to a lake that they needed to cross or take the long way around. Parts of the lake were thawed, but they found a path that they thought might work. Rønneberg edged his way out on the ice on his hands and knees, ax in hand. He inched forward, ears tuned to any snap or pop, testing the surface with the ax head. The ice was weak, but should hold their weight.

20

The Hunt

IN MØLLERGATA 19, Olav Skogen lay quietly on his bed, the rainbow of bruises across his body throbbing with pain. Even though nine days and nights had passed since his torturers last came, he was on constant alert for their return. He knew their routine: They came at night. A door opening. The shuffle of footsteps down the corridor. The rattle of keys. Then the light in his cell would flicker on, the door thrown open, and they would be upon him.

On their last visit, March 1, four of them had taken him to Victoria Terrasse. The shouting began immediately: What did he know about the underground resistance in Rjukan? "What are you trying to hide?" They were in a murderous mood, and Skogen knew that something significant was at play. When he told them he knew nothing, their leader, a burly bear of a man, pulled back his fist and punched him in the face. The blow knocked him from his stool onto the floor, where he lay half-conscious. As they pulled him back to his feet, he promised himself yet again: *Never a word.*

Then they fitted a screw clamp to his right shin. When it dug into his bone and still he did not answer their questions, they tightened another on his left shin. He tried not to scream as the clamps tore into his flesh but could not help the guttural sounds that escaped his mouth. Silently to himself, he repeated the words he'd heard Churchill deliver in a radio address before the German Blitz, as if the British prime minister were speaking directly to him: *This is your finest hour.* When the punishment of his legs did not persuade him to talk, his torturers inflicted the same on his arms, which swelled like balloons. Then he passed out. A bucket of water poured over his head had the intended effect of waking him up. Then they kicked him in his sides until he passed out again. As he crept back to consciousness, he overheard the four Germans speaking about Vemork, about how the plant had been blown up and the saboteurs had yet to be found. A faint smile flickered across Skogen's face before another kick lifted him from the floor. When he next awoke, he was back in his cell.

That night, March 10, still suffering from his injuries — one eye half-shut and his limbs swollen — he heard the familiar footfall. His torturers reeked of alcohol. "Until now you had first-degree and second-degree torture," one told him. If he did not cooperate, the third degree would commence: ripping off his fingernails and breaking his bones. Continued resistance would see him hanging from the hook on the wall until he spoke or died, whichever happened first. Skogen was silent. "You're not fit to be examined tonight," another said. "But soon you will be well enough, and then we'll come for you one last time."

Never a word.

"Operation carried out with 100 percent success. High-concentration plant completely destroyed. Shots not exchanged since Germans did not realize anything. Germans do not appear to know whence the party came or whither they disappeared."

On March 10, Tronstad had his long-awaited confirmation from Swallow. More messages followed to Home Station over the course of the day: news of Falkenhorst's inspection at Vemork; a multitude of arrests in Rjukan; the information that only three of the party had been sighted — and only three were being hunted; and a note from those Swallow men who wanted a firm mandate to continue their work in Norway, Haukelid among them.

Tronstad was moved by this last request. Here were men who had already risked so much, and they wanted to stay and do more, no matter the inevitable crackdown after Gunnerside. He and Wilson sent a message of their own: "Heartiest congratulations on excellent work done. Decision to continue your work approved. Greetings from and to all." Two days later he delivered his report to the SOE at Chiltern Court. His best estimate was that between six hundred and seven hundred kilograms of heavy water had been destroyed (four months' worth of production) — and without an aerial bombing that would have caused considerable collateral damage.

The Germans would need at least six months to reconstruct the high-concentration cells and an additional four to six months to return production to previous levels. In total, this would delay German heavy-water supplies by ten to fourteen months.

Tronstad sent the same report to Eric Welsh at SIS, writing in an attached note: "It's justified to say the Germans have suffered a very serious setback of their project in utilizing the atomic energy for war or other purposes." Sir John Anderson, his team at the British atomic program, and Winston Churchill were all informed of the same. The operation gave both the SOE and Kompani Linge a major victory, elevating their reputations.

This success was in stark contrast to the fortunes of Operation Carhampton. The *Tromøsund* never made it to Aberdeen. Odd Starheim had escaped some tight spots in the past, but this time Tronstad knew that he was probably dead. The RAF swept the North Sea but found no sign of the steamer. German newspapers celebrated its sinking: "England's once-proud Navy is so depleted that she needs to steal ships from little Norway . . . but just as the pirates were jingling their ill-gotten golden reward in criminal pockets, the Nazi fighter planes came upon the scene and sent the gangsters and their prey to the bottom of the ocean."

In his diary, Tronstad blamed the British for providing insufficient cover for the escaping ship. In other entries, he wrote, "We must accept these losses" — they showed the brave fight being put forward by Norway — "grim in our loneliness," and that Norwegians had "sacrificed enough for a while." He was most affected by the loss of Starheim, and kept a photograph of him, shoulders draped in his country's flag, on his mantelpiece.

But Tronstad knew, above all, that they had to carry on. He arranged the dispatch by sea and air of several teams of Kompani Linge agents to establish resistance networks with wireless radio stations in Oslo, Trondheim, Ålesund, and elsewhere — all in the expectation of a future Allied invasion. To provide for their security, he was also fighting a political war, pushing the high command and the SOE to keep these resistance cells independent instead of having a central Milorg command that would risk the entire network if it were infiltrated.

Using intelligence provided by the cells run by Tronstad, the RAF destroyed key structures at the Knaben mines, which provided molybdenum to the Germans, used in armor plating. Operation Granard saw the sinking of a cargo ship loaded with pyrite. And a mission named Mardonius was underway that planned to use limpet mines to blow up enemy troop and cargo ships in the Oslofjord.

All the while, Tronstad continued to develop his atomic-intelligence network. On March 15, he and Eric Welsh sat down with Victor Goldschmidt, a Swiss-born but Norwegian-educated professor who had recently fled Oslo for Britain. Goldschmidt offered some limited insights into the Nazi program and urged that Niels Bohr, who was Jewish, be brought to London as soon as possible. As one of the fathers of atomic physics, Bohr was too important to be left in occupied Denmark.

In fact, Goldschmidt's proposal had been rebuffed by Bohr himself, but he'd insisted something be done to convince him. The Danish physicist, who believed he could best serve his country by remaining in Copenhagen, had also turned down Welsh and Tronstad when they had approached him earlier that year.

In the midst of this work, Tronstad hoped every day to hear news of Rønneberg and his men — to hear their "heartbeats" by wireless once they arrived in Sweden. With a hunt for the saboteurs now underway, they needed to hurry, just as those who had stayed behind needed to keep out of sight.

On March 13, the tenth day of their trek, Rønneberg and his men were just north of Lillehammer, still roughly a hundred miles from the Swedish border. Despite averaging twenty miles a day, they had moved more slowly than Rønneberg anticipated. The snow, the need to travel at night, and the difficulty crossing valleys and fording rivers had all cut at their pace. They were struggling to stay nourished, stealing food from cabins along the way to supplement the ten-day supply of rations they had brought for the journey.

Their muscles were exhausted from the persistent strain, and their skin chafed from the constant damp. Now they faced the heavily trafficked Gudbrandsdalen Valley that lay between them and the Swedish border. They set out before dawn to dodge any weekend skiers who might be staying in one of the several hotels in the area. As the sky grew light, two German Junkers shot past overhead. The men hoped they were mail planes traveling between Oslo and Trondheim. They could not be certain. A few hours later, they settled down in their sleeping bags, not quite yet out of the valley, and took turns keeping watch.

That evening, when they were preparing to get going again, Idland asked to speak with Rønneberg alone. Idland had been struggling to stay with the team, and there had indeed been some stretches of rough terrain that they would have traversed on skis if Idland had had the skill. "You must all speed up and get yourself to Sweden," Idland said. "I'll follow." Rønneberg, who felt that Idland's workhorse attitude more than compensated for any lapse in athletic ability, dismissed the idea. "Stop it now," Rønneberg said. "You're imagining things." Idland tried to protest, but Rønneberg cut him off. They would arrive in Sweden together.

Under the light of the moon, they headed across Gudbrandsdalen. The roads were sheer ice so they kept to the fields, and there was little wind, but it was very cold. The thin straps of their rucksacks cut into their shoulders, and their legs ached after climbing the valley. After midnight, Rønneberg signaled them to stop, and they prepared beds in the woods, laying pine needles and heather on top of the snow. Then they wriggled into their custom sleeping bags, which were proving to be lifesavers.

Over the next seventy-two hours, the men tramped in a southeasterly direction, through woods and fields, often battling driving snow and sudden winds. They crossed a number of ski tracks, evidence that there was a lot of movement in the area, possibly German. They often struggled to navigate, unable to find points in the distance or in the dark to orient themselves on their maps. At these times, Rønneberg advanced on instinct, buttressed by his experience orienteering and the months of studying their escape. Perilously low on rations, they also sometimes ventured deliberately off course, desperate to find cabins with food stores. Often they found little or nothing.

Late on March 16, after a mistaken diversion down the wrong valley, they came to the Glomma, the largest river in Norway. To their shock, it was clear of ice. Rønneberg sent Storhaug, a native of the region, off to find a boat. The rest of the team took cover in a hay shed to wait for him.

After a few hours, Storhaug returned: he had located a boat they could steal. In the dark morning hours of the seventeenth, they rowed across the Glomma, then sent the boat drifting downstream. A miserable, frost-ridden rest in their sleeping bags followed. In the morning, they continued their journey, giving a group of lumberjacks a wide berth, then crossing through a confusing tangle of woods, roads, and streams that for hours left them at a loss as to their position. The heavy snow made each step a labor.

They broke into the cabin of someone Storhaug knew to be a Nazi sympathizer, convinced he would have a rich supply of food. He was wrong; the cabin was unsupplied. Again, they slept outside in damp sleeping bags and sweat-soaked clothes, but they were too ruined by exhaustion and hunger to care. Rønneberg dreamed of tables groaning under the weight of platters of food.

They woke up in a blanket of fog. It was fifteen days since they had set out from Lake Skrykken, and they were fewer than twenty miles from the Swedish border. After a time, they approached a road that cut through a long, open field. Crossing it now would expose them in broad daylight. They would have waited until dark, but they had too few rations and they were so close to reaching Sweden and safety. Rønneberg instructed the men to stay low and move quickly. Then, "All right, let's go."

The team skied as fast as their legs would carry them across the field, feeling like they were in the middle of an assault, eyes darting left and right, watching out for any trucks or cars. They reached the road, their breath heavy in the air. After checking that nobody was coming from either direction, they crossed over. Then they hurried through the other side of the field, their backs exposed to the road. Within minutes, hearts thumping in their chests, they were into the woods and able to slow down as they passed through a half-frozen marsh.

In the midafternoon, sun blazing down, they finally took a long rest. They shed their shirts and boots, spread out their sleeping bags to dry, and ate what was left of their food.

"Guys," Idland said. "When we get to London, I don't want to see your bloody faces for fourteen days. I'm completely sick of you." The others smiled and laughed, at ease for the first time in weeks. They were so close to safety now.

As night fell, they moved through some low country pocked with stone-ridden ravines, thickets, and gnarled, twisted trees. It was tough going, and orientation was difficult as well, but there were no Germans in sight. At 8:15 p.m., March 18, they finally passed Border Marker No. 106 into Sweden. Then they built an open fire, settled themselves down around it, and burned the final map. Afterward, they crawled into their sleeping bags, shattered with relief and exhaustion.

In the morning, they buried everything that marked them as soldiers, including their guns. Wearing civilian clothes, they had to hike twelve miles over the border before they found a patrol to surrender to. Their cover story was that they had escaped from a German prison where they had been held for underground activity. If the Swedes believed their story, the five would be brought to a refugee camp, from where they could reach out to Norwegian officials connected to SOE.

When Knut Haukelid was a little boy, he believed that there were trolls living in the Norwegian countryside, far from prying eyes. Now, barely a day's journey from the mountains and lakes where he'd spent so much of his youth, he and Kjelstrup were just like those trolls, hiding high above the treeline in a thin-walled hut.

But it was not a carefree life. In truth, they were starving. It was not just hunger pangs they suffered but the kind of deprivation that left the body weak and the mind empty of purpose. They knew that they needed to hunt but also that the attempt, if unsuccessful, would sap what little strength they had left. Scraping his plate of crumbs after another paltry meal one night, Kjelstrup said, "When this war is over, I'll spend all my money on food." Haukelid looked at his friend's cheeks sunken under his red beard. Badly weakened after months on the Vidda, Kjelstrup was in a bad way.

Over two weeks had passed since they left Skårbu. They had trudged west across fifty miles, hauling a sled laden with weapons, radio equipment, and other gear, until they arrived at a lakeside cabin on the grounds that encompassed the Haukeliseter mountain lodge. Haukelid's cousins lived three hours away in the small farming village of Vågslid, and they had stocked the place with canned food and oats. From it, Haukelid and Kjelstrup planned on launching an underground resistance cell in the area.

Soon after their arrival, they heard that the local magistrate was on the lookout for the Vemork saboteurs, aided by several German patrols, so they headed southwest to a mountain hut in the neighboring district to lie low. They ran through their rations quickly, and their early hunts for reindeer were unsuccessful. With his ski pole, Haukelid had skewered a scrawny squirrel stuck in a snowdrift, but it took almost as much effort to skin and cook as it provided in nourishment. They took to eating raw any small animals they trapped or shot. When they were not dreaming of food, they were dreaming of firewood. The best they could find were juniper bushes buried in the snow or small birch trees, which took half the day to collect and get back to their hut.

By the last week of March, Haukelid and Kjelstrup were in parlous shape and knew they had to get back to Vågslid for food. Then they would travel onward to Lake Møs. After leaving Skårbu, Skinnarland and Haugland had intended to move on to Nilsbu. There, Haugland planned on teaching Skinnarland how to code and transmit wireless messages so he could operate his own radio station in the area.

Haukelid and Kjelstrup left for Vågslid on a gray, foggy morning. They skied down from their hut, then headed east through the mountains until they reached Haukeli Road, which had been hewn out of the rough rock with pickaxes some decades before.

In the late afternoon they reached Haukelid's uncle's farm outside Vågslid. Haukelid waited by the side of the road while Kjelstrup went up to get bread and other provisions. Haukelid was too well known in the area to risk exposure. As he waited, an ominous feeling came over him. Trusting his instincts, he retreated behind a mountain birch. Moments later, two German soldiers carrying rifles came around the hill and walked toward his position. Crouched low, Haukelid gripped the handle of the pistol tucked into his belt. The soldiers passed within five feet of him, but did not detect his presence. Nor did they notice his ski tracks on the road.

At the farm, Kjelstrup did not receive a warm welcome. Haukelid's cousin passed him four loaves of bread and urged him to get away. A huge sweep of troops had descended on the district. "It's not safe in the village," the cousin said. "It's forbidden to go between farms, and the Germans are patrolling the road every hour." Kjelstrup left and descended toward the road on his skis. As he rounded the bend, he spotted several soldiers on patrol. He braked abruptly and ducked down in the snow, hoping that Haukelid was out of sight. They had unwittingly entered a trap.

After reports pointed to the Hardangervidda as the base for the Vemork saboteurs, a manhunt was launched on March 24, and Fehlis himself established temporary headquarters south of Rjukan. No effort was to be spared in rounding up those responsible for the attack and anyone who supported them. Fehlis sent an army to accomplish the task: thousands of Wehrmacht infantry, hundreds of German and Norwegian police, Gestapo investigators, and SS shock troops, and, finally, dozens of Jagdkommando platoons. These were elite soldiers who specialized in destroying guerrilla groups where they lived and operated. Numbering close to eight thousand men in total, Fehlis's army was aided by locals who knew the countryside and was supported by roving patrols of Fi-156 Storch spotter planes.

A Norwegian hunter named Kristiansen had, after an outing on the plateau, returned to his village with English chocolate and a tale about the well-armed soldiers who had taken him hostage on the Vidda. The police chief had arrested him and passed him on to the Gestapo for interrogation. Patrols had uncovered evidence to corroborate his account. According to their interviews, "Seven men were seen on skis on the Hardangervidda, going toward Rjukan. Two were in civilian clothes; five in uniform; and carrying, among other things, submachine guns. All had white camouflage clothing on." Reports stated that a cabin on Lake Skrykken had been broken into, and that "tracks of five pairs of skis and a sled were seen running from Rjukan and avoiding inhabited areas" after the attack.

If enemy commandos and the Norwegian resistance were persuaded that the Vidda was an ideal base for their operations, then Fehlis intended to prove otherwise. His troops circled the barren plateau like a noose, then he launched his raid. Troops scoured the countryside, searching every farmhouse and cabin for the fugitives or their supporters. They were also on the lookout for illegal weapons, explosives, radios, newspapers, and other contraband. Travel within the Vidda was banned, and anyone found wandering in the region would be arrested immediately. Any dwellings used for resistance purposes would be immediately burned. In his operational orders, Fehlis warned his men that the agents were heavily armed and would use any means to escape. They should search buildings in force and be prepared for ambushes. Every effort should be made to bring them in alive so they could be interrogated, but if they refused to surrender then they should be shot.

Although the Vidda was his key target, Fehlis knew that he could not limit his search to the plateau. In Rjukan and surrounding towns, roadblocks were established. Travel, even on foot, was restricted to those with passes. Curfews were imposed, and posters informed residents that anyone who violated the new restrictions would be "shot without warning." Further, Fehlis ordered intensive searches in the neighboring regions to the south and west of Lake Møs. Intelligence had revealed that these areas were hotbeds of resistance. He ordered his troops to flush out those hiding in the mountains so that they would be snared on the roads, which were easily patrolled. Another force was sent to the Swedish border, in case the fugitives were headed in that direction.

It took weeks to assemble his army, but Fehlis used the men to full effect. The Vemork sabotage and the burgeoning resistance movement must be dealt with, and he was the one to deal with them. Terboven and Falkenhorst were paying attention, and Berlin surely awaited news of his success.

21

Phantoms of the Vidda

As soon as the German soldiers passed out of sight, Kjelstrup swept down the slope, his skis leaving a telltale trail in the snow. When he reached Haukelid, he told him, "The whole district is lousy with Huns."

"Two of them went by just now. They're really asking for it."

Haukelid knew that they should get rid of the bread—a sure sign that someone in the village had helped them—but they desperately needed the food. They raced down to Lake Vågslid, then across its still-frozen surface. There were more Germans on the road north, but the two men spotted them first. They hid in some brush until the threat passed.

They could either rejoin Haugland and Skinnarland at Nilsbu, a long journey to the east, or go back the roughly twenty miles they had come and hope that the German action did not extend that far. They decided to head back into the mountains, hoping that the rising winds would erase all trace of their movements before they were noticed by a patrol. A good tracker could tell if a ski track was a minute, a day, or a week old. Whenever they could, they maneuvered onto trails and summits that had been blown clear of snow to reduce the chance of leaving a mark.

They continued their flight into the next day, until the winds picked up such force that they had to stop. They bedded down under a slight overhang of rock on a hillside, certain now that the gales were strong enough to obliterate their tracks. Overnight, a storm hit, leaving them two snowbound dimples in the landscape. For the next few days, they remained in the hideout, their sleeping bags soaked through. They ate their loaves of bread, melted snow with a candle for drinking water, and saved their last sliver of pemmican until they could bear waiting no longer.

When the storm abated, they headed west. As they journeyed farther away from the ongoing manhunt, another of the saboteurs was heading straight into the heart of the danger.

In the late afternoon of March 25, Claus Helberg was skiing across Lake Skrykken to Jansbu. He needed to fetch some weapons and explosives, which were buried in a depot near the cabin. Then he planned to head down to Notodden, southeast of Rjukan, to connect with the underground cell there. When he reached Jansbu, he saw that the door was ajar, yet there were no skis outside or trails leading up to the cabin. He unfastened his skis, stuck his pistol in his pocket, and entered, his rucksack in his hand.

The place had been ransacked, the furniture upended, mattresses ripped open, cupboards broken. The thought of the enemy intruding so far into the Vidda — territory Helberg considered his own — left him deeply unsettled. Then the fear struck that the Germans might still be close at hand, perhaps even hiding in wait. He moved to the window to scan the surrounding area.

After separating from his Gunnerside and Swallow team members, Helberg had headed back to Fjøsbudalen to retrieve his civilian clothes and his papers. These were in the name of a clerk from Oslo. He left to meet up with the others, but when the storm had hit, he'd lost his map to the wind. His choice was stark: retreat back to the cabin or get lost in the storm. When the blizzard calmed, he set out again, but by then his compatriots had already scattered in different directions on the Vidda.

He finally made it to Oslo. On March 8, he went to the Majorstua Café, the prearranged spot for a meeting with Poulsson. Being in the capital was a dizzying experience, the shuffle of the crowds, the trams screeching past, the German soldiers milling around. In the café, Helberg drank his coffee, trying to act like he did not have a care in the world.

Minutes later, Poulsson arrived. They were overjoyed to see one another but masked their emotions with a casual hello. Poulsson told Helberg that he would soon leave for Stockholm. There was too much of an uproar going on around Rjukan for them to risk working together. Helberg was determined to stay, and his first order of business was to move the Skrykken depot. The two men said their goodbyes and departed in opposite directions.

On March 22, after spending a couple of weeks in a safe house, Helberg received a note from Rolf Sørlie informing him that things had quieted down in Rjukan and it was safe to return to the area. What Sørlie did not know, and could not have known, was that trains and buses filled with soldiers were streaming into the area from Oslo that very day. Helberg returned to the Vidda at the worst possible moment.

Peering out through the window at Jansbu, Helberg spotted not a soul. Still uneasy, he ventured outside. Then, from the direction of the lake, he saw three Wehrmacht soldiers skiing toward the cabin. They were roughly four hundred yards away and coming fast. All Helberg had on him was his Colt .32 pistol. Outnumbered and almost certainly outgunned, he knew that he had no choice but to run. He raced inside, grabbed his rucksack, returned to his skis, and then dashed away.

A soldier shouted out to him in German to halt; the crack of gunfire followed. All around him, the snow mushroomed up as the shots missed their mark. Looking over his shoulder, he gauged his pursuers to be skilled skiers. He would have to be a better one. He veered west, straight into the setting sun — its piercing light was sure to make him a harder target to hit.

For the next hour, Helberg cut around hills, down into ravines, up short valleys, and past rocky outcroppings. He hoped to find some way to mask a change in direction, but the Germans were too close on his tail. He knew the terrain better than they did, but he had already skied many miles that day. Ten miles, maybe more, from Lake Skrykken, he finally began to distance himself from all but one of his pursuers, a giant of a man. No matter how hard Helberg pushed, the soldier maintained a distance of about a hundred yards behind him. For another hour, Helberg kept at it, glancing over his shoulder now and again to see if he had finally broken free. The hound stuck on his trail, his job made easier by the tracks Helberg was plowing through the unbroken snow.

On the uphill slopes, Helberg managed to outpace him. On the downhills, his pursuer closed the gap again. Eventually, probably soon, he would catch up. Either Helberg's legs would give out or his skis, with their poor wax and cumbersome metal-lined sides, would disadvantage him. In a Trojan effort, he aimed for every hill within reach, climbing higher and higher, gathering distance from his pursuer until he had nowhere to go but down. But every time the terrain flattened, the soldier drew close again. And then he was almost in range.

"Halt! Arms up!" the German shouted.

In that instant, Helberg made his decision. Pulling the Colt from his pocket, he stopped and turned. The soldier came to a sharp halt as Helberg fired a single shot from forty yards.

He missed.

The soldier drew a Luger. If it had been a submachine gun, it would be all over. Now Helberg knew how this would play out: whoever emptied his magazine first would lose unless he managed to kill the other.

Helberg calculated that the soldier was not in the best position to aim true. The setting sun was in his face, he would have sweat in his eyes, and his muscles would be burning. Helberg stood his ground.

The soldier fired his shots in quick succession, eight in total. They all missed. Knowing that he would not have time to reload, the soldier spun around and skied off. His poles struck the snow fast and forcefully as he speeded up the hill.

Helberg followed him, Colt in one hand, both poles in the other. He could not allow the soldier to get clear, reload, and come after him again. As the man approached the top of the hill, Helberg slowed. He was within twenty-five yards. It was close enough. He leveled his Colt and fired. The soldier stumbled forward slightly, then hung over his poles in the snow. It looked like he was taking a much-needed rest.

Helberg turned straightaway and raced downhill. It would be dark within the hour, but he knew that the next day his pursuers would try to follow his tracks. He needed to get as far away as possible and cut across some lakes with bare ice to throw them off his trail.

For at least two more hours, he journeyed south. He could see little, but the terrain was mostly flat, and his instincts guided him well. Then, suddenly, he felt himself falling. He had skied straight off a cliff.

He crash-landed in a hard-packed bank of snow. Once he caught his breath and realized he was still alive, a rush of pain enveloped him. He rolled over, his left shoulder and arm useless. Looking up at the precipice outlined by the starry sky, he figured that he had fallen a hundred feet or more. He inspected his upper left arm and felt sure that it was broken and his shoulder mangled too. He knew that he could not remain long in the mountains in such a state. He brought himself to his feet. At least his skis were intact. With one pole, he pushed off.

Since that morning, he had already journeyed some sixty or seventy miles. He had still farther to go, now crippled, exhausted, and hungry. At a slow, steady pace, his left arm hugging his side, he continued down to the tail end of Lake Møs. If he could reach the farm of Jon and Birgit Hamaren, he knew that they would help. At 8:00 a.m. he finally staggered to their door. Birgit answered. She gave him some food but warned that some fifty Hirdsmen and Gestapo were billeted at a neighboring farm — a five-minute walk away. Her brother was there too, conscripted by the Germans as a guide. "You have to get out of here," she told him.

Leaving quickly, Helberg skied along the shoreline of Lake Møs, then toward Rauland, a village twenty miles to the south, where he had another contact. Thirty-six hours had passed since he last slept. Given his exhaustion and injuries, he knew that if he met any Germans, he would have little fight left in him.

A mile outside Rauland, he ran straight into a patrol. The German soldiers asked for his papers, and he presented them: "Sverre Haugen." They told him that nobody was meant to be traveling about the area. Concealing his wounded arm, Helberg pleaded ignorance, saying that he was only a postal clerk out to visit a friend. The soldiers allowed him to pass.

At 9:00 p.m. he reached the house of his contact. When the door opened, it was to a pair of Germans flanking the owner. Helberg knew there was only one course open to him: to talk his way out of the situation. He smiled and lied, explaining that he had been injured while guiding the Germans in the mountains and now needed medical treatment. When one of them offered to put his arm in a sling, Helberg took off his coat, revealing his pistol. Coolly, he explained that the company he was with had allowed him to carry a gun in case of trouble. The soldiers accepted his story. They played cards with him, and even offered to take him in a medical truck to the neighboring town from where he could go on to an Oslo hospital. Helberg smiled and thanked them. True to their word, the next day they drove Helberg twenty-two miles south, past one checkpoint after the next, to Dalen. "Auf Wiedersehen," he said, waving to the soldiers before they drove off.

There were scores of Germans in the waterside town, but it was beyond the restricted zone and Helberg felt he would be safe there. The boat for Oslo left the following morning, so he checked in to the Dalen Hotel, an architectural confection in wood: carved dragonheads, rounded balconies, and elaborate turrets. After an early dinner of fried trout, seasoned carrots, and thick bread spread with strawberry jelly, he retired, well sated, to his second-floor room. He hid his pistol on the outside windowsill and crawled gingerly into bed, his arm and shoulder throbbing.

Shortly after he fell asleep, he was awakened by the sound of pounding on doors, heavy footsteps down the hallway, and commands barked in German. SS soldiers emptied every room and sent the occupants to the lobby. The sleepy guests were informed that Reichskommissar Terboven, accompanied by his security chief, Heinrich Fehlis, was taking control of the hotel for their temporary headquarters. With soldiers surrounding the hotel and guarding every entrance, Helberg knew there was no escape. He presented his false papers, lied again about his injury, and was one of the few who were allowed to return to their rooms instead of staying in the lobby all night. Once there, he dared not leave.

While Helberg rested uneasily in his room, the Reichskommissar and his high officials sat down at two long tables by the fire. They ordered dinner and some bottles of wine. The conversation was about how they should reposition security forces throughout Norway to better defend the country from an Allied invasion.

Upon learning that among those who had been turned out of their rooms as a result of his arrival were two young attractive Norwegian women, Terboven invited them to join their table. One of them, Aase Hassel, spoke fluent German and won Terboven's unwelcome attentions. Later in the wine-soaked evening, Terboven asked Hassel about her family. She told him that her father was a Norwegian Army officer. Then he must be happy, Terboven said, to be safe and part of the compulsory workforce recently instituted by the Germans. "No," Hassel said. "He's in Britain and I'm proud of it."

Everyone at the table went stock still. Seething from her remark, Terboven turned his attention to her friend. Before long he started in again, criticizing university students who mistakenly thought themselves "patriots."

Hassel could not resist. "All good Norwegians are patriots."

Again the table went still. Again Terboven kept silent. He would deal with her later.

At 10:30 a.m. the next morning, his arm in a sling, Helberg slowly descended the stairs from his room, a soldier trailing behind him. The Gestapo was sending all but a few of the hotel's guests to Grini in return for some "bad behavior" shown to Terboven the night before. Once at the camp, Helberg knew he would no longer be able to talk himself clear. His identification papers would be double-checked and determined false. Then the interrogations would begin — if he did not swallow his cyanide capsule first.

The soldier behind him kicked him in the back for not moving fast enough. He pitched down the steps, and his Colt sailed out of his belt and clattered to the floor. It came to a halt between the black boots of another soldier. Helberg could barely believe his bad luck. The soldier picked up the gun. "As you can see, it's not loaded," Helberg said in his pidgin German, as he struggled to his feet. He was sure that he was as good as dead now.

A ruckus followed, several soldiers speaking quickly to each other about what to do. Since Terboven and his entourage had already left, there was nobody there to countermand the order to bring Helberg to Grini.

Let the officers at the prison camp sort it out, they decided. Helberg was pushed into the line of prisoners filing out of the hotel to a rickety bus with blacked-out windows. He was one of the last to climb onboard, and found a spot on the floor at the back of the bus. A single SS soldier in a steel helmet, armed with a rifle and grenades, watched over them from the front.

The bus rolled out of Dalen for the 140-mile ride to Oslo, escorted in front and behind by SS riding in motorcycles with sidecars. Helberg was resolved to escape, somewhere along the way, somehow.

The afternoon passed in silence, the bus straining to get through the mountains, its occupants shivering from the cold. Two young women were in the seat beside Helberg. One of them scolded him for putting all their lives at risk by trying to sneak a pistol onto the bus. Hungry, and a little eager to tease her, Helberg took her notebook, tore off some pieces of paper, and ate them. Her response was to give him a throat lozenge to help him swallow. She introduced herself as Aase Hassel and spoke proudly of her father and uncle who were in Britain.

In the middle of their conversation, the guard came down the aisle. "You sit there," he said to Helberg, pointing to the front of the bus. Helberg shuffled down toward the driver. If the guard wanted to flirt with the young women, then fine — it would give him his chance. He sat down by the door and eyed the pull handle that operated it. From the passing landmarks, they must be thirty miles from Oslo. With some luck, he could reach the woods. The bus started up a hill and slowed down to a crawl. Helberg rose, grabbed the handle, pulled, and jumped.

He tumbled onto the road, slamming into his broken arm. The guard inside the bus started screaming for the driver to stop. Before he could obey, Helberg was already scrambling through the snow-covered field toward the woods. He fell several times, each time certain that the Germans were about to reach him.

The field ended in a thick, tall hedge that stopped him in his tracks. He couldn't get past it. "Stop!" a German guard yelled.

Helberg knew what he had to do. He was sure to get shot, but he saw no other choice. He turned around and barreled back across the field toward the soldier on the road. A grenade exploded in the snow behind him. Unhurt, he continued. Several gunshots sounded. Nothing hit him. Not that he could feel, anyway. He dashed across the road between the bus and a motorcycle, zigzagging to avoid being tackled by the German soldiers, momentarily confused by his head-on approach. Then he ran across the field on the other side of the road. Another grenade exploded behind him, too far away to cause harm. Then something hit him in the back, hard: a third grenade. He would never get clear of it.

The explosion never came. The grenade was a dud.

He sprinted headlong into the woods and darkness. There were several more gunshots, but the Germans had to be aiming blindly. Helberg slowed to recover his breath, then threaded through the trees, his arm ablaze in sheer agony. A soft rain was falling, but he knew that his tracks would still be evident in the snow come morning. Finally, after a long hike through the forest, he came upon a long rectangular building lit up from inside. He climbed over the barbed-wire fence surrounding it and staggered to the front door.

An old man answered his knock. Helberg was out of lies. His arm was shattered. He was bloodied and dazed. His clothes were in tatters. He surely could not go any further this night. If this man was a good Norwegian, he would help. If not, Helberg was lost. The man welcomed Helberg in and told him he had arrived at a psychiatric hospital. They had food, doctors, clothing, and beds. Helberg was safe.

At 9:00 a.m. sharp on April 8, Heinrich Fehlis stood at the steps of the Hotel Dalen in front of four battalions. After sixteen days combing the

Vidda, his men needed to be relieved. They were all exhausted. Some suffered frostbite on their hands and feet; some had broken bones; all were weather-beaten, their faces blistered. They were in terrible shape. They had trekked hundreds of miles through the mountains and the surrounds of Lake Møs. They had struggled through storms to penetrate the plateau, searching cabins as they went. One of their number had been shot by a Norwegian during a chase from Lake Skrykken.

Fehlis had visited their quarters in the days before, making sure that they had cognac and vermouth to put in their hot drinks. Now he thanked them for their effort. "Every day, you boys have endured long marches and still you assemble in high spirits, without complaint. Among my troops, you have distinguished yourselves." Then he discharged them. Other battalions would take their place.

To date, there was little to show for the action. Some stores of explosives and weapons had been found, and the huts in which they had been hidden had been torched. There had been some arrests — one of a wireless operator — but nothing of note and certainly none of the Vemork saboteurs. If his manhunt failed to make progress soon in finding those responsible — his intelligence reports indicated they were likely members of the so-called Norwegian Independent Company No. 1 — then Fehlis would have to call off the search.

High in the Hamrefjell mountains, Skinnarland and Haugland settled down for their second night in a cave they had dug out of the snow. The southwest wind howled outside the narrow opening, and the cold burrowed down into their bones like a sickness. Having spent nearly a month in hiding since the German razzia began, they were used to such conditions.

They had been staying at Nilsbu when, on March 24, Jon Hamaren hurried up from his farm to warn them that a raid was underway. Skinnarland's brother Olav had been one of the first arrested, Hamaren told them, and the soldiers who had taken Olav away now occupied his hotel by the dam, where his wife, Ingeleiv, was forced to wait on them at all hours while also tending to her young son and newborn daughter.

Skinnarland and Haugland had cleared Nilsbu of weapons, radio equipment, and other gear, and buried the stash away from the cabin. Then they skied and hiked up the narrow gorges and steep cliffs of Hamrefjell to a spot over five thousand feet up. There they stayed, for ten days and nights, with only a tent, a kerosene stove, and their sleeping bags to keep warm. Through binoculars, they watched the German patrols moving about Lake Møs and the surrounding hills. On occasion a Storch search plane shot overhead. Though exposed on the mountainside, they knew it was unlikely they would be found. If anyone approached their tent they would see them from a long way off. Given the precipitous terrain, their pursuers would struggle even to reach their position. Best of all, the Norwegians who had been conscripted as local guides included Hamaren and other local farmers, and they knew to keep the Germans away from their hiding spot.

On April 1, when the Nazis' search had moved away from Lake Møs, Skinnarland and Haugland returned to Nilsbu. Although the Germans set up a mobile D/F station down by the lake to sniff out any transmissions, they positioned it in such a low-lying place that it was unable to pick up the Nilsbu signal. Haugland continued to train Skinnarland as a radio operator, and he was sufficiently adept to send his first message to Home Station a week later, describing the razzia, the lack of news from Haukelid, Kjelstrup, or Helberg, and the pressing need for a drop of supplies. Tronstad answered with the news that Poulsson had made it to Sweden along with Rønneberg, Strømsheim, Idland, Kayser, and Storhaug. Those members of the team were safe at least.

On April 16, Hamaren had warned Skinnarland and Haugland about renewed enemy activity around the lake. The two fled back up into the Hamrefjell mountains at speed and built their snow cave. The next day, Skinnarland went down to Nilsbu to investigate and discovered two sets of skis leaning against the cabin wall. Fearing they belonged to Germans, he retreated back into the mountains to stay again burrowed in the snow.

The following morning, no sign of any patrols down below, he and Haugland skied back down to Nilsbu. As they edged their way carefully to the cabin, they sighted the trespassers: It was Haukelid and Kjelstrup. They enjoyed a warm and happy reunion, and the four men shared the stories of their narrow escapes with each other. They also discussed the cruel waste of herds of reindeer destroyed by German machine guns while the razzia was underway. Later that day, the farmer Hamaren came to the cabin to tell them that Claus Helberg had been shot and killed while trying to flee a German patrol. There was little hope the report was false. They forwarded the news to London and mourned the loss of their friend.

The time had come to launch their resistance work. Skinnarland was transmitting at a fast-enough clip to run his own wireless station. Haugland was headed for Notodden, then on to Oslo, to start a network of radio operators for Milorg. Haukelid and Kjelstrup would build up the resistance cells in the district. For all four men, their original mission had been accomplished. As far as they knew, they would have nothing more to do with Vemork.

22

A National Sport

I N M I D - A P R I L 1943, a military truck drove across the suspension
bridge into Vemork. Secured in the truck bed was what looked like
a steel drum of ordinary potash lye, an ingredient in the electrolysis
process. What the drum actually contained was 116 kilograms of almost
pure heavy water from Berlin. It had been originally produced at Vemork.

Soon after the sabotage on February 28, a stream of Norsk Hydro company men and German officials had come to the plant to decide its fate.
Some argued that all the salvageable equipment be shipped to Germany,
since destroying the plant had virtually become a "national Norwegian
sport." Others, including Bjarne Eriksen, the Norsk Hydro director general, wanted to start up again at Vemork. Esau and the Army Ordnance
Office were asked for a "swift decision," which they delivered: the cells
should be repaired and the plant expanded as soon as possible. Heavy
water facilities at Såheim and Notodden should also be completed. The
German command provided any materials and manpower required (including slave labor from abroad) and warned that if the work was not
completed quickly enough, there would be severe reprisals.

By the time the secret shipment of heavy water from Berlin arrived at
Vemork, the round-the-clock work on the plant was almost complete. The
shipment was used to fill the new high-concentration cells, overriding the
slow process of accumulating the precious substance drop by drop, and
accelerating the return to production by several months. With three new
stages and a number of cells added to the preliminary electrolysis process

as well, the Germans projected that daily output would soon reach 9.75 kilograms. Given the plans to double the size of the high-concentration plant yet again, the daily yield might reach almost 20 kilograms within the year.

While this was going on, SS officer Muggenthaler and Lieutenant Wirtz, the new head of guard at Vemork, finalized the security measures. Another guard was placed at the suspension bridge and two more at the railway gate. Others patrolled the grounds with Alsatian dogs night and day. Sappers laid more mines on every approach. Barbed-wire fences were raised. To defend against an air attack, additional wires were strung across the valley, and fog-producing machines were placed about the area. The pipelines were camouflaged, and torpedo nets were set up to protect the Lake Møs dam. A permanent guard was posted outside the rebuilt high-concentration plant. All the doors but one, which was reinforced with steel, were bricked up or sealed with wooden planks. The windows were similarly blocked off or fixed with iron bars and wire mesh. Within the plant, a team of guards was armed with submachine guns.

Muggenthaler weeded out any employees perceived to be a threat, and German technicians took on roles within the facility to spy on any illicit activity. Vemork may have been a fortress before, but now its high-concentration plant was a fortress within a fortress. And on April 17, 1943, at 2:00 p.m., heavy water began to flow securely through the cascade of cells.

Three weeks later, on May 7, the Uranium Club met at the Reich Physical and Technical Institute in Berlin. The scientists were under pressure for results like never before. With German fortunes in the war deteriorating, the Allies set to retake all of North Africa, and the Soviets continuing to defeat the Wehrmacht on the Eastern Front, the Nazi brass felt a desperate desire for something that would quickly turn the tide in their favor. One report stated that "rumors abound in the general German population about a new-fangled bomb. Twelve such bombs, designed on the principle of demolishing atoms, are supposedly enough to destroy a city of millions." Worse yet, Abwehr intelligence had revealed that the Americans were on the path to creating "uranium bombs." The German atomic program was riven by factions, and its scientists and research centers now exposed to attack by Allied bombing. They needed a breakthrough to focus their efforts.

The first item on the agenda for the May 7 meeting was heavy water. Only the previous day, at a German Academy of Aeronautical Research conference, Abraham Esau had placed part of the blame for their slow atomic progress on the recent lack of heavy water. He wanted to press ahead with production in Germany, the plans for which had long been stalled because of the cheap supply from Norway. Paul Harteck advised that after a few more experiments, the Leuna pilot plant, using his catalytic exchange process, could likely be expanded to produce five tons a year, if they fed it with slightly enriched water from Vemork or a couple of Italian electrolysis plants. He also suggested that they try another method, invented by Klaus Clusius, which capitalized on the slightly higher boiling point of heavy water to produce enriched amounts. Esau gave the go-ahead on the preliminary work for Leuna and charged Harteck with determining whether the Clusius method made sense on an industrial scale. But with Vemork back online, and these additional projects, Esau felt confident they would soon have all the heavy water they needed.

Diebner was less sure there was enough to go around. He made it clear that he needed every drop for his next two experiments. His team's most recent uranium machine (G-II), which used uranium metal cubes suspended in frozen heavy water, showed neutron production at a level one and a half times greater than any German experiment so far. The machine proved that a cube design was far superior to any other in fostering a chain reaction, he said, and at the right size, it would likely be self-sustaining.

Heisenberg disagreed. In his mind, the best design was still an open question. He claimed the dimensions of his latest machine, a sphere with alternating layers of uranium metal and heavy water, were "too small to yield absolutely certain values." But he added that the company producing uranium metal for them was already casting it in plates, like he needed, rather than cubes, as Diebner demanded. Thus his experiments — and the heavy water they required — should be first in line. "This doesn't rule out a subsequent cube experiment, if one is needed," Heisenberg offered condescendingly.

The tension in the room was palpable. Diebner had his supporters, including Harteck, who believed that Heisenberg was blind to the value of any experiment that did not originate from his own brain. Esau, who had appropriated heavy water from Heisenberg for Diebner's latest experiment, said he needed to think further about whose work should be given precedence. Eventually both men received a share of heavy water and uranium to allow them to proceed with small-scale experiments, but both were left dissatisfied.

General Leslie Groves was worried. And he was not a man to leave a worry to fester for long. The graduate of West Point and MIT was just shy of six feet tall, with a blocky head, a thick sweep of brown hair, and a barrel of a chest to accommodate his medals. Groves was known as a "doer, a driver, and a stickler for duty." Charged with running the Manhattan Project, he was also regarded by some of his staff as the "biggest sonuvabitch" they had ever worked for: critical, abrasive, and egotistical. Those same people would have wanted nobody else to lead the American project to beat Germany to the bomb.

As Groves saw it, there were two complementary ways to achieve that end: first, to accelerate the U.S. effort and, second, to slow down the enemy. For the former, he led a full-throttle campaign that employed tens of thousands of scientists, engineers, and workers and drew on hundreds of millions of dollars. In the hills of Tennessee, monumental plants were being built to separate the rare isotope U-235 from U-238 using two different methods. Beside the Columbia River in Washington State, construction had commenced on reactors that used two hundred tons of uranium moderated by twelve hundred tons of graphite. Working with their Canadian ally, the Americans were building a massive heavy water plant at a hydropower station in Trail, British Columbia. At the Los Alamos Ranch School in New Mexico, a small city of physicists was working to build a functioning fission bomb. All these efforts brought worry, but at least they were under Groves's direct control.

Slowing the enemy down was not. First, he had limited intelligence about German advances. Second, he did not have authority over operational forces or bombers to direct them to enemy targets. At the end of March, he learned — from a Swedish newspaper, no less — of the success of the British sabotage operation against Vemork in Norway, a plant that he had long known provided the German program with critical resources.

Through the Army chief of staff, General George Marshall, and Field Marshal John Dill, the lead British military representative in Washington, Groves demanded to know the details of the operation. In April he was told that the plant would be inoperable for two years. Then, only days later, he was informed that this period had been reduced to one year.

A short while after, Michael Perrin told him that the Germans would not realize the bomb before war's end. Perrin, one of the leaders of the British Tube Alloys Committee, was on a visit to Washington at the time. "You might be right," Groves replied. "But I don't believe it." Even if Perrin were correct, Groves knew that there were other dangers — one of which was a radioactive attack.

From a detailed report produced by his scientists, he learned that if the Germans succeeded in starting a heavy water reactor, they could easily produce "colossal amounts" of radioactive substances that could be dropped over a city. Although the report concluded that there were challenges to creating an effective radioactive bomb, the Germans could, at the very least, "completely incapacitate" a city like London, requiring large sections of it to be evacuated.

On the morning of June 24, 1943, Groves convened with Vannevar Bush, the blue-eyed, beanpole New Englander who served on the committee that oversaw the Manhattan Project. The two men went over the progress of the program, including the likelihood that they would have at least one bomb ready to be deployed by early 1945. They also reviewed a list of targets to be attacked in order to slow down the German project, targets pointed out with the help of British intelligence — and Leif Tronstad. These included Vemork and several German research centers. Groves and Bush agreed there was no sense in spending half a billion dollars to produce a bomb if they did not "strain every nerve on the countermeasure side."

A few hours later, Bush lunched at the White House with President Roosevelt. He outlined how they were "going very aggressively" and were on schedule for a January 1, 1945, delivery of a bomb. Roosevelt wanted to know where the Germans were. Bush answered that the Nazi scientists were "doing serious work on this before we were and that they might therefore be ahead of us." However, "arrangements were under way" to hit the few German targets they had on their list.

When Roosevelt gave his assent, Groves put his wrecking-ball force of will into seeing this was done.

Earlier in the summer, Knut Haukelid was in the Oslo apartment of Trond Five, resistance leader and an old family friend, when there was a knock on the door. Trond went to open it, and Haukelid heard the distinctive voice of his own father, Bjørgulf. Quickly, he hid in the next room, sticking to one of the fundamental rules of illegal work: never make contact with family. As far as any of them knew, he was still in Britain. In almost two years, he had neither seen nor spoken with anybody in his family, including his wife, Bodil. Her involvement in the resistance had forced her to leave for Sweden in mid-March.

Now that his father was only a few feet away, separated only by a door, Haukelid wanted to break the rules, to step into the room, to embrace his dad. They may not have always seen the world in the same way, but they were still father and son. Resisting the temptation, Haukelid remained hidden. After some brief words, Trond sent the surprise visitor away.

Haukelid and Kjelstrup spent several weeks in the capital city, resting up and waiting for new identity cards in the names of Nasjonal Samling members. While there, they learned that Helberg had not in fact been killed while attempting to escape, and the happy news of his survival was sent to London. Their friend's travails reinforced with them how careful they must be at all times — and that there should be no overnight stays in hotels.

In June, they left Oslo to start their underground organization in earnest. They headed back to Vågslid on bikes bought on the black market, acting the part of tourists out for a summer ride and sleeping out in the woods, avoiding hotels.

Their plan was to create a safe base of operations, then Haukelid would recruit several district commanders to lead their own cells and assemble resistance fighters. As these commanders would be the only ones to know of his existence, they had to be absolutely trustworthy. Once trained and armed, the underground organization would lie in wait, ready to carry out guerrilla attacks that would sap their enemy's strength and inhibit their movement through the district's important east–west corridor from Oslo to the North Sea.

Haukelid and Kjelstrup built their base high in the mountains southwest of Vågslid, up from Lake Holme. They gathered moss-covered stones to make a double-walled cabin that would look like just another pile of stones from a distance, packing dirt and peat between the walls, making them impermeable to the wind. For the door and roof beams, they salvaged wood from an abandoned mine to the south. When they were hungry, they fished for trout in the lake. When tired, they lounged in the sun, glad to be free of the winter at last. Haukelid bought an elkhound puppy from a local farmer; they called him Bamse (Little Bear) and named their rising cabin after him, Bamsebu.

Haukelid took a break from the construction to go and meet one of his new platoon commanders. Then he traveled on to a hamlet a few miles northeast of Dalen. On June 18, as planned, he met Skinnarland at the farmhouse of one of their contacts. Skinnarland was in a terrible state. Two weeks before, his baby niece had died, suffocated in her crib. Her father Olav, who was in Grini because of his connection to Einar, could not even attend her funeral. The loss was one tragedy too many for their elderly father, who was now on the verge of death himself, and there was no way Skinnarland could visit him with the Germans stationed so close to the dam. He had been living alone at Nilsbu, suffering from a cracked front tooth that caused him constant pain. Traveling as "Einar Hansson," a life-insurance inspector, he had emerged from hiding to undergo several procedures with a dentist located outside Dalen. The one in Rjukan was sure to identify him — and potentially turn him in to the Gestapo.

Haukelid managed to comfort Skinnarland, and for the next few nights they tried to forget that they were men on the run. "Bonzo was waiting— very nice meeting," Skinnarland wrote on the eighteenth in his abbreviated diary. "Big party with eggnog cream — very stately," he dashed off the next day. "Began at the dentist — fun and games," was his entry for the day following that. On one of the days they spent together, Haukelid showed Skinnarland a list of Norwegian girls' names he had found in a book. The names had gone out of fashion long before, but Haukelid liked one in particular: Kirvil. If he ever had a daughter, he said, he would name her that. For a moment, Skinnarland thought of the future too, one in which he might have a wife, children, a whole life beyond this one. He promised Haukelid that if he had a daughter, he would name her Kirvil as well. Shortly after, Haukelid returned to Bamsebu. Skinnarland remained for almost two weeks and several more procedures at the dentist's.

After one such appointment, as he came back to his hut up from Lake Møs, Jon Hovden, a farmer who was one of his main pillars of support, visited him. Lillian Syverstad had delivered a note from her brother Gunnar for Hovden to give to him. The news contained in the note, Skinnarland knew upon reading it, would have to be sent to London straightaway.

· · ·

On July 8, Tronstad received the disturbing message from Skinnarland: "Vemork reckons on delivering heavy water from about August 15." Wilson asked him to get confirmation on the report. John Anderson and the War Cabinet would need to be alerted.

Until that point, Gunnerside had been an unqualified success. Their target had been destroyed. Not a shot was fired. There had been no major reprisals. Every single member of the team had escaped to safety, their identities unknown to the Germans. Tronstad could not have dreamed of a better outcome. Praise had come from every quarter, from the Norwegian high command to Churchill himself, who had asked plainly, "What rewards are to be given to these heroic men?" The success had raised the profile of the SOE and its Kompani Linge, giving them more opportunities for future missions in Norway.

If Skinnarland's report were true, the Germans would have full production back online much sooner than Tronstad had originally thought. Gunnerside may have set back supplies of German heavy water by two tons, half of what they needed for a working reactor, but Tronstad knew any renewed deliveries to Berlin would not be tolerated, particularly after recent statements secreted out from Niels Bohr. After two German physicists visited his lab in Copenhagen, the Danish physicist stated that he believed atom bombs were practicable in the immediate future, particularly if there was enough heavy water on hand to manufacture the necessary ingredients. When asked if heavy water production was "war-important" and whether such plants should be destroyed, Bohr answered yes to both questions. Coming from Bohr, one of the fathers of atomic physics who was soon to be secreted out of Denmark at last, this declaration put the bull's-eye back on Vemork.

Tronstad sent orders to Skinnarland to investigate progress at the plant. On July 19, Tronstad wrote a lengthy report to SOE on how to "tackle the juice issue" again, as he described it in his diary. More than anything, he wanted to prevent a massive bombing run on the plant like the Americans were urging. He doubted such a bombardment would destroy the basement-level high-concentration plant, which was protected by tons of steel and concrete overhead. Also, such an attack would almost certainly inflict enormous collateral damage, both in terms of the lives of the everyday Norwegians living around the plant, and on the postwar Norwegian economy. Further, he had doubts that the Germans were pursuing the bomb with the fervor suggested by Bohr's statement. From what Brun had gleaned from his time at Vemork, not to mention recent intelligence Tronstad and Welsh had received from their moles in Sweden, Norway, and Germany, the chances of the Nazis developing a "devil machine" were limited.

In his report, Tronstad advised caution but also gave various options to stop production at Vemork and slow the Nazi atomic effort. They could destroy one of the dams that provided water to its power generators. They could target the transport of heavy water from Vemork to Germany. They could sabotage the high-concentration plant from within. They could hit sites in Berlin where experimental work was taking place or from where it was managed. He even provided the addresses. He did not include a bombing run on Vemork or on the other two plants at Såheim and Notodden, which were either already producing heavy water or were soon to be.

By coincidence, as Tronstad's list of options for a second attack on Vemork's heavy water was being circulated within the SOE, he and his men were presented with British awards recognizing their service in the initial attack. On July 21, at Chiltern Court, they assembled in uniform. On behalf of King George VI, Lord Selborne, the minister of economic warfare who oversaw the SOE, awarded Rønneberg and Poulsson the Distinguished Service Order, and the others present (Helberg, Idland, Kayser, Storhaug, and Strømsheim) were given the Military Cross or Military Medal. Tronstad received the Order of the British Empire.

Later, Selborne hosted a dinner for them all at the Ritz Hotel. Appropriately, they were served grouse, gnawing at the bones until they were picked clean. There was much to reflect on — their struggles on the Vidda, the tense moments during and after the sabotage — but still more to laugh over. They recalled their time at a Stockholm hospital, where a pair of Swedish nurses had deloused and scrubbed them clean. Then, once freed from the camp, they had attended *La Traviata* at the Stockholm opera house like proper civilized people. After the dinner, Tronstad took the men out on the town, the merry group singing songs as they made their way up Piccadilly. Tronstad said nothing of what he had learned from Skinnarland. He did not want to sour the evening.

Not seventy-two hours later, an armada of American heavy bombers from the Eighth Air Force division roared through the clear blue sky over southern Norway. Because of fog over Germany, the bombers had been diverted from a run over Hamburg. Instead they set out to hit several industrial targets in Norway, including the massive new Norsk Hydro aluminum plant at Herøya. They dropped over 1,650 bombs, leveling the area and killing fifty-five people, mostly local workers. Tronstad feared Vemork might be next.

23

Target List

O N AUGUST 4 Bjarne Eriksen rushed to Rjukan after his visit to the rubble-and-casualty-strewn site of the Allied bombing raid on Herøya. Subsequent to the attack, Vemork had been shut down. The raid had also destroyed the Norsk Hydro fertilizer plant adjacent to the aluminum plant. No more fertilizer production meant no need for ammonia and, therefore, no need to run Vemork's hydrogen plant. At the administrative building in Rjukan, Eriksen faced a host of Nazi officials who demanded he restart the plant solely in order to continue its heavy water production. Eriksen refused, and like the lawyer he was, laid out his objections point by point.

First, heavy water had minor commercial significance for Norsk Hydro — not least because the Germans had never paid one krone for the supply since the invasion. Second, if they continued to run Vemork's hydrogen plant to feed the cascades of the heavy water, the volatile hydrogen gas produced would have to be released into the air, which was clearly dangerous. Third, he could not understand the need to restart, since he had repeatedly been told that heavy water was not important for German war purposes but only for scientific investigations. Fourth, and principally, deuterium production increased the risk of an Allied aerial bombing, putting a great number of lives in danger, not to mention an industrial site of inestimable value to his company as well as to Norwegian agriculture and exports.

Despite Eriksen's arguments, Terboven's representative at the meeting, a Dr. Albrecht, was unmoved. He insisted that the production of deuterium be brought up to full speed, and then accelerated further. In June 199 kilograms of heavy water, a record for Vemork, had been tapped from its rebuilt high-concentration plant. July had missed that mark, not least because of the Herøya bombing. Albrecht made clear that Terboven expected August to yield record production levels. Eriksen replied that "personal conviction" led him to insist not only that heavy water production not be restarted any time soon, but that production be permanently halted. He would recommend the same to his board of directors.

Eriksen had a mixed history as a patriot. Before the invasion, he had leaked to the French spy Jacques Allier that there was a German interest in Vemork. On the other hand, he had encouraged King Haakon to resign after the April 1940 invasion because of "practical politics" and had signed a twenty-five-thousand-kroner check to the Nasjonal Samling Party on behalf of his company soon after. Now he was standing up to the Germans again.

Albrecht was not impressed. He had orders to see that "the greatest possible output of SH-200 be maintained regardless of the risk involved." One last time, he asked Eriksen if he was sure he wanted to continue to object. Eriksen said that he would and that he was "fully prepared to accept the consequences." So be it, Albrecht responded. His country's demands would be enforced, whether Eriksen was in charge or not.

Within a few days, Fehlis received notice to have the Gestapo arrest Eriksen. A former Norwegian Army officer, he was rounded up with over a thousand other officers, ostensibly for being a subversive threat within Norway. However, Eriksen was dragged from his home and sent to a German concentration camp for a very different reason: nobody refused a direct order from Terboven.

At Nilsbu, on the very same morning that Eriksen demanded Vemork be mothballed, Skinnarland tapped out a series of messages to Home Station. They detailed how, by June 1, the Germans had begun producing the "usual output" of heavy water again. The Såheim plant was now operational as well, with concentrations reaching 15.7 percent purity. Its engineers expected to begin delivering supplies in "finished condition" by October. The Notodden facility was also approaching startup. Skinnarland concluded: "We can arrange to reduce the output at Vemork, while Såheim and possibly Notodden can be hindered. Our men are willing but would like orders from you."

By return message, Tronstad wanted to know how Skinnarland planned to slow the production and made it clear that no action should be taken without prior approval. Gunnar Syverstad was already contaminating the high-concentration cells with drops of cod-liver oil, reducing output by 1.5 kilograms a day. "With care," Skinnarland answered by cipher message, "that can be reduced to one half without direct danger for our people." Over the next two weeks, he suggested other ways to prevent the Germans from obtaining any more heavy water. Vemork was too well protected, but a small force, five men at most, could attack shipments to Oslo, whether they were by train or motorcar. His orders, however, were to watch and wait.

To sustain himself until the longer fall nights when the SOE could again drop supplies, Skinnarland hunted, fished, and collected wood. It was marvelous to spend so much time free in the mountains, but eventually the isolation began to press in on him. Although he needed to travel down to the Skindalen, Hamaren, and Hovden farms to obtain scarce extra food, recharge his wireless batteries, and collect any messages brought by Lillian Syverstad, he also went down for the company.

Some nights he slept over, then spent the next morning working in the fields to extend his visits with the families. They were his lifeline and his protection, too. Far from civilization, subsisting off the land, those secluded mountain farmers may not have known much about the world, but they were worldly wise when it came to underground work. Whenever a German patrol approached their lands, they always seemed to know far in advance and sent warning to Skinnarland at Nilsbu. Once all was clear, they brought word to him in the mountains that he could return to his cabin, and often stayed for a while with him. On occasion he would hunt with them; skilled as he was, they were far better. As he noted in his diary one night, "Olav killed a small buck. I shot a stone."

During this period, Skinnarland also risked seeing his Bergen girlfriend, Gudveig, for the first time since he went underground. Over the past year, he had sent a few gifts by post: some reindeer meat, a knife whose reindeer-bone handle he had carved with an intricate design. In mid-August he hiked to the Haukeliseter mountain lodge, where, through a contact, he had invited Gudveig to meet him. In his diary, he simply wrote, "Gudveig came. Big fuss!" They spent almost a week together before Skinnarland had to return to his hideout.

He was still completely cut off from his family, and there was nothing he could do to help them when, on July 11, his father died. Einar dared not attend the funeral that took place a week later. The Gestapo was still looking for him, and it would have been an obvious opportunity to catch him. He did learn that his father had been comforted in his final hours by the news that his son was safe in the mountains, but it would have been better by far to have been at his side. Moreover, though Skinnarland knew his big brothers Olav and Torstein and his best friend Skogen, who had survived his Møllergata tortures, would never have accepted his surrender in exchange for their freedom from Grini, that did not soften his guilt.

None of them, including Skinnarland himself, could have known that what was transmitted from the old reindeer hunter's cabin deep in the Norwegian wilds was being heard in the corridors of power in London and Washington, D.C.

On August 9, A. R. Boyle, chief of the Air Ministry's Directorate of Intelligence, sent the SOE a note suggesting that Vemork be targeted again. He was receiving reports that the Germans were doing "everything required" to produce a "secret weapon" that used Vemork's heavy water to win the war.

Wilson informed Tronstad that the plant was once again being considered for attack. Tronstad drafted a message to Skinnarland that gave precise instructions on where, how much, and at what rate to insert castor oil into the high-concentration plant cells to contaminate the heavy water in a way that could be naturally explained. Before the message could be sent by the Grendon Hall operators, however, Tronstad was informed that it would be placed in a file until further notice. Anderson was forbidding any "local action" until he achieved consensus between British and American military authorities on the best course forward. Now more than ever, Tronstad feared Vemork was on the "target list."

Late in 1942, in a meeting between Norway's General Wilhelm Hansteen and U.S. General Dwight Eisenhower, the two men had come to an understanding that any Norwegian sites slated for destruction, whether by the RAF or by the U.S. Air Force, were to be sanctioned by the British Air Ministry and placed on a target list so that the Norwegian high command could advise for or against them.

As one of those who reviewed this list for his government, Tronstad knew that some targets were best hit by air. He himself had suggested that the Knaben molybdenum mines be put on the list for attack by the RAF, and this went ahead in March 1943. Herøya was different. Although the aluminum and magnesium plant was essential to the German war effort, Tronstad always considered the safest and most efficient way to cripple its production was by sabotage, not by a daylight bombing raid. On the target list, it was flagged as low priority, first because it was not yet running at full production and second because there was a commando operation in the works against it. Nonetheless, without warning, the Americans had launched their bombing raid in midsummer.

In the uproar that followed, Hansteen and Tronstad pushed for a monthly review of the target list and for more direct consultation on what sites should be designated for sabotage or bombing. Most pointedly, Tronstad used all his influence to push for Vemork's exclusion from the list. According to his sources, it was unlikely the Nazis would realize a bomb within the next several years. Njål Hole, Tronstad's young spy in Sweden, had shared recent conversations between Lise Meitner and some visiting German scientists that indicated their uranium machines were focused on producing energy, not bombs. A Norsk Hydro executive who had recently escaped to Britain confirmed this. From his many contacts with those same German scientists, he believed it unlikely that the Nazis considered heavy water to be "of vital importance to the prosecution of the war." Tronstad thus believed that a bombing strike was incommensurate with the level of the threat. Instead, a straightforward contamination of the high-concentration cells by insiders at the plant was the most prudent way forward.

While Tronstad awaited a response to his lobbying efforts, another German program — on which he also provided intelligence — was struck by the Allies. On the night of August 18, bombers hit Peenemünde, on the Baltic Coast, the site of the Nazi V-1/V-2 rocket center. That same evening, Tronstad was at his desk, writing his daughter Sidsel a letter, a photograph of the ten-year-old in front of him. He asked how she liked the watch he sent, praised the writing of her recent letter (though admonishing her for some spelling mistakes), and asked after her younger brother Leif. But Tronstad dedicated most of the letter to explaining why he needed to be away from her: "We have to do everything for our land to make it free again. When we say 'Our Fatherland,' we don't just mean the land, which is beautiful and we also love, but also everything else we love at home: mother, little boy and you, and all the other fathers and mothers and children. I also mean all the wonderful memories from the time we ourselves were small, and from later when we had children of our own. Our home villages with the hills, mountains and forests, the lakes and ponds, rivers and streams, waterfall and fjords. The smell of new hay in summer, of birches in spring, of the sea, and the big forest, and even the biting winter cold. Everything . . . Norwegian songs and music and so much, much more. That's our Fatherland and that's what we have to struggle to get back." His efforts to protect Vemork were one way Tronstad was fighting for the future of Norway, when the Germans were driven out. Two days after writing to Sidsel, he met with the Air Ministry, and as a result of that meeting, he believed he had won an agreement to shield Vemork and other sites in Norway and to prevent another Herøya from happening again.

Later that afternoon, however, Michael Perrin distributed a secret report stating that he and his fellow Tube Alloys members believed that the production at Vemork needed to be stopped, and in the opinion of British intelligence, a bombing attack was the only way to eliminate production in the long term. "I would propose," Perrin concluded, "the Americans be informed of the position and asked whether they would take the matter up with the USAAF. In that case, it would probably be advisable that no decisions should, at present, be communicated by us to the Norwegian authorities."

"My name is Knut. That's all you will learn. I have come from England and represent the King." This is what Haukelid told yet another potential recruit. Governments may come and go, but King Haakon was the symbol of Norway's struggle for freedom. "He wants you to set up and lead a military resistance group in your county . . . I shall instruct you about the organization. I shall give you what you need in the way of money, and I shall arrange for weapons to arrive by airplanes or by sea, at the time and place you want them." Throughout September, their Bamsebu headquarters now built, he and Kjelstrup continued to develop underground cells throughout western Telemark. They stressed security above all else to their roughly seventy-five recruits: Use dead drops. Never put your commander into contact with a stranger. And "Remember: Keep your mouth shut."

Traveling by bicycle, with high-quality fake papers and travel passes, they only rarely ran into trouble. Once, by a lakeside boathouse, Kjelstrup was confronted by a local sheriff and his deputy. The sheriff was a known Nazi sympathizer. When they demanded his papers, Haukelid, who had stayed out of sight, crept around the boathouse and readied to shoot the two policemen with his Colt pistol. Before it came to that, however, the officers let Kjelstrup pass.

They established a wireless station on a mountain farm, and the farmer agreed to the operator living there undercover — just another farmhand working the land. They also scouted German positions throughout the area and along the Haukeli Road, in preparation for the time when the Allies invaded Norway — or when their cells were called on for sabotage operations. British supplies were their most pressing need: their boots had holes in the soles and their clothes were threadbare. They could also do with tents, maps, compasses, rain gear, matches, shaving blades, cigarettes, rucksacks, and, as their drop list noted, "Six housewives." Given that winter was approaching, they also requested food. Above all, they were desperate for weapons — sniper rifles, Sten and Bren guns, and Colt pistols — to train and arm their cells.

After midnight on September 22, the tail end of the first moonlit period in which the RAF could send a supply drop, Haukelid was alone at Bamsebu when he heard a plane. Barefoot and wearing only trousers, he ran out of the cabin with nothing but a single flashlight to signal the plane. When the sound of its engines faded, having sighted no parachutes, he hurried back into the cabin to dress. The plane came around again, and again he signaled frantically to the pilots. Still he saw no drop. Twice more he heard the hum of the aircraft's engines. Then nothing. Frustrated, he returned to the cabin, fearing he would have to wait several more weeks for supplies.

A few days later, he was out at the edge of Lake Holme, pulling in fishing nets, when Kjelstrup returned from a visit to Skinnarland. "How'd it go?" he asked Haukelid.

"Go? Nothing's happened here as far as I know."

"London reported that a plane had been here and made a drop."

Haukelid and Kjelstrup searched far and wide in the area around the cabin until they came upon two containers on the banks of the lake. Five others were half-submerged or had sunk completely in the water. After stripping down, they dove into the frigid water and began hauling them to shore.

On inspection, the weapons and most of the contents were fine, and only a share of the ammunition was ruined—a thousand rounds for their Krag rifles. That night, the men celebrated with a big dinner. For the first time in months they had more on their plates than they could ever eat: crackers, corned beef, chocolate, tinned fruit, jam, and raisins. The next morning, they feasted on more of the same.

But the full meals did nothing for the stiffness and swelling that had plagued Kjelstrup since the previous winter. A doctor in Oslo had diagnosed beriberi—a disease caused by the chronic lack of the vitamin thiamine—as a result of his eating only reindeer meat for such a long stretch of time. His teeth were a mess too, cracked and breaking from the same bad diet. Kjelstrup knew that he would not be able to endure another winter in the countryside. He decided to cross the Swedish border and move on to Britain.

The two said their farewells at the end of September, and Kjelstrup cycled away. A couple of weeks and some snowfalls later, Haukelid shuttered up Bamsebu. A heavy rucksack on his back, he set off on foot toward Lake Møs with his elkhound by his side. On the way, he realized that the snow was much deeper than it had been around Bamsebu, and he found himself trudging through waist-deep banks. He was forced to spend one night out in the mountains, sleeping in a clearing under a boulder with Bamse. The next day, it took twelve hours to hike a distance that should have taken a quarter of that time. As he approached Nilsbu, half hidden in fog, he was especially relieved to see that there was no shovel on the cabin roof, which would have been a warning signal telling him to keep away.

Skinnarland welcomed him—and Bamse—into the cabin. The two men planned to spend the winter together. It was good to have someone to rely on—not to mention the company—through the long, dark months. They settled into their new routine and continued to supply London with intelligence on heavy water production at Vemork. They were ready to slow or stop it if they were called upon to do so. But a call to action did not come.

"The powers that be wish us to consider whether we can have another go at the Vemork plant," wrote one SOE deputy to another on October 5, 1943. Those "powers" were Sir John Anderson and his American counterpart, General Groves, who believed his scientists would have a working bomb in twelve to eighteen months, and the Germans might not be far behind. And even if they were not close to having a bomb, they might still be in a position to inflict a radioactive attack on London or another major city. Whatever "another go" entailed, the two SOE deputies decided over a series of correspondence, the Norwegians must be excluded from any involvement because of their clear opposition to any attack.

Using intelligence provided by Skinnarland, Wilson and his British staff drew up a report with three options: (1) internal interference with production; (2) coup-de-main attacks along the lines of Gunnerside; and (3) aerial bombing. The first was deemed only a temporary solution. The second was a long shot given the new defenses at the plant. The third, if carried out during a daylight precision attack, likely offered "the best and most effective course." Armed with the report, Anderson proposed a bombing raid on Vemork to the chief of the air staff, Sir Charles Portal. Portal passed the target dossier to the commander of the American Eighth Air Force, General Ira Eaker, who was already well aware of the target.

From his base at the Wycombe Abbey, an hour's drive west of London, Eaker governed a force of 185,000 men and four thousand planes. His mandate was to pummel Germany into submission, chiefly by destroying its ability to fight. The cigar-chomping, soft-spoken American, who early in his command told the RAF, "We'll bomb them by day. You bomb them by night. We'll hit them right around the clock," doubted the importance of the target and resisted the mission.

Groves continued to press for the attack through General George Marshall, the Army chief of staff, and the senior British representative in Washington, Field Marshal Sir John Greer Dill. On October 22, Eaker complied, informing his men, "When the weather favors attacks on Norway, [Vemork] should be destroyed."

Throughout all these machinations, Tronstad and the Norwegian high command were left outside the circle. They were told that their efforts to influence the target list had been successful, that there was a star next to Vemork's name, indicating it was to be attacked "in a special way, and therefore should not be bombed." Everyone from Eric Welsh to the RAF reassured Tronstad that nothing would be done without telling him first.

On November 11, 1943, at a joint meeting between the British and Norwegian high command, Tronstad pitched a new policy toward industrial targets in his country, one that, given the state of the war, "should be changed as far as possible from the offensive and destructive to the defensive and preservative." The same deputy who had delivered to Anderson the recommendation to bomb Vemork without a mind to the Norwegians now promised Tronstad that he would take up the case with the chiefs of staff. This was quite simply a lie.

24

Cowboy Run

A T 3:00 A.M. on November 16, 1943, when the duty sergeant roused pilot Owen Roane from his bed, Station 139, the massive U.S. airbase by the North Sea coast, was already alive with the preparations for an impending mission. Orders from the Eighth Air Force command, over a hundred miles to the southwest, had come into the base by Teletype, identifying the target, the weather prospects, the force needed, and the plan of attack. The operations officer, Major John "Jack" Kidd, and his staff had been working since the field order arrived. They determined bomb tonnages, fuel loads, routes, zero hour for launch, and which groups and squadrons would participate in the assault.

While First Lieutenant Roane shaved his boyish, bright-eyed face — the better to improve the fit of his oxygen mask — armament crews were at the bomb dump, loading their trailers with explosives weighing a thousand pounds each. At the same time, fuel tankers rumbled across the tarmac to fill a row of B-17 bombers, while mechanics checked out the engines and bomb bays. At the mess hall, the cooks and kitchen staff were preparing the morning's pancakes, powdered eggs, and oatmeal. And at the Group Operations buildings, maps and photographs of the target were being assembled for the crews.

Dressed for the minus-thirty-degree temperatures at high altitude (wool underwear, two pairs of wool socks, a wool sweater, a brown leather jacket lined with sheep's wool, and heavy trousers), Roane crossed the cold, fog-ridden airfield and gathered with the other pilots and aircrews in the huge Nissen hut used for briefings. A curtain covered the map showing their route and target. Once the doors were closed, Major Kidd, the operations officer, stood up in front of them, and a duty clerk pulled aside the curtain. They were headed to Norway, to a place called Rjukan.

Given the distance and the short November day, they would need to depart soon after 6:00 a.m. The target was Vemork, a power station and hydrogen plant, where the Germans made some "special explosive." To limit civilian casualties, they would hit the site during the lunch hour. Major Kidd did not expect much enemy resistance, either from antiaircraft batteries or from fighter planes, and called the attack nothing more than a "milk run." It was never said, but Roane was left with the decided impression that this Vemork place was a priority target.

Although Roane had just celebrated his twenty-second birthday, he was something of an "old-timer" — only two missions away from joining the "Lucky Bastards Club." Membership was earned by beating the odds and making it through a twenty-five-mission tour alive. Nicknamed the Cowboy, with the hat to match, Roane was from Valley View, Texas, population 640, a town north of Dallas and little more than a dirt strip bordered by a few buildings. Roane was one of nine children (eight of them boys). His family owned a small farm, growing cotton, wheat, and corn, and his father also herded cattle on a nearby ranch. Owen loved to help when he escaped from school each day. On graduation, he joined the Army Air Corps, his aim being to become a mechanic. A few plane rides later, he was hooked and enrolled in flight school. Soon he was assigned to fly the B-17. The four-engine, long-range bomber had an arsenal of machine guns and could take punch after punch and still deliver its bomb load — over ninety-six hundred pounds. Crewed by ten men, the B-17 was known as the Flying Fortress and was a giant in the sky.

In June 1943, after ten months of flight training, Roane arrived in Britain, where he was assigned to the One Hundredth Bombardment Group. Over the coming months, they would earn their own sobriquet: the Bloody Hundredth. They hit submarine bases, airfields, and factories across occupied Europe and far into Germany. Over that time, Roane saw B-17s that were flying next to him end their war, either shredded by enemy fighters, exploding midair, barreling down into the sea, banking sharply into the ground, or simply falling helplessly from the sky, their engines dead, their pilot and crew parachuting out over enemy territory.

The average lifespan in the Eighth Air Force was eleven missions; his fellow crewmen, many of them friends, were killed or went missing at a sobering rate. With equal measures of luck and skill, Roane always made it back. On one mission to Stuttgart, a fire raging across his wing and hounded by Messerschmitt fighters, Roane sent his plane into a spinning nosedive at three hundred miles per hour to extinguish the flames and throw off the enemy. Over Bremen and Schweinfurt, he waded through storms of flak, hell-bent on dropping his bombs on the target. In August 1943, after a run against a Messerschmitt factory, he was forced to land his plane, riddled with 212 bullet and flak holes, in North Africa. While there, he adopted a twenty-five-pound black donkey he named Mo, short for Mohammed. Mo came back to Britain with him, spending the flight hooked up to an oxygen mask in the radio room, covered with a sheep-skin jacket. Approaching base, Roane messaged the control tower: "I'm coming in with a frozen ass."

At 5:00 a.m. Roane made his way onto the hardstand to check out his plane. Circling it, he inspected everything from the tires and the fuel vents to the propellers and wing-deicing boots. The ground-crew chief advised that the plane had been loaded with six-thousand-pound bombs and an overload of high-octane gas, bringing its total weight to sixty-five thousand pounds (twelve thousand pounds over its rated maximum). Takeoff in the dark through a low cloud ceiling would be a neat trick.

"Ready to go," Roane told the ground chief, then he headed to the food wagon, where he milled around with his men and had some tea, then a last cigarette.

They were joined by Major John Bennett, their new squadron commander. The hard-nosed thirty-six-year-old was coming along on the mission aboard Roane's plane, the *Bigassbird II*.

After checking the safety straps on each other's Mae West life preservers and parachutes, the ten-member crew entered the plane through the rear fuselage and took their positions. Two of the crew, the flight engineer and the radio operator, were a little more tense than usual, it being their twenty-fifth mission—a sortie infamous for bringing bad luck. Roane ran through his checklist again, and on seeing a green flare fire into the morning sky, started the engines. Their roar coursed throughout the plane, turning any conversation not conducted via the interphone into a shouting match. The ground crew removed the wheel chocks, and Roane taxied the plane onto the runway. All about him, lights flashed, brakes whined, and the air reverberated with the growl of engines.

A minute behind schedule, *Bigassbird II* was in position for takeoff.

Roane released the brakes and headed down the runway. Three thousand feet down the pavement, the throttles at maximum power, Roane lifted the plane into the dark sky at 120 miles per hour and retracted the wheels.

Almost immediately the plane was shrouded in clouds. Departures were spaced out every thirty seconds, but if the plane ahead had engine trouble or if its pilot misdirected his course, the plane behind could find itself flying straight into it, and the crew wouldn't know a thing about it until it was too late.

At three thousand feet, they emerged from the clouds. A half-moon hung overhead. To assemble with the twenty other B-17s in his group, Roane made a wide left-hand circle around Station 139. He kept his eyes peeled on the swirling beehive of planes in the sky, both to prevent a mid-air collision with the other planes stacked at various altitudes and also to spot the colored flares corresponding to his own formation.

Three hundred and eighty-eight B-17 Flying Fortresses and B-24 Liberators from three divisions of the Eighth Air Force were headed to Norway that morning. Roughly half of them were set on Vemork; the others were assigned to destroy an airfield just north of Oslo and mining operations in Knaben.

After some time spent circling, Roane rendezvoused with the other planes in the One Hundredth. Since Bennett was onboard, *Bigassbird II* was the lead in the group. There was a fair bit of mayhem, pilots barking into their radios as B-17s and B-24s scrambled to find their places in the moonlight. Once they were all together, Roane told his crew to "go on oxygen" and climbed to an altitude of fourteen thousand feet for the journey across the North Sea. The armada expected limited fighters, and so did not assemble in combat wings but rather hung together in groups of about twenty planes.

They crossed the North Sea on a northeastern course at a steady cruising speed of 150 miles per hour. The sun rose over the horizon to their right, illuminating a mesmerizing view: drifts of feathered clouds hanging below them, pure blue skies above, and hundreds of bombers surrounding them. The cockpit heater kept Roane and Bennett toasty warm.

When they neared the coast, Roane and his crew donned their flak jackets and steel helmets. They would have to lower to twelve thousand feet to drop their bombs, and given that Vemork was roughly two thousand feet above sea level, the difference would put them at a prime distance for antiaircraft fire.

When they sighted Norway, Roane checked his watch and found that they were twenty-two minutes ahead of schedule. The first bombs were not supposed to be dropped until 11:45, when the plant's workers would be eating lunch in the basement-level canteens. They had a choice: drop the bombs early and risk more civilian casualties, or make a 360-degree turn at the coast to delay the run, which would give the Germans time to muster a defense.

"Make a large circle over the North Sea," Bennett decided.

When the bombers came around again, the Germans were ready for them. Two coastal patrol boats fitted with antiaircraft guns fired away. One B-17 went down. The rest of the bombers continued through the flak, most of it meager and inaccurate. Then German fighter planes scrambled into the sky, Messerschmitts and Focke-Wulfs. They attacked sporadically but were too few in number and the Flying Fortresses too well armed with machine guns for them to press the attack home. A B-17 in another group was hit. Its crew parachuted out, then the unpiloted plane performed a series of sharp turns, whipstalls, and corkscrews before slamming into the sea. Still, there was nothing the enemy could do to stop the force of bombers.

Having survived missions over Germany where hundreds of fighters attacked for hours on end and where the Eighth Air Force lost sixty bombers in a single day, Roane was aware that this journey to Norway remained very much a milk run. As they crossed the coastline, the temperature in the cockpit registering minus-forty-five degrees, he peered down on a monotonous landscape of snowbound mountains, steep canyons, and frozen lakes. The view felt ominous. The navigator, Captain Joseph "Bubbles" Payne, had few landmarks — cities, rail lines, or roads — to guide him to the target. Nonetheless, he charted a true course to Vemork.

The Ninety-Fifth Bomb Group was ahead of the One Hundredth on the approach. Roane had to descend slightly to get out of their contrails, and *Bigassbird II* rocked and shuddered in the prop wash. Minutes later, the bomb bay doors opened. At twelve thousand feet, free of the turbulence at last, the crew readied to drop their load. Some low clouds hung in the sky ahead, but they would not interfere with an accurate run. In total, 176 Flying Fortresses and Liberators soared on toward Vemork.

On a farm west of Vemork, Einar Skinnarland had just finished transmitting a message to London by wireless. While he waited in the barn for a return message to come through, he heard a distant rumble. Stepping outside, he found Haukelid staring up into the sky. Far above, an endless parade of bombers was heading east, and neither German fighters nor antiaircraft guns harried their course. The two men could only guess at their target. There was a good chance it was the power station and hydrogen plant at Vemork. The bombers made for an awesome sight.

At 11:33 a.m., air-raid sirens blared throughout Vemork. Transport manager Kjell Nielsen ran down the steps of the hydrogen plant to the basement shelter. Only months before, he had been working at Herøya, in magnesium production, when American bombers had attacked. At that time, Nielsen had been supplying intelligence and photographs of the industrial site to the resistance — and, by extension, to the Norwegian high command.

Down in the shelter, the chief engineer, Fredriksen, received a phone call from the operator at Våer, the hamlet across the bridge. She reported twenty aircraft above the valley, then another fifty, then cried out, "There are even more planes!" Fredriksen had no doubt as to their purpose.

The panicked families of the workers and engineers who lived on the Vemork side of Vestfjord Valley were shepherded into the air-raid shelter near their homes. The concrete structure, built above ground, was to be used as a garage when the war was over. Considering the limited protection it provided its occupants, it would have been better had this already been the case.

Down in Rjukan, citizen volunteers directed the townspeople into a range of structures prepared for such an assault. At the local school, teachers hurried some sixty pupils into a tubular concrete shelter, which had a layer of sand on the floor. Four Germans who were living on the first floor of the school building joined them. They all heard the thunder of airplanes overhead. Fearing what was to come, one of the teachers ventured outside. Seeing the formation of bombers directly overhead, he shouted, "We're in the center of the circle! Run to your homes!" The schoolchildren dashed from the shelter and scattered in all directions.

The Ninety-Fifth, the lead group in the attack, swept directly over Vemork and held on to its bombs. Roane figured their crews could not see the plant through the bank of low clouds hanging over the target area. No doubt they would come around for a second pass. He wanted *Bigassbird II* to win bragging rights for the One Hundredth by being the first to hit the target. Whether they did or not would be down to the skills of his bombardier, Captain Robert Peel, who was now in charge of the plane's flight controls.

At 11:43 Peel spotted the plant through a slight break in the cloud cover. The Germans had started to generate smoke screens over the valley, but they were not enough to obscure the target. "Bombs away," Peel called, releasing his ordnance. With the sudden loss of weight, *Bigassbird II* bucked upward in the sky. Peel watched his four thousand-pound "eggs" strike the target. The concussion rocked the plane as it continued on a straight course for ten seconds, giving the Fortresses behind time to release their loads along the line.

Squadron after squadron followed. Over the next twenty minutes, the planes, with names like *Hang the Expense*, *Raunchy Wolf*, and *Slow Joe*, poured destruction down onto the plant. Those who had missed their drops on the first pass circled back to try again through a haze of contrails and billowing smoke. There was occasional gunfire from the ground. It made little more impression than the few fighter planes that continued to nip at the edges of the armada.

In total, 711 explosions ripped across Vemork and the surrounding area. Some bombs fell in the valley and woods, causing no harm. Others struck the penstocks, severing nine pipelines and spewing tons of water down the hillside. The suspension bridge was torn in half and hung over the southern cliffside. Three direct hits on the power station ripped away part of its roof, destroying two of the generators and damaging others. Bombs sheared off the top two floors on the western corner of the hydrogen plant. Several houses at Vemork and Våer were leveled, and the homes not eviscerated by explosion were destroyed by flying stones and splinters and by the fires that followed. Flames — red, green, and orange — rose throughout the area.

Just as the main body of bombers banked away from the hydrogen plant, a pack of twenty-nine B-24 Liberators flew down the Vestfjord Valley. The pack had been assigned to the bomb run outside Oslo, but they found their target covered in clouds and so had come the hundred miles to Vemork. At 12:03 p.m., these B-24s mistook the nitrate plant in Rjukan for the target and released their five-hundred-pound bombs. Most of the cluster hit the plant, bringing down a pair of brick towers and demolishing a number of small buildings. Some of them struck the town's populated center, a few hundred yards away.

Roane and the others directed their bombers back toward the Norwegian coast at twelve thousand feet, Roane one run closer to joining the Lucky Bastards Club. As they made their way safely home, the residents and workers of Vemork and Rjukan were emerging from hiding to reckon with what they had left behind.

"My God, what's happened to my family?" One engineer, covered in concrete dust, gave voice to everyone's fears as he stepped toward the door of the hydrogen-plant shelter. Nielsen tried to calm the man next to him, who was frantic about whether his wife and children had reached the air-raid shelter in time.

A former member of the Norwegian Red Cross, Nielsen had cared for wounded soldiers during the Finnish war against Russia. Now he headed straight to the shelter to see if anybody there needed his help. The air-raid sirens were still wailing, and all of Vemork was choked with smoke. People ran through the rubble putting out fires and carrying the wounded out of buildings on the verge of collapse. Some workers managed to shut down the flow of water from the penstocks and also closed off the valves to the severed hydrogen and oxygen pipes that ran across the valley. Screams and moans sounded from every direction.

There were no survivors at the Vemork air-raid shelter. There was nobody in need of Nielsen's help. Where it had stood there were two craters, the result of two direct hits from the bombers. The concrete walls and roof had been pulverized, and the sixteen people who had huddled inside were dead: eleven women, two children, three men. Their bodies were all but irrecoverable: an arm, a head devoid of its features, a dismembered torso. Flesh and bone littered the broken concrete and twisted steel bars in gruesome chunks. Fathers, husbands, and friends knelt down in the open holes, their cries joining together into a macabre song of grief.

In Rjukan, four miles away, plumes of heavy dark smoke filled the sky. The nitrate plant and a number of houses were in ruins. As fate would have it, the teachers and students who ran from the bunker during the attack had saved themselves. The shelter had been leveled, just like the one at Vemork. When the smoldering fires had been put out and the wounded treated, the dead were counted: in total, twenty-one Norwegians had lost their lives.

SS officer Muggenthaler took cover in Rjukan during the bombing, but still received lacerations to his face from flying debris. In his first dispatches to Oslo and Berlin, he painted a stark portrait of the devastation he had witnessed in the attack and its aftermath. At Vemork, there was much that needed urgent repair: the pipelines, the suspension bridge, and the generators, as well as the equipment and the hydrogen plant. But after he and others carefully surveyed the site, investigations revealed that the "SH-200 high-concentration plant" was undamaged. Only a brief period of time and a limited amount of material would be needed to get things running again. In summary, the bombing run was a lot of storm and fury for what was, in effect, a limited blow to the German war machine.

Part V

25

Nothing Without Sacrifice

O N T H E D A Y of the American bomb attack, Leif Tronstad was where he felt most at home in his exile: in Scotland. He was staying at STS 26, checking in on his men and taking part in a week of commando training. It was a nice reprieve from the corridors of the high command, where he sometimes felt he spent more time navigating bureaucracy than fighting for Norway. He passed the afternoon setting off small explosive charges, the sun shining brilliantly on the hard-packed snow. For the first time in a long time, he felt carefree and happy.

Back at Drumintoul Lodge in the early evening, he heard a radio report about a U.S. Air Force run over Norway. There were no specifics. In his diary that night, he simply wrote, "Hope that the Norwegian losses have been small and that this devilry will end soon." Over the next few days, he learned the full story.

The Allies had broken their word. Without consulting the Norwegian government, they had sent a fleet of bombers to strike Vemork. Many civilians had died. Much needless destruction had been wrought, especially on the nitrate plant in Rjukan. That site had never appeared on any target list and only produced fertilizer for Norwegian agriculture. Hardest to accept was the fact that the primary target, the heavy water plant, had not even been damaged, just as Tronstad had warned it would not be. The whole situation — the betrayal, the needless death, and his own powerlessness to stop it — left Tronstad deeply embittered. But he was not one to wallow in such an emotion. He wanted reassurance that nothing like this would happen again.

In London, Jomar Brun took the news personally. Furious, heartbroken, and ridden with guilt that he had somehow played a part in the raid, he protested to the Norwegian and British governments and tendered a letter of resignation to the Directorate of Tube Alloys. "On leaving Norway about one year ago," he wrote, "I thought I would be able to contribute to the Allied war effort, at the same time helping to spare Norwegian lives and property. I now understand my mission was in vain."

When Tronstad returned from Scotland, he calmed Brun and persuaded him not to resign. Then he got to work, rallying General Hansteen and the Norwegian foreign minister to protest to the British and American authorities. Investigations were started and apologies given, but no offer was made to seek Norwegian approval in advance of any future strikes. The Allies made clear that they needed a free hand to wage war against the Germans.

Tronstad learned that the decision to bomb Vemork came from outside the typical target-list channels; in fact, the highest Allied authorities had mandated the attack. They felt that even if the Germans were unable to manage a fission bomb like the ones the Allies were developing, they might still produce radioactive weapons that could ruin entire cities. Unless and until they had clear, unequivocal intelligence that the Germans were no longer pursuing fission research, then the Allies had to act with unflinching resolve.

From messages received in December from Njål Hole, Tronstad knew the Germans had not abandoned the atomic field. Their studies and experiments continued, and they were still eager to produce heavy water. There were further reports from Germany that Hitler was heralding "secret weapons," soon to be loosed on the world. One of them, from a Reuters correspondent who interviewed several fugitives from Germany, warned of a Nazi bomb "filled with explosive gases of fantastically high destructive power . . . that will be used against Britain soon." To remind himself of what he was fighting, Tronstad tucked the article into his diary.

Allied intelligence soon learned of reports that Norsk Hydro had "decided to abandon entirely the production of heavy water." This unconfirmed intelligence failed to appease Tronstad. The Americans had demonstrated that they were willing to launch huge raids on Vemork to halt production. Now, he warned, even if Vemork was shuttered, the Germans might dismantle the high-concentration plant and recommence production elsewhere — somewhere unknown to the Allies. If he had been allowed to move forward with the castor-oil sabotage, they could have slowed production almost to a standstill without arousing suspicion.

Struggling emotionally with the aftermath of the raid, Tronstad also had to suffer through another Christmas away from his family. Although he knew that Bassa was holding strong in his absence, her letters throughout the year spoke about the effect of that absence on all of them. In one, she wrote, "It was a heavy time for me after you were gone, and I still have the most terrible nightmares in my sleep, so it has probably left a deep mark." She feared for him too: "It's as if you are living only half a life in this way. The children are growing up without you, it's so sad, our little boy is so funny, he will be a big boy when you get home."

Tronstad feared the same, and wondered in what other ways the war had changed him. In one letter to Bassa, he wrote, "It is my fervent wish that the two of us will find our way through this darkness and come together again in life in an open and trusting way as before." All the suffering was the price they had to pay. Just before Christmas, he wrote, "We get nothing without sacrifice in this world . . . and perhaps least of all freedom and independence." It was the nature of his own contribution that troubled Tronstad the most. While he sent his boys, one after another, into danger — and often to their deaths — he himself always stayed behind. "It is hard to spend my time quietly here to determine other people's life and death," he lamented in his diary. He wanted to be in Norway, bringing the fight to the Germans on the ground.

Throughout November and December 1943, as Allied bombers pummeled Germany night and day, Kurt Diebner and his crew of young physicists continued doggedly to work on the assembly of uranium machines. For their G-III experiment, his team constructed a hollow sphere out of tons of paraffin wax and poured 592 kilograms of heavy water at room temperature into it. Then they fitted 106 uranium cubes to thin metal wires (eight or nine cubes to each wire) and secured the wires to a lid, each cube in the lattice an equal distance apart. The whole arrangement was lowered into the heavy water by a winch. The machine showed impressive results.

The Gottow team then repeated the experiment, using the same sphere and the same heavy water, but this time with 240 uranium cubes. The design was efficient enough to be assembled in a day, and the results showed a neutron increase of 106 percent. As Diebner reported to his fellow physicists, "Given the relatively small size of this apparatus, these neutron in-crease values are extremely large." Straightaway his team began building out the third iteration of the strung-cube design.

His success came at an opportune moment: the constellation of offi-cials involved in overseeing the atomic group — Speer and Göring among them — were about to oust Abraham Esau as head of the program. It was true that under Esau there had been a number of advancements, from Diebner's cube design to Harteck's ultracentrifuge work, to Erich Bagge's U-235 enrichment using an "isotope sluice," to numerous theoretical pa-pers by a host of physicists that laid the groundwork for practical prog-ress. But there was still no working reactor, nor any prospect of a bomb.

Furthermore, the program continued to lack a steady supply of that essential moderator, heavy water, despite much experimentation and the establishment of pilot plants to produce it. Building a full-scale plant at Leuna was dogged by high costs and protracted negotiations with IG Far-ben. The recent Allied bombing on Vemork made clear the mistake they had made by relying on the Norwegian supply.

Göring sent Esau his official marching orders on December 2. His replacement was Walther Gerlach, a scientist who had made his name studying subatomic magnetic fields. Gerlach taught experimental phys-ics at the University of Munich and had spent most of the war working on torpedo fuse design for the navy. Tall, with a narrow, beak-nosed face, Gerlach was well liked among German physicists and was considered to be a soft, yet deft, hand at navigating the channels of power. Some thought him a curious choice, most likely to push the program only as a way of saving the scientists involved from the frontlines.

They were soon to discover that he had ambition. Indeed, he planned to be "the emperor of physics." Although he was not a member of the Nazi Party, Gerlach was a militarist who wanted to see Hitler remain in power. By late 1943, he felt doubtful that Germany would be victorious in the war, but he was sure that if they possessed a working reactor or a bomb, they could secure whatever peace-treaty terms they wanted. As he told one of Speer's deputies, "In my opinion, any politician in possession of such a de-vice can get anything he likes."

Prior to his official appointment on January 1, 1944, Gerlach met with Diebner several times. He recognized that the Army Ordnance physicist was showing the most progress in his experiments and reassured him that he would be given whatever resources he needed to succeed in his efforts.

In addition, Gerlach promised to appoint Diebner as his administrative assistant and return him to his office at Harnack House, the headquarters of the Kaiser Wilhelm Institute. Gerlach was not interested in appeasing the feelings of all parties as Esau had been. Heisenberg, who continued to resist the superiority of Diebner's cube design over his plates, would simply have to adjust.

On December 11, 1943, Diebner headed to Norway to remedy the continued roadblocks with heavy water. At Norsk Hydro headquarters in Oslo, he called a meeting with Axel Aubert, the seventy-year-old director general, who had resumed his position after his successor's arrest. Although the same stand led to the arrest of Eriksen, Aubert now told Diebner that Norsk Hydro was discontinuing heavy water production completely, as they could not "expose the company's workers to further attack, nor invest another fortune in rebuilding a plant that would be lost in the event of a new air raid."

Diebner agreed with him. If production continued, the Allies would likely attack Vemork again. He wanted to move the plant's high-concentration equipment — including all existing stocks of heavy water at every level of concentration — to Germany, where a new plant would be constructed. Aubert tried to argue against this — surely it was better they store everything in Norway until after the war. However, Diebner was determined. He needed approval from Berlin but was confident it would be granted. Then, when a new plant was built, he would have all the moderator he required for a reactor to produce plutonium.

At the same time, Diebner was forging a second path toward a completely different type of atomic explosive. Through December and into the new year, he devoted a team of scientists and engineers at the Army Ordnance Research Department to it. Almost a decade before, physicists had shown that when two deuterium atoms collided at high speeds, pulses of energy were released. At Kummersdorf, some German scientists had perfected the shaped charges that could bring about these collisions at very high temperatures. Diebner and his team began putting together a series of experiments that would squeeze deuterium atoms together through the use of explosive shock waves inside a hollow silver ball, their goal being to trigger a fusion reaction — and create a bomb. One way or another, Diebner aimed to give Germany the weapon it needed to reverse its failing fortunes against the Allies.

· · ·

"We are sending all our friends our best wishes for a Happy Christmas," Skinnarland tapped out in Morse, following with special greetings to Kjelstrup, Poulsson, and Helberg. They soon received a reply: "All your friends here send you best wishes for the new year with the hope that we can soon meet again." Bunkered down at Bamsebu, they tried to find what cheer they could. They made a little Christmas tree from some spruce branches, ate reindeer, and played the handful of records they had nabbed along with a gramophone, Haukelid's gray elkhound at their feet. They tried to get a signal to find some holiday music, draining a little of their precious battery supply, but the best they could do was a BBC broadcast of Big Ben striking its chimes at midnight on New Year's Eve, heralding 1944.

The two men had much to reflect upon as 1943 came to an end. Skinnarland had lost his father and niece that year. Although the Gestapo had finally released his brother Olav, Torstein remained at Grini. Skogen had been sent to a prison camp in Germany in early October. His fate was unknown.

Haukelid was somber too. He had learned through Milorg that his father had been arrested in late September and taken to the torture chambers at Møllergata 19. Bjørgulf Haukelid knew a lot about his son's activity before he had left for Britain, but he knew nothing about the wireless sets secreted in one of his company's storerooms in Oslo, which were discovered during a raid. Haukelid hid the sets there in 1941, thinking it best not to say anything to his father. Now he feared the Gestapo would inflict even worse torture on him, assuming that he was lying when he said he didn't know anything about the radios.

After the New Year, Haukelid continued to build his underground resistance network and Skinnarland to collect intelligence on Vemork. The two men, though now as close as brothers, made an odd couple. Emotional and intemperate, Haukelid was quick to act. Skinnarland was reserved by nature, patient and deliberative, always planning things out before making a move.

Their differences, and their living so closely together in a ten-by-ten-foot cabin, bred tension. One evening, tired of Skinnarland playing the same record for the hundredth time, Haukelid took it off the gramophone and broke it into a dozen pieces. Skinnarland's response was to do the same with Haukelid's favorite. Another day, Haukelid returned after a long, unsuccessful hunt, hungry and vexed. He'd thought the conditions perfect for stalking a herd and yet still he came home empty-handed. "I don't understand it, it's supposed to be best in thick fog," Haukelid complained. "Then you can creep up quite close to them without their seeing anything."

"Indeed — and without your seeing them," Skinnarland said, unsympathetic. "Everyone knows it's impossible to find deer in a fog. You only scare them away."

"Do you think I don't know where they are?" Haukelid said.

Skinnarland gave him a quizzical look. "Only a fool goes out shooting in a fog."

"Oh, go to hell!" Haukelid shouted. He stormed out of the cabin and spent most of the night fuming in the snow.

Despite these squabbles, both were committed patriots who loved life in the mountains and were tough enough and skilled enough to endure the worst of conditions. While Haukelid could be volatile, Skinnarland's equanimity served as a balance. More than anything, they trusted each other with their survival.

On January 29, 1944, Skinnarland received an urgent message from Tronstad. He had just received some troubling intelligence from Stockholm, and he needed Skinnarland to confirm its accuracy. "It is reported that the heavy-water appliances at Vemork are to be dismantled and transported to Germany. If it is true, are there any possibilities of preventing the transport? It is a matter of great importance."

That same day at Vemork, construction engineer Rolf Sørlie was helping the war effort by dragging his feet with respect to the plant's rebuilding. In the afternoon, he was visited by Thor Viten, a Milorg leader in Rjukan. Sørlie felt excited by the visit: maybe his chance had come at last.

Although the American raid had cost a number of lives, Sørlie and many others in Rjukan understood the need for it. Clearly, the Germans were keenly invested in what Vemork was producing, and in wartime, one hit the enemy where it hurt most. There was also talk that Nazi scientists were using heavy water as a catalyst for splitting the atom and, potentially, creating a bomb. These rumors were fed by a number of sensational articles in Swedish newspapers smuggled into the country, which described Vemork as a "secret weapon forge" for the creation of "a single atomic bomb to lay London flat."

Like Sørlie, most of the workers tasked with the reconstruction believed that the sooner they completed the job, the sooner the Allies would send another wave of bombers. The next run might target the dam at Lake Møs, flooding Rjukan in less than an hour, which would be a catastrophe. The new German antiaircraft batteries and torpedo nets at the dam only reinforced this fear. As a result, none of the local workers were interested in finishing the rebuilding. While such passive resistance was important, Sørlie wanted to be part of the bigger struggle, like his friend Helberg. Editing a few illegal newspapers, hiding some radios, passing along a little intelligence — these were no longer enough for him. He wanted to be trained. He wanted to fight. He had never allowed his physical disability to keep him from adventures out in the woods when he was a boy. It wouldn't stop him now.

Viten told Sørlie to come up to Lake Møs with him and bring his skis. After work, they took a bus up to the lake, then started across its frozen surface. The wind blew with such force that Sørlie had to turn his head to the side to draw breath. Keen to show Viten that he could handle himself as well as anybody in the mountains, Sørlie fought to keep up.

After a few miles, they reached the Hamarens' farm. There they enjoyed a satisfying meal before being brought to a cabin a couple of hundred yards away to spend the night. Neither the Hamarens, nor Viten, explained to Sørlie why he was there, but he suspected it was to meet Einar Skinnarland, whom he knew to be in hiding in the area. In the early-morning hours, his suspicions proved correct when Skinnarland arrived at the cabin. He wanted to know if a recent message he had received from London was true: Were the Germans planning on disassembling the high-concentration plant and moving it out of the country? Sørlie confirmed that this was the case, but that first they would be shipping out the existing stocks of heavy water. Skinnarland urged Sørlie to find out as much as he could about the impending transport: they might need to stop it. Later that morning, Sørlie headed back to Rjukan, the journey made easier by his excitement at finally being part of an important mission. Little did he know that he would be returning to the farm much sooner than anticipated.

26

Five Kilos of Fish

ON FEBRUARY 1, 1944, a Milorg courier arrived in Rjukan with the news that Minister President Quisling wanted to mobilize seventy-five thousand Norwegians to fight for the Germans on the Eastern Front. Milorg called for all its members in the area, young and old, to go into hiding in the mountains. Already in the past three months, a spate of Gestapo arrests in Rjukan, Notodden, and Kongsberg had decimated the underground organization. It could not afford to lose more fighters.

That night, Sørlie, who had just returned from the Hamarens' farm, met with others in the resistance at a house in town. It was decided: in the morning, they would head to cabins in the mountains, bringing enough food with them to last at least a week. The danger was clear. The Germans would not fail to note the disappearance of some two hundred to three hundred men, many of whom worked for Norsk Hydro. Patrols were sure to be sent out to find them. If it came to a fight, they were armed with only a scattering of guns and homemade grenades filled with glass and nails.

The orders from Oslo were plain, but there was a chance it was a Nazi trick to draw out the underground cells. Viten asked Sørlie to seek out Skinnarland and obtain confirmation of the order from London. Sørlie was happy to go; he had more information to provide to Skinnarland on Vemork. Some workmen had discovered fat floating in the high-concentration cells — the result of Gunnar Syverstad's doctoring them with cod-liver oil. Any further such sabotage would pose too great a risk. Further, Sørlie did not yet know when, or how, the drums of heavy water would be moved from the plant, but Syverstad and Kjell Nielsen, the transport manager, whom Sørlie had recruited into the local Rjukan resistance, were going to find out soon.

The next day, Sørlie returned to the Hamaren farm, arriving late in the afternoon. As before, Jon Hamaren invited him in for a meal, then led Sørlie through the darkness into the hills. The terrain grew steeper and rougher with every passing minute. Eventually they reached a farmhouse. Sørlie thought he had arrived at Skinnarland's, but he was greeted by Olav Skindalen. Skindalen offered them coffee and, after a short break, they continued into the wilderness.

At last they arrived at an old hunter's cabin shrouded in darkness. Hamaren knocked firmly and a light flickered from inside. Skinnarland appeared in the doorway and welcomed Sørlie and Hamaren to Nilsbu. The floor was covered with reindeer pelts; there was a line of rifles on the wall, and a wireless set on the table in the center of the room. Sørlie delivered his report on Milorg and the heavy water. Soon Skinnarland was sending a message to Tronstad and Wilson. Hopefully, they would respond soon.

Hamaren left them and headed back down the hillside, and Sørlie was soon asleep, exhausted from the journey. When Sørlie awoke the next morning, Skinnarland was shaved and in fresh clothes. "Do you always keep the place this nice?" his visitor asked, admiring the orderly cabin.

"Yes," Skinnarland replied. "If it were a pigsty, I'd have gone mad. If you're careless and disorganized, you lose your ability to see things clearly." Skinnarland then spoke about the Hamarens and the Skindalens, how essential they had been to his survival. "Through generations," he said, "these farmers have learned how small people are in the face of the forces of nature. Being helpful, without the thought of betrayal, is necessary when you live in a place like this. It's in their blood."

At last, an answer came in about the mobilization order. No one in London knew of any order from the Oslo Milorg leadership to retreat into the mountains. There must be some mistake; either that or the Germans were trying to draw them out. Tronstad instructed Skinnarland to have the order reversed immediately. A second message stressed how important it was to find out everything about the heavy water shipment.

Sørlie returned the twenty miles to Rjukan as fast as he could travel.

Three days later, on Sunday, February 6, Knut Haukelid was returning to Nilsbu from the mountains to the west, furious that a herd of reindeer had escaped him after his rifle failed to fire. As he made his way down toward Lake Møs, he spotted a number of ski tracks in the snow. There were far more than there should have been at that time of year. Being cautious, he returned up the slope and followed a ridgeline until he was close to the Hamaren farm. Jon Hamaren answered the door with, "Did you meet them?

"We had fifteen men here not long ago," Hamaren explained. "German soldiers, combing the hills."

Haukelid skied away quickly for Nilsbu, keeping a close eye out for Germans. When he reached the cabin, his dark beard was half-frozen from the cold. Inside, Skinnarland was with a stranger, a slight-looking man he introduced as Rolf Sørlie, a construction engineer from Rjukan. He had helped Helberg before the Gunnerside operation.

Skinnarland dished up a meal and some coffee as the men spoke. Sørlie reported that the Milorg order had been reversed, but not before many in the resistance had taken to the hills. The initial communication had been valid, but the Milorg courier had failed to relay that the evacuation should be undertaken only if the mobilization order was put into effect. The German patrol that Haukelid had so narrowly avoided was likely a response to this flight. As far as Sørlie knew, however, the Germans were of the mind that it was not an organized retreat but rather a rash escape into the mountains by the men of Rjukan who were afraid of being sent to the Eastern Front.

Haukelid was less concerned about Nazi patrols on the Vidda than he was about Sørlie's intelligence on the recent activity at Vemork. The Nazis intended to move all the stocks of heavy water, at every level of concentration, within the week. The drums would be shipped from Vemork by train, then ferry. Haukelid and Skinnarland knew they had to do whatever they could to stop it. With limited time and no commando team at the ready, an operation would not be easy. Skinnarland sent a message to Tronstad: "We will probably be able to blow up the transport, but as time is short, I must be told soonest what to do."

Sørlie returned to Rjukan to glean more information about the transport. Meanwhile, Haukelid and Skinnarland started collecting explosives and selecting a team of men they knew could help. What they needed more than anything else was the order to proceed.

• • •

After a week of long days at Kingston House followed by long nights in his bedroom listening to the German bombers thundering over London, Leif Tronstad just wanted to spend his Sunday afternoon at rest. Maybe he would catch up on some reading, take a walk in Hampstead Heath, or write a letter to his family. Then he got a telephone call from a member of his staff. It was urgent he come into the office. There, he received the cipher message from Skinnarland, asking for the go-ahead to sabotage the Vemork transport.

Over the past week, he had sent many questions for Skinnarland to answer: What equipment was to be dismantled? How much heavy water was to be shipped? Of what concentration? Might they be able to contaminate it first? When was the transport planned? Could it be prevented? And if so, how?

Now Tronstad had all his answers. The time for contamination was past, and any opportunities to tamper with the heavy water had been taken. Transport was imminent and included even the most dilute concentrations. The shipment would weigh many tons and would require roughly forty drums. Immediately Tronstad alerted Colonel Wilson and Eric Welsh. An attack on the shipment needed approval at the highest level.

In the end, everyone from General Hansteen at the Norwegian high command to the British War Cabinet agreed that a strike should be made. When Welsh brought the news to Michael Perrin and his boss John Anderson, they were dismayed at the amounts of heavy water remaining at Vemork and thought it vital to the war effort to stop the Germans obtaining either the stocks or the equipment required to produce more. The order was given, ultimately by Anderson himself, to intercept the shipment, no matter what it took.

On February 8 Tronstad drafted a note to be sent by Home Station operators on the next scheduled transmission to Swallow: "We are interested in destroying as much of the heavy water as possible. The demolition or perforation of the drums, especially drums containing high concentrations is of the greatest importance . . . Leave British effects where the action takes place and if possible use uniforms as before . . . Try to make the action to cause the least harm to the civil populations."

In his diary, Tronstad wondered what the outcome would be: "We will do the best we can, but with a heavy heart for the consequences at home. I fear it will result in a lot of suffering, but we have to hope that it will save us from worse things. The guys are outstanding. They are happy to give everything."

"Einar, you awake?" Haukelid asked late in the night of February 7. They were still waiting for the green light from London on the attack, and the thought of it kept him from sleeping.

"I wasn't until you began bellowing," Skinnarland replied.

"You're as fast as anybody in Norway on the British sets now."

Skinnarland grunted.

"You'll have to stay in the hills as the radio link-up with London."

"Not bloody likely," Skinnarland said. It was true that he did not have as much commando training as Haukelid, but it was still a sight more than anybody else they would find to help with the job.

"If we 'buy it' — as they say so charmingly in England," Haukelid said, "you'll have to go down and see that the heavy water never makes it to Germany."

"I'll sleep on it," Skinnarland said.

The next morning, they received the critical message from Tronstad to "make the action" against the shipment.

Sørlie arrived at the cabin again soon after, with detailed information from Nielsen about the transport route. What was more, Syverstad had recruited his boss, the plant's lead engineer Alf Larsen, into the fold. From what Larsen said, the Gestapo was already aware that the shipment might be targeted. Security for the transport would be high. Syverstad and Larsen would tap the heavy water into the drums as slowly as possible in order to allow more time to prepare for an operation.

Seated at the table in Nilsbu, Haukelid, Skinnarland, and Sørlie began what they called a "council of war." From their intelligence, they knew that a train would leave Vemork with roughly forty iron drums onboard. The drums would be labeled "potash lye" but would contain heavy water at various levels of concentration (from 3.5 to 99.5 percent). Taking into account the German security preparations and the inside efforts to delay filling the drums, this train was unlikely to depart before February 16. When it did leave, it would travel down to Rjukan, then on to Mæl at the north west tip of Lake Tinnsjø. Then a ferry would bring the railcars down the length of the long, narrow lake. On the opposite shore, eighteen miles away, another train would haul them the short distance to Notodden, then on to the port of Menstad to meet a ship bound for Germany.

The three men talked over their options. First they could try to blow up the drums while they were still at Vemork. With all the additional defenses put up after Gunnerside — minefields, steel doors, bricked-out windows, and soldiers at every entrance — and more guards expected soon, it was unlikely the commandos could get into the plant. Next, they could hit the train while it wound its way down the cliffside to Rjukan. There was a shack along the route where Norsk Hydro kept explosives it used for construction. As the train passed, they could use pressure switches on the track to set off a huge detonation that would send the railcars pitching down into the gorge. The heavy water drums were thick, however, and some of them might survive the fall intact. Such an attack would also kill any Germans guarding the transport — deaths that would be avenged on the people of Rjukan.

Alternatively, they could wait until the railcars were loaded onto the ferry, then sink it. Given the depth of Lake Tinnsjø, it was unlikely that any drums would be recovered. But the ferry also carried civilian passengers, and some of them were sure to drown, along with the Germans guarding the railcars. They could attack the train as it traveled from the far shore toward Notodden, but this carried the same risks as hitting it between Vemork and Rjukan. Or an operation could be staged near Menstad, or at sea, en route to Hamburg, Germany. Both of these last options, however, were far from their home base and involved many unknowns.

By the end of their war council, they had decided that sinking the ferry was the best way to stop the shipment, despite the potential loss of life. They were also agreed that Skinnarland would stay at Nilsbu to keep up contact with London and as backup in case the operation unraveled. Haukelid and Sørlie would set up in a cabin near Rjukan to organize and carry out the operation.

The following morning, Sørlie left for town, charged with learning as much as he could about the ferry: its schedules, schematics, and any known security measures. Back at Nilsbu, Skinnarland sent a message seeking approval for an operation against the ferry. He felt oppressed by the thought that their decision would mean the deaths of innocent Norwegians. While they were waiting for an answer, Haukelid headed away to collect the explosives stored at Bamsebu, and Skinnarland went to recruit some local men to haul supplies and perhaps participate in the operation.

When the three men reconvened the following evening, they had received their permission from Tronstad. "Agree to sinking of ferry . . . If the seacocks are opened, this must be combined with an explosion to indicate a limpet attack from outside . . . The engine must be put out of order so that the ferry cannot be driven further to shallow water . . . The sinking must not fail . . . Good luck."

Thanks to Syverstad and Larsen's efforts to slow down the tapping of the heavy water cells for transport, the ferry would definitely not leave until at least the sixteenth, probably later. Now that they had some time, Haukelid decided to give Sørlie a crash course in commando training. Over two bright, clear days, he taught him how to fire a pistol and a machine gun. Sørlie practiced until it was dark, using snowmen as targets. Haukelid also showed him how to throw grenades and the basics of hand-to-hand combat. "You have to know it all," he told him whenever he wavered. "You have to be tough." He would punctuate this by throwing Sørlie into a bank of snow.

The training helped Haukelid keep his mind off his wife, Bodil, who was in Sweden. Sørlie had brought him a letter from her. Apparently she had been trying to get in touch with him for months, but no one in the Norwegian government or in the British Embassy would help her. Believing he was part of the action in Rjukan, she sent her letter via Milorg. Too much time had passed without contact, she said. The absence was too great, and their wedding had been a rushed affair in the first place. She was seeing another man in Stockholm—and she wanted a divorce.

The letter was a blow to Haukelid. He never once regretted his decision to fight against the Germans, but his soldier's life had come at a price. He had faced many dangers. He had starved, he had almost frozen to death, he had nearly shot off his own foot. Many of his friends—some of them close ones—were dead. His own mother had been brought in for questioning by the Gestapo; his father was in their hands and, as far as Haukelid knew, he was still being tortured. And now this.

When he received the letter, he asked Skinnarland to send a message to London, requesting three weeks of leave after the ferry mission so that he could travel to Stockholm and attempt a reconciliation with Bodil. Even if they agreed, he would still have to survive the operation, a long shot at best.

During the Gunnerside mission, now almost a year before, Haukelid had been on a team with nine other hardened commandos. They had trained for months and knew almost every detail of their target and its defenses. In this new action against the ferry, he was going in with the brave but inexperienced Sørlie and whatever ragtag collection of men he could assemble. The Germans were on the highest possible alert, and he would probably have to improvise a plan as events developed. There would be only a short window of time to destroy the target while limiting casualties, and if innocent people died he would carry that burden for the rest of his life — if, that is, he himself survived.

On February 13 Haukelid and Sørlie skied away from Nilsbu into a bitter southwest wind. Skinnarland waved them goodbye from the door. He wished he was going with them, but they all knew that he was more valuable as their lifeline to London during the operation — and after, if things went sour.

Olav Skindalen met them at his farm with two local men: Karl Fehn, who had helped Poulsson and his Swallow team haul batteries and equipment throughout the long winter of 1943, and Aslak Neset, an unmarried farm owner, strong as a mule. Fearing German patrols, the four waited until after midnight to cross Lake Møs and climb up into the Vidda. Their heavy load of supplies included explosives, detonators, and enough food for ten days.

Sørlie led them through the pitch black. He stared alternately at his compass and at his watch to measure where they were on their route. Four hours later, he ran into the wall of a building and knew they had arrived at their destination: Ditlev Diseth's cabin by Lake Langesjå. The door was locked, and Haukelid used a hacksaw to break in. An icy sleet was starting to fall. They woke up at noon the next day and looked out at the sun reflecting sharply off newly fallen snow.

Haukelid believed that the operation required three men: one to be lookout and two to place the timed explosive charges on the ferry. Fehn volunteered to be their third man, and Neset returned to his farm. At dusk, with Sørlie again leading the way, they moved to a hut near the top station of the Krossobanen, nestled in the narrow, heavily forested ravine. It was an ideal hideout from which to move back and forth to Rjukan.

The next day, February 15, Haukelid and Sørlie climbed down the northern wall of the Vestfjord Valley into town, where they met Diseth. The pensioner had been in the resistance since the start and had spent time in a Gestapo prison in the aftermath of the failed glider attack in November 1942. He offered to help them in any way he could.

In the early evening, Haukelid and Sørlie went to an apartment where they met Gunnar Syverstad and Kjell Nielsen. The four decided that the best day to attack the ferry would be Sunday, February 20. There was only one crossing that day, which meant that they would know exactly where the heavy water was, and when. Further, there would be fewer passengers onboard. Nielsen and Syverstad could not guarantee that they would be able to delay the transport until that date, but they would try.

They also warned Haukelid and Sørlie that a number of Gestapo agents had come to town, and a battalion of elite assault troops was expected. A host of other soldiers had just arrived in Rjukan, ostensibly for some mountain training, and two German planes would sweep the area every day in advance of the shipment.

Both Nielsen and Syverstad pushed for the operation to be called off. Norwegians were likely to die, and reprisals would be inflicted on the local population. Despite the thousands of liters of heavy water being shipped, neither believed the Germans would make any headway in purifying them without electrolysis. Haukelid listened to the men and agreed to communicate their reservations to his bosses in London. They would have the final word.

When the meeting finished, Haukelid and Sørlie sent a coded letter to Skinnarland via a messenger. It contained a brief account of the meeting and once again asked Tronstad to confirm that the "effect of the operation" was worth the many risks. They returned to their hideout in the ravine only to find that Fehn was missing. Sørlie set out to find him but soon concluded that their third man had abandoned them, whether out of fear or an unwillingness to take part in the mission. He and Haukelid both knew they would have a hard time finding someone else suited to the task.

After dinner that night, Sørlie produced a bottle of aquavit. Diseth had given it to him earlier, and he knew that Haukelid could surely use a drink. They sat by the fire and drank glass after glass. Haukelid told Sørlie the story of his love affair with Bodil, the long absences, and his belief that if only he could see her, he would be able to save their marriage. By the end of the night the bottle was empty. Haukelid's torment was somewhat

eased; and he and Sørlie had grown to understand each other and the determination they both had to see this mission through.

A heavy snow was falling outside Nilsbu on February 16 when Skinnarland made contact with Home Station and sent Haukelid's message. At the
next scheduled hookup, he got the answer: "The matter has been considered and it is decided that it is very important to destroy the juice. Hope
it can be done without too great misfortune. We send our best wishes for
success in the work." Skinnarland did not like the order: it meant Norwegians would die. With a troubled conscience, he passed a coded message
to Hamaren, who brought it to Jon Hovden, who conveyed it to Haukelid's
hideout above Rjukan. The phrase "Ten kilos of fish" meant there would
be no operation. "Five kilos" gave the green light.

Having no other choice, Skinnarland sent five kilos of fish.

27

The Man with the Violin

THE FOLLOWING DAY after dusk, Haukelid and Sørlie came down from their mountain hideout. They kept their eyes peeled for the German soldiers who were practicing their winter maneuvers on the plateau less than a quarter mile away from their cabin. Both of them knew the steep woods inside out, so many times had they taken the journey into Rjukan.

They went to a stately guesthouse owned by Norsk Hydro. Sørlie rang the bell, and a maid answered. She brought them up to Alf Larsen's room — he had been living there ever since being bombed out of his own home beside Vemork. Larsen, who was suffering from the flu, stayed in bed, but reassured the two men that he was committed to assisting the sabotage, namely by making sure that the Germans didn't ship anything until Sunday.

Soon after, Syverstad and Nielsen joined them in the room, and Haukelid informed the assembled group that London had sent the order to proceed with the attack on the ferry. "I know it's tough," he said. "Sure it's tough — but London says there's no other way."

It was one thing to send a platoon of soldiers toward an enemy position, knowing that some of them would die. But this mission, no matter how well they executed it, put their own compatriots in jeopardy. They must accept that risk against the hope that their actions would save many more lives. They all accepted the responsibility grimly.

Haukelid continued. They had three options: One was to persuade the engine-room staff to open the seacocks and shut off the engine. With the ferry disabled and unable to reach the shore, a small explosion would cause it to sink slowly in the water, allowing time for more passengers to get away safely. A second option was for him to get onboard himself, to disable the engine when they were in the middle of the lake and then use a limited charge to sink the ship. Their third option was to "place a time bomb onboard and sink the vessel quickly before it could be run ashore."

The first option was not possible. The others, all Rjukan locals, admitted that they did not know any of the ferry staff well enough to trust them. The second option was not practicable. Given that the ship would be so closely guarded, it was unlikely that they would escape the notice of the German sentries when they stopped the engine and placed an explosive charge. The third option was by far the best, so long as Haukelid could sneak onboard to set the timed explosives, and so long as the ship stayed on schedule.

The operation still needed another man. Besides Sørlie, whose training was limited, to say the least, none of the others had any experience with commando work. Things could go wrong. They might have to fight their way out. Furthermore, it would raise too many questions if Larsen, Syverstad, or Nielsen were seen lurking about the ferry on the night before its departure. Sørlie promised he would recruit someone they could depend on soon.

Next they needed to plan for what happened after the operation. Haukelid urged Larsen to escape with him to Sweden. Otherwise the Germans would continue to avail themselves of his heavy water expertise. Larsen agreed. Sørlie would move out to Nilsbu to live with Skinnarland and continue the resistance work. Syverstad wanted to remain in Rjukan. He had a wife and children he did not want to leave, and he thought it unlikely that he would come under suspicion. Nielsen was keen to stay as well, but because he was the transport manager, it was clear to all that he needed to arrange a foolproof alibi for the day of the action. He promised that he was working on one.

After the meeting, Haukelid and Sørlie went to see Diseth. The pensioner had a small workshop crammed with tools, disassembled radios, and boxes of wires, hinges, springs, and screws. Haukelid needed some kind of accurately timed detonator, as he had no easy way to set an explosion on a long delay. Diseth proposed an alarm clock. Instead of hitting the bell on top of the clock, the strike hammer would close an electrical circuit attached to a detonator and set off the explosive. Diseth offered to have the device ready within twenty-four hours.

The next day saw Haukelid strolling through Rjukan, wearing a borrowed blue suit and dress shoes and carrying a violin case. He looked like any member of the orchestra that was visiting the town, scheduled to play that night. Their conductor was the well-known composer and violinist Arvid Fladmoe. It seemed like there were Germans everywhere: standing on street corners, sitting in restaurants, driving past in cars. Time and again, Haukelid witnessed residents called aside to show their IDs. Neither the Sten gun in his violin case nor the hand grenades and pistol in his rucksack would be of much help to him if he were stopped. At the railway station, he bought a ticket for Mæl and waited for the train.

Now that the plan was in place, Haukelid wanted to check out everything for himself. He knew the schedule for the Sunday ferry, and Nielsen had provided him with a diagram of the vessel, but he wanted to determine exactly when the ferry would reach the deepest point of the lake and the best place to lay the explosive charge. If anybody asked about his presence on the return ferry, he would simply say that he was out for some sightseeing before the evening's concert.

The train arrived, and Haukelid took it the eight miles down the Vestfjord Valley to the ferry terminal at Mæl. With only a surrounding fence and a single attendant at the ticket booth, the terminal was far from being a fortress on a par with Vemork. In two days, however, when the shipment came down, Haukelid knew a heavy guard would accompany it.

The ferry, the D/F *Hydro,* was in port, readying to depart. One of three that worked the crossing, it was also the vessel scheduled for the Sunday run. Once, years before the German invasion, Haukelid had taken the ferry to Rjukan to buy fingerling trout before heading up to his family's mountain farm. He remembered now that there was nothing beautiful about the *Hydro.* Launched in 1914, the 174-foot flat-bottomed ferry had a broad angled bow that could break ice. A set of tracks ran along either side of the main deck. Together they would fit a dozen railcars of whatever was to be transported across the lake — usually fertilizer and potassium nitrate. Under this deck there was room for 120 passengers. The captain's bridge, flanked by two tall black funnels, stood over the deck. When Haukelid boarded, these funnels were belching steam into the dreary, overcast sky. He watched some railcars being shunted down a ramp from the quay onto the deck. Shackles secured them in position. Checking his watch and taking notes in his head, Haukelid timed everything, from the boarding of passengers, to the ship's actual moment of departure, to when they cleared the terminal. From his study of maps of Lake Tinnsjø, and using the landmarks on both shores, he knew that they reached the deepest spot in the lake thirty minutes into the two-hour crossing. The lake ran thirteen hundred feet deep at that point. Recovering anything sunk at that depth would be all but impossible.

For the rest of the four-hour round trip, Haukelid kept busy. He headed up to the bridge and chatted with the helmsman about navigating the lake. He walked the length and breadth of the ferry, searching through any compartments to which he could gain access. He managed to drop his pipe through a metal grate above the engine room, giving him an excuse to go down and get a look at the two 250-horsepower engines driving the ferry. After retrieving his pipe, he offered the chief engineer a pinch of tobacco. They discussed the building of the *Hydro,* and the engineer even gave him a little tour. All the while, Haukelid was looking for the best place to put his explosives. He tried to force from his mind the thought that this same engineer might die in the sinking.

By the time they neared Mæl again, he was fairly sure of his plan. He could blow a hole in the ferry with a couple of charges attached to its bow. Water would pour through the front holds, and its weight would pitch the ferry's front end down into the lake. This might cause the railcars to roll forward, speeding up the sinking of the ship. Even if the cars remained in place, the rudders and screws in the stern would rise out of the water, immobilizing the vessel.

The key question was how big a hole he should make. Lake Tinnsjø was narrow enough that it would take only five minutes to steam from its center to either of its banks. Haukelid needed to balance the risk of the captain bringing the crippled ferry to shore against allowing enough time for as many passengers as possible to escape.

Thinking on it still, Haukelid returned by train to Rjukan. By chance, he passed Nielsen on the street. Neither gave any sign that they knew each other. Anyway, Haukelid was in a hurry to meet Sørlie's new recruit.

It turned out that Knut Lier-Hansen, the twenty-seven-year-old former Norwegian Army sergeant with a boulder of a chin and a level gaze, was as much of a maverick as Haukelid himself. A local Rjukan boy, his father worked as a Norsk Hydro electrician. Lier-Hansen had graduated from infantry school and was studying at a technical school in Oslo when the Germans invaded. After fighting in a few skirmishes, he was captured, but when the truck bringing him to a prisoner camp stalled, he leaped out of the back and escaped into the woods. From there, he shed his uniform and joined in the struggle to resist the German advance.

When Norway surrendered, Lier-Hansen hitched a ride on a milk truck toward the Swedish border. Since then, he had been back and forth between Stockholm, Oslo, and Rjukan, working for Milorg as a weapons instructor, radioman, and spy. In the past few weeks he returned to his hometown to work in maintenance for Norsk Hydro. Most people in Rjukan knew he was in the resistance, which made him something of a target for the Gestapo, yet he continued to evade arrest.

Haukelid liked Lier-Hansen straightaway. He was eager for the job. He knew how to handle a gun. He had seen action. And he could tap connections of his own if they were needed. Less than forty-eight hours before the ferry was due to be sunk, Haukelid had his team in place.

Every few weeks, Fehlis and his staff sent a "Mood Report from Norway" to the military and security leaders in Berlin, chronicling their success in "smashing" and "rolling up" one underground Milorg or communist cell after another. From late 1943 into February 1944, his men broke up a huge organization led by Bergen firefighters. In Trondheim, they uncovered a large cache of weapons and explosives. In Oslo, bands of "radical socialists," mostly students, were rounded up. In Kongsberg and Notodden, they arrested many of the top leaders of the underground, including the brother of Knut Haugland. Overall, arrests were up. Executions were, too.

In these reports, Fehlis also bluntly reported the many continued acts of sabotage and resistance, both active and passive. The Norwegians, who Fehlis and his superiors had hoped would come to appreciate their place in the Nazi Reich, were more recalcitrant than ever. Fehlis reported a quote from an illegal newspaper that encapsulated the attitude of many Norwegians: "Anything that can be aligned with resistance against the Germans must necessarily be done. No one may voluntarily provide his labor, his expertise, or his business to the Germans. Take the Norwegian farmer: each liter of milk, each piece of butter or bacon that fails to come into German hands is a loss for them and a win for us . . . Not the slightest attempt of Nazification, not a single thought or sound, must penetrate into the soul of our people. We are fighting not only for the destruction of Nazism but for the reconstruction of our democratic country. This idea is to serve all of our actions."

The efforts to destroy Vemork were a further example of Norwegian intransigence. The American bombing raid had pushed Berlin to dismantle the plant and ship its stores to Germany. However, in the past few weeks, wireless transmissions hinting at sabotage had been intercepted, and Muggenthaler, Fehlis's man in Rjukan, was making last-minute changes to the transport plan. Fehlis had sent his elite Seventh SS Police Regiment to Rjukan to protect the heavy water on its route to Menstad, and Himmler himself sent a pair of Fieseler Storch airplanes to guard the train and identify any move to attack it.

On February 18, Tronstad was hiking through the Scottish Highlands outside the village of Spean Bridge. He gazed out at Ben Nevis and Aonach Mòr rising in the distance, heavy water and the impending attack on the *Hydro* never far from his mind. According to a recent Swallow report, the Germans were tapping almost 15,000 kilograms of heavy water of various concentrations from Vemork, and 100 kilograms of that was 97 to 99.5 percent pure. He calculated that if the German scientists managed to concentrate this whole supply, they would have 633 kilograms of pure moderator. Added to what they had already received from the plant, and any production they had managed on their own, they might have enough to start a working reactor.

When Einar Skinnarland had communicated that the plan was to blow up the ferry, Tronstad recognized that it was a drastic but unavoidable solution to their problem. He consoled himself with the thought that there was no way to destroy the supply of heavy water without loss of life. Brun pleaded with him to stop it, but Tronstad knew that if he refused to dispatch Haukelid to take care of the ferry, the Allies would bomb Vemork again before the shipment left the plant or while the train or ferry were in motion. Many more innocent civilians would die in these scenarios.

Nonetheless, Tronstad and the British arranged backup plans in case the shipment did make it off the ferry. Wilson sent instructions to a two-man Kompani Linge team working in the Oslo area. They would prepare a limpet-mine attack on the cargo ship in Menstad. They would sink the ship before the heavy water left the harbor. If that failed, the RAF Bomber

Command would launch an aerial attack on the ship while it was at sea. A lot could go wrong with these operations. Attacking the *Hydro* was still the best option.

After his hike, Tronstad returned into Spean Bridge to oversee a training exercise for a new crop of Linge commandos. They were in fine form as they attacked a building kitted out as a Gestapo headquarters riddled with traps. Tronstad hoped that Haukelid and his hastily assembled team would be equally ready for surprises.

Two pops ripped Haukelid from his dreams. He was sure they were rifle shots. On his feet in an instant, gun in hand, he moved toward the door of the cabin high above Rjukan. Sørlie was already at the window, his Sten gun pointing out toward any threat that might be approaching. It took a moment for the fog of sleep to lift before they both realized the source of the sounds. "At least they work," Sørlie said, referring to the two alarm clocks on the floor.

The explosive time delays Diseth had made were ingenious. He had removed the alarm bells from both clocks, then screwed strips of Bakelite in their place, taken from a broken telephone he had in his shop. On top of these strips, he fixed copper plates, which were then connected by wires to detonator caps pilfered from Norsk Hydro. The caps would go off when triggered by an electrical current, supplied by four nine-volt flashlight batteries. Diseth welded their terminals together to keep everything secure.

Haukelid and Sørlie had set the clocks to "ring" at noon to wake them up after their early Saturday morning visit to Diseth's workshop. At the right minute, the strike hammers on the alarm clocks closed the electric circuit and off went the caps. These same type of caps would trigger the detonator fuses connected to the long, rust-colored sausage of Nobel 808 explosive they had rolled — almost nineteen pounds of it. Once the explosive was set in the forepeak of the *Hydro,* Haukelid figured it would blow a hole eleven feet square that should sink the ferry within four minutes. He based his calculations on his own engineering experience and on what he had learned at STS 17. Four minutes would give the passengers enough time to escape, either into lifeboats or into the water with life belts, but would not allow the ferry to reach shore.

Settling back down into their beds, Haukelid and Sørlie went over again what was ahead. In a few short hours, when darkness fell, they would head down to Rjukan to meet the others. Then, after midnight, they would travel to Mæl to plant the explosives.

Over in Vemork, Larsen and Syverstad were overseeing the hitching of a locomotive to two railcars loaded with forty-three four-hundred-liter drums and five fifty-liter flasks of "potash lye" (roughly four thousand gallons total). Dozens of soldiers with machine guns encircled the train, scanning the surrounding hillsides and valley for any threat. They had guarded the cars overnight under the glare of spotlights. In a couple of hours, the train would move down to Rjukan, where it would remain overnight at the station under similar guard until Sunday's journey to Mæl.

Before the train left Vemork, the local Norsk Hydro director, Bjarne Nilssen, came up from Rjukan to ensure that everything was in order. After he had inspected the shipment, Larsen asked him back to his office to fill out the necessary paperwork for the consignment. It had to be done today, Larsen explained, because he needed to take the day off tomorrow.

While in Larsen's office, Nilssen made a phone call to Norsk Hydro's Notodden plant, requesting that a truck be waiting at Notodden railway station when the train arrived the next day. The man on the other end of the line asked where the truck was headed, and Nilssen replied that it was an "unnamed destination." When he hung up, Larsen asked his boss why he had ordered the truck. Nilssen told him that the Germans wanted to send the highest concentrations of heavy water, which were in the five fifty-liter flasks, on a separate transport.

The Nazis were clearly expecting an attack. If the ferry reached its destination at the southern end of Lake Tinnsjø and if the railcars made it the twenty miles to Notodden railway station, then the most valuable part of its shipment would disappear out of reach.

28

A 10:45 Alarm

SHORTLY BEFORE MIDNIGHT, Haukelid and Sørlie were in Rjukan, looking down from a bridge over the Måna River toward the town's railway station. They could see two flat railcars laden with drums and flasks and drenched in light. Soldiers with machine guns were perched on top of the cargo, their breath steaming in the frigid cold, clearly following, to the letter, Falkenhorst's advice after the first Vemork sabotage: "When you have a chest of jewels, you don't walk around it. You plant yourself on the lid with a weapon in your hand."

The two saboteurs left their vantage spot on the bridge and threaded through the backstreets of Rjukan to a detached garage behind a house. The garage door was locked. A practiced thief by now, Haukelid jimmied it open easily. Inside they found an old car. The owner had offered the car to Lier-Hansen for the night and had been told that it would be best for him if it looked like it had been stolen.

Lier-Hansen and a man named Olav, whom he had recruited as their driver, met them at the garage soon after. Then Larsen showed up, carrying a rucksack and skis for the journey to Sweden and looking altogether nervous. Syverstad was at Vemork, and Nielsen in an Oslo hospital, awaiting an appendectomy that his sister, a nurse there, had arranged for him to have on Sunday—the perfect alibi. Olav opened the choke of the car and then rotated the crank to turn over the engine. It sputtered and died. He tried again. This time, the engine did not even feign life. "You dumb brute!" Haukelid kicked the side of the car.

It would be a tough eight-mile slog to Mæl in the cold and dark if they were not able to get the car started. And they could not be late: the closer it was to the train's arrival, the more security they expected at Mæl station. They opened the hood and inspected the engine. The battery was fine. They had enough gas. The fuel pumps were clear of ice. Everything else looked in order. They tried a third time. Again, nothing. "I'm sorry," Olav said, "but it won't go."

They had already attempted to obtain cars from the two doctors in the town, who were the only ones allowed to operate vehicles during wartime. Neither car was in working order. The Germans had taken most of the rest.

Once again, Haukelid and Olav combed the engine. This time they loosened the carburetor and found it clogged with soot. After a thorough cleaning, their hands and faces black, they tried again. Olav yanked on the choke, spun the handle, and at last the engine started. It was almost 1:00 a.m., an hour later than their planned departure time.

The lone car moved east from Rjukan along the snow-covered road, its tire chains churning up a trail of murky white behind them. If they ran into any checkpoints they would have a tricky time talking their way free. Five men in a stolen car without travel permits in the middle of the night was bad enough; their weapons and bundle of explosives would see them shot, but only after they were tortured. The short ride to Mæl was made long by the dread they all felt. Haukelid knew that the responsibility of seeing that the charges were set rested with him, as the operation's leader. From what Larsen had told him, they would not have another chance to hit the whole shipment once it was on the other side of Lake Tinnsjø.

Roughly a mile from the ferry terminal, Haukelid told their driver to pull over on the side of the road and shut off the headlights. When the car stopped by a bank of trees, Haukelid handed Larsen a pistol. It did not settle easily in the engineer's hands. "Wait for us," Haukelid told him. If they were not back in two hours, he and Olav were to drive off, and he would have to get himself to Sweden. Then Haukelid turned to the driver. If shots were heard, he was to "get the hell out of there as quickly as possible." All agreed. Haukelid, Sørlie, and Lier-Hansen got out of the car.

"Good luck to you," Larsen said before the doors closed. They would need luck, and plenty of it. Overhead, a half-moon hung in the clear sky, providing barely enough light to see their way to the station. The winter chill stung their skin, and the ice under their feet cracked and snapped with every step. If there was a keen-eared guard ahead, he would hear them coming long before they saw him. Indeed, for all the popping of ice, he might think an army was on the way.

Moving slightly ahead of the others, Haukelid kept his eyes trained on the station, watching out for any movement. A single hanging lamp lit the gangway leading to the shadowed hulk of the *Hydro*. Otherwise there were no other lights. From what he could see, no soldiers patrolled the ground around the ferry. A few nights before, on reconnaissance, he had seen between fifteen and twenty men in the station house. They had rarely ventured out in the cold, but he figured they would be significantly more careful tonight, the night before the shipment arrived. If they were indeed keeping a closer watch, they were doing it discreetly.

One hundred yards from the station, Haukelid waved for the others to stop and provide him with cover as he ventured closer. His pockets stuffed with grenades and alarm clocks, a Sten gun hidden under his parka, and a nineteen-pound pack of explosives wrapped around his neck and waist, he felt like some kind of lumbering giant. He spotted a handful of soldiers in the station house, but again, none outside. For all their defensive measures, the Germans had apparently not thought to protect the ferry the night before the transport.

Gesturing for Sørlie and Lier-Hansen to follow, Haukelid crept along the dock toward the ferry. Again, no guards or lookouts. Fear snaked up his spine. He did not trust that the Germans would be so foolish.

Unchallenged, the three moved onto the ferry at last. Haukelid heard some faint voices below the main deck. Slowly, he continued down into the companionway. Near the door leading to the crew's quarters, he listened closely. It sounded like the crew was in the middle of a heated game of poker. He stole forward, briefly catching sight of the Norwegian crew sitting around a long table, playing cards.

He then came to the third-class passenger compartment. Now he needed to find a hatch below decks where they would place the explosives. While he and Sørlie searched for the hatch, Lier-Hansen provided cover. Just then, they heard the shuffle of feet down the passageway. Before they could hide, the watchman appeared in the compartment. If he called the alarm, they were lost. "Is that you, Knut?" the watchman asked.

"Yes," Lier-Hansen said coolly, recognizing a man, John Berg, whom he knew from the Mæl Athletic Club. "With some friends." Haukelid and Sørlie stepped forward. There was another tense moment as the watchman tried to reconcile why they were onboard in the middle of the night.

"Hell, John," Lier-Hansen said. Berg knew he was in the resistance. "We're expecting a raid, and we have something to hide. Something illegal. It's as simple as that."

"Why didn't you say so?" Berg said. "No problem." He pointed out a hatch in the floor of the passageway that led below deck. "This won't be the first time something's been hidden below."

Lier-Hansen kept up the conversation with the watchman as Haukelid and Sørlie climbed down into the bilge. On their hands and knees, a flashlight to guide them, they sloshed through the ice-cold water until they reached the bow of the ferry. The low, dark space resembled a tomb. If the Germans came onboard and found them there, it might well be theirs. Sørlie held the light as Haukelid got to work.

The minutes passed quickly as he delicately prepared the charges. He secured the explosives under the water on the corrugated-iron floor. The sausage of Nobel 808 was curled into an almost complete circle. Haukelid fixed two detonator fuses to both ends of the sausage. He drew the other ends out of the water and taped them together. Then he connected the electric detonator caps to these fuses and fixed them to the ribs of the ship. After connecting the two alarm clocks with their battery packets, he made sure that the wires leading from the clocks were not already electrified, then wound the alarm clocks, setting them to go off at 10:45 a.m. These were taped to the ribs of the ferry as well.

Hands numb, eyes stinging from sweat, Haukelid began the most dangerous part of setting the charges, taping the wires from the alarm clocks to the detonator caps. With only a third of an inch separating the strike hammers on each alarm clock from the plates that would complete the circuit and trigger the detonator caps, Haukelid took great care. If he jarred the alarm clocks, if his hands slipped, if he lost his balance on the slick floor, disaster would strike—at nine thousand meters per second, the burn rate of the fuses. He and Sørlie would be dead before the thought so much as flashed across their minds.

His hands remained steady, and he finished his work. Haukelid and Sørlie wriggled their way back over to the ladder and emerged through the hatch, dirty and soaked. Lier-Hansen and Berg were still talking away. Berg did not ask any questions about what had taken them so long—and Haukelid was not about to answer any. He simply shook Berg's hand and thanked him for being a good Norwegian. Then the three men sneaked off the ferry and away into the night.

They reached the car just a few minutes before their two-hour deadline and drove away a short distance before Olav stopped on the roadside. Sørlie got out. He was heading back into the mountains to reconnect with Skinnarland. After putting on his skis, he said his goodbyes to the others. "I'll be back before long," Haukelid promised. Then Sørlie disappeared into the woods. Olav drove the rest of them south toward Kongsberg, where they planned to board a train to the capital. Back on the ferry, the clocks ticked.

On Sunday morning, February 20, at 8:00 a.m. sharp, whistles blew across the train station at Rjukan. The soldiers guarding the containers of heavy water settled into position on the two flatbed cars. The train was also hauling seven tanks of ammonia and two wagons of luggage, mostly Wehrmacht supplies, to the ferry.

Never had a shipment from the town been so well protected. Soldiers were stationed all along the railway line as well to make sure there was no attack on the route to the docks.

In his childhood home in Rjukan, Gunnar Syverstad was tending to his sick mother. She was supposed to be taking the 9:00 a.m. passenger train to Mæl, then the ferry to Notodden, to attend a hospital appointment. But the night before she had suddenly come down with terrible stomach pain and was now too weak to travel. She did not know that her son had added a generous amount of laxative to her dinner the night before.

But dozens of other Rjukan residents — men, women, and children — as well as several others, including the visiting composer Arvid Fladmoe, were not hindered in boarding the passenger train and beginning their journey to Mæl and the ferry that waited there.

At the Kongsberg station, a sixty-mile drive from Rjukan, Haukelid and Larsen bought two train tickets to Oslo. Olav had dropped them off roughly ten miles from the town, and they had skied the rest of the way through the forest. Lier-Hansen was supposed to be with them, but he decided at the last minute not to leave for Sweden. As they were waiting for the Oslo train to depart, a German troop train arrived from the east. Within minutes, the station was overflowing with soldiers.

Haukelid knew they would be fine if they stayed calm. Then Larsen seized his arm. "That's the chief of the Rjukan Gestapo," he said, watching Muggenthaler step from the train. "And I'm not supposed to leave town." The two men hurried to the lavatories, and Larsen locked himself inside a stall, where he remained until the train with Muggenthaler onboard departed and they were able to board their own train.

Up on the hillside overlooking Mæl, Lier-Hansen watched the freight train come down the line on time. Over the next hour at the dock, the train wagons and flat cars were shifted onto the *Hydro* and locked into place. Then the passengers began to arrive. The clerk checked their tickets before they crossed the gangway and took their seats. Those with the cheapest seats were seated farthest from the main deck. Lier-Hansen checked his watch frequently. A few minutes after ten o'clock, the scheduled departure time, the dockworkers cast free the mooring lines, and the *Hydro* moved away from the pier, its propellers stirring up the placid waters of Lake Tinnsjø behind it.

It was a cold but clear morning, and the sun shone brightly overhead. Captain Erling Sørensen had made this crossing hundreds of times. He came from a family of ship captains. His brother had been torpedoed — twice — while plying the North Atlantic. On Lake Tinnsjø, one did not worry about such things. Up on the bridge, Sørensen directed the *Hydro* toward the center of the lake.

Just before 10:45 a.m., Sørensen stepped out of the wheelhouse to make a note in the logbook. Below deck, the thirty-eight passengers were passing their time in the saloon and other compartments. Some played cards, others were engaged in conversation or were quietly reading their books. An elderly woman thumbed through a photo album. In the engine room, three crew members were eating a late breakfast, taking a break from the cold. Except for the presence of eight German soldiers keeping careful watch over two of the railcars, all was normal.

As Sørensen descended from the bridge, a sharp crack sounded below deck, and the ferry shuddered. If they had not been in the middle of the deep lake, he might have figured they had run aground. But no, this was something different altogether. Running back up the steps, he saw smoke blanketing the deck. "Steer toward land!" he shouted to the helmsman. Before the ferry could change course, it began to keel violently. In the compartments below, terror struck the passengers. The lights went out, and water poured across the floor. Hot steam hissed through cracked pipes. The third-class saloon lacked portholes and was almost pitch black. "A bomb!" one of the passengers shrieked. "We've been bombed!" Everyone struggled to find the door.

On the bridge, Sørensen knew the ferry was lost. The bow was under water, and they were nowhere near the shore. He shouted for the passengers to get into the lifeboats and, with a member of his crew, managed to release one of the boats. Some of the passengers were already hurling themselves into the water. Then he told his helmsman to abandon ship and took over the bridge. He swung the wheel to direct the *Hydro* to the right, but it continued to list to port. One crew member was almost crushed between two tilting railway cars.

Passengers fled from their compartments onto the main deck. Some managed to find life belts before jumping into the water. Others simply took off their thick coats and leaped overboard, Arvid Fladmoe among them. Those who could not swim faced a Hobson's choice between the sinking ship and the treacherously cold water. Down in third class, the passengers finally found the door and their way out but were confused in the darkness about which way to go. Water poured down the passageway in a torrent.

One young woman named Eva Gulbrandsen struck a porthole with her fists, hoping to crawl through it to the light above. The glass was too strong to break. A man ran past, and she asked him for help, but he was in a panic. "I can't help myself. I can't even swim." Screams and cries for help rang out in every direction. The floor underneath their feet steepened into a slide. At last Eva made it to the upper deck. She stripped off her wool coat and climbed over the railing. Still in her heavy boots, she sprang from the ship.

Sørensen scrambled out of the wheelhouse. The bow was now completely submerged, and at the stern, the propellers spun higher and higher out of the water. The ferry now tilted so far to one side that he could almost crawl across the starboard side of the hull. At that moment, there was another huge bang, and he saw the eleven railcars break free and pitch into the lake. He had only seconds now before the whole ferry sank, and he would go down with it if he did not move quickly. He jumped.

Four minutes after the explosion, the *Hydro* was lost to the lake.

The one lifeboat quickly became inundated with passengers. Fladmoe was drawn onboard, his soaked violin case in hand. So too was Eva Gulbrandsen. Others held on to the spread of debris left in the wake of the sinking ship, including suitcases and four half-empty drums of "potash lye."

Those who had managed to free themselves from the ferry now struggled to get to shore before the ice-cold water overwhelmed them. Some local fishermen and farmers who had witnessed the disaster hurried out in rowboats, pulling so hard at their oars that their hands bled. Of the fifty-three people who boarded the ferry, twenty-seven survived, including the captain and four German soldiers. All eleven of the railcars sank to the bottom of Lake Tinnsjø, and the drums and flasks of heavy water went with them.

Rolf Sørlie spent most of Sunday bunkered down in a small hut on the Vidda. He did not dare rejoin Skinnarland at Nilsbu while it was daylight. Although exhausted from the night before, he stayed awake, fearing that German soldiers out on exercises might come across him. Ill at ease, he wondered what had become of the ferry: Had it sunk? Had there been loss of life? When the sun set, he started out. A fierce wind was blowing, and the cold seeped into his bones.

After several hours of skiing, his arms and legs burning from the strain, he reached the Hamaren farm. He had hoped to rest there, but it looked like there were visitors at the house, so he pushed on. The wind was now howling, but he could not stop. His arms were so weak, and he wasn't sure he could continue on. He was plagued by the thought, *What have I done?* running over and over in his mind, knowing that if the sabotage was successful, Norwegians had paid with their lives. Just when there was no way he could move another step, the wind suddenly died down. For the first time he no longer felt he was being beaten by the gusts, and the momentary reprieve gave him renewed vigor.

At last, he looked ahead to see Nilsbu framed in the moonlight. As he neared the cabin, the door opened, and Skinnarland came out in the snow to greet him. Sørlie felt like he had come home. Skinnarland made him coffee and set out food. While he was eating, Sørlie told him about the night before. Skinnarland promised to go down to the Hamarens' the next day to see if there was any news. Then Sørlie lay down. He was asleep in moments.

When he woke up the next morning, the cabin was empty. At the Hamarens', Skinnarland learned that the *Hydro* had sunk, with all its precious cargo, and that the first reports were that fourteen Norwegians and four Germans had died. As soon as he returned to Nilsbu, he transmitted this information to Home Station. Now he and Sørlie needed to go into hiding farther away from Lake Møs. They carried with them the burden of what they had been asked to do.

On Monday morning, soon after Gunnar Syverstad reported to work, Bjarne Nilssen asked him to come down to his offices in Rjukan. He arrived there to find soldiers and Gestapo milling about the hallways. Nilssen wanted to know where Chief Engineer Larsen had gone: He was not at his home. He had not come into the office. Syverstad pleaded ignorance. Nilssen warned him that the Gestapo would soon be giving him a thorough interrogation.

Syverstad knew immediately that he needed to flee. After leaving Nilssen's office, he ran into another engineer from Vemork who had already been questioned by an enraged Muggenthaler. Face flushed, the German had placed his gun on the table in front of the engineer and threatened him: "If you disappear, I will blow up your home with your wife in it." Syverstad returned to his house, gathered his belongings, and said goodbye to his wife and two young children. He would have to get to Sweden before Muggenthaler came for him. When the Gestapo arrived, they found him already gone.

Knut Lier-Hansen remained in the town, few any the wiser that he had played a part in the ferry's sinking. Two Gestapo officers showed up in Kjell Nielsen's hospital room in Oslo. He was still recovering from his appendectomy, and told them that he knew nothing about the sabotage. After all, he had been in the hospital since Saturday. They did not question him any further. When the Gestapo interrogated John Berg, the watchman, he admitted that he had let three men onboard the night before the ferry's departure. This was a common practice, he pleaded, since some passengers arrived early from the mountains and needed a warm place to rest. Berg said that he did not know the men, and his descriptions of them were indistinct at best.

Another hunt on the Vidda began, focusing yet again on Lake Møs and the Skinnarland home, but the search for the saboteurs ended as fruitlessly as the one almost a year before. It seemed like they were chasing shadows.

Knut Haukelid was somewhere north of Oslo, in a cabin owned by a resistance member, when he read the headlines in the Monday-evening edi-

tion of the newspaper: "Railway Ferry Hydro Sunk in the Tinnsjø." At
10:45 a.m. on Sunday morning, on the train from Kongsberg to Oslo, he
had stared at his watch, picturing in his mind's eye what was unfolding on
the lake. The explosion. The ship keeling. More Norwegian lives added to
the butcher's bill to bring an end to Vemork's heavy water. So much death
and sacrifice. So much endured. Now he held the news in his hands: The
Hydro was sunk with all its cargo. He had followed his orders, tough as
they were to bear, to the very end.

With the help of the underground escape network, he and Alf Larsen
crossed the border a few days later and made their way to the Swedish
capital. In a hotel, he bathed and changed into new clothes. After many
months in the Norwegian wilderness, Haukelid found it strange to sit in
a restaurant and eat his fill or to pass shop windows piled high with mer-
chandise. Soon after his arrival, he met with Bodil in an attempt to reconcile with her. But much — too much — separated them now, and he feared
that their marriage was yet another casualty of the German invasion. After two weeks in Stockholm, Haukelid was ready to return to his resistance work. It was the only life that made sense to him.

On February 26, 1944, Tronstad returned to London from Scotland, arriving into Euston Station. On the way to his office, he passed a large apartment building cordoned off by police. It had suffered a direct hit twelve
hours before during a renewed blitz by German bombers. Over the past
few evenings, incendiary and explosive bombs had knocked the sides off
buildings and leveled houses and shops. A school was hit in Tavistock
Crescent, and a convent destroyed in Wimbledon — the nuns had to pick
through the rubble to find their sisters. Across the city, hundreds were
dead, and many more were left without homes.

That evening, Tronstad returned to his house, and through the night he
listened to the fighters overhead and the crack of gunfire. He thought further about the sabotage of the *Hydro*, consoling himself with the thought
that the operation had at least prevented another Allied bombing run
against Vemork. The death toll from such an attack would have been far
worse than that of the ferry sabotage.

The truth was that many more were dying in London, every night. If
the Germans had managed to build an atomic bomb, they would leave the
British capital — and perhaps other cities too — a scorched ruin littered
with innumerable dead. Tronstad understood that in war, leaders had to

measure their decisions against such comparisons, whether on the field of battle or at the planning table. Still, when he read the names and ages of those who perished on the ferry, he felt terribly diminished.

Within days, Tronstad received final confirmation from Skinnarland's spies that the entire shipment of Vemork's heavy water — except for a few drums of nearly worthless concentrate — was at the bottom of Lake Tinnsjø. In his diary, he noted the closing of this "brave chapter" in the fight against the Germans. He would ask Winston Churchill to reward Haukelid and the others involved in the ferry mission. As for Vemork, when the war was over he hoped to rebuild the plant, make it better than before. Until then, he would content himself with the knowledge that his men had succeeded in destroying the Nazis' source of heavy water — and potentially stopped them from realizing a weapon unlike any known before.

29

Victory

B Y T H E E N D of March 1944, Walther Gerlach, the head of the
Uranium Club, and Kurt Diebner, his administrative head, were
under relentless assault, chiefly from the Allied air raids that were
leveling one research center after the next. Only days before the sinking
of the *Hydro*, waves of bombers had targeted the Kaiser Wilhelm Institute,
destroying many departments, save, by chance, the one devoted to physics. With no end to the raids in sight, Gerlach and Diebner began to evac-
uate their scientists and equipment to the south.

In a report sent to Göring on March 30, Gerlach detailed the state of his
program, from advances in the ultracentrifuges that separated U-235, to
their successes in uranium-machine design, to the new methods of concentrating heavy water. Because of the Allied attacks, supplies of the precious moderator were in a "precarious position," but Gerlach was hopeful
that significant investments in German plants would bring about a steady
flow in the near future. Further, he stated, the attacks on Vemork's supply
made it clear that the Allies themselves placed the "utmost importance"
on fission research as a path to obtaining new explosives. It was essential
that his teams do the same.

Diebner embarked on a new uranium-machine experiment while also
pursuing his design of shaped charges for a fusion bomb. Harteck prod-
ded IG Farben to move ahead with heavy water plants at Leuna. Heisenberg was still attempting a large new reactor, and other German scientists
in the program continued with their research as well, all the while fleeing

bombings, evading draft call-ups, and hauling their laboratories to hidden bunkers.

In mid-1944 Hitler, increasingly deluded and desperate, proclaimed Axis victory was imminent. "Very soon I shall use my triumphal weapons, and then the war will end gloriously . . . Then those gentlemen won't know what hit them. This is the weapon of the future, and with it Germany's future is likewise assured." Few believed him. Germany was under assault from land, air, and sea. Allied forces drove their way to Berlin from the west, and the Russians pushed from the east. In July, a bomb run of 567 Flying Fortresses destroyed the Leuna works and with them the possibility of a renewed heavy water supply. Other attacks halted uranium production and U-235 separation. By the end of 1944, the best Diebner — or anybody he had called to nuclear research in 1939 — could aim to build was a self-sustaining uranium machine.

The Allies knew it. In August, Colonel Boris Pash, a U.S. Army intelligence officer, found Frédéric Joliot-Curie at the Collège de France in Paris. Joliot-Curie told Pash about his dealings with a number of German physicists, including one Dr. Kurt Diebner. Joliot-Curie believed the German program was far from advanced. That November, Pash and his boss, Dr. Samuel Goudsmit, discovered a bounty of secret papers in a Strasbourg hospital commandeered by the German atomic program. While American forces battled the Wehrmacht on the outskirts of the city, Goudsmit had soldiers haul away the files, and for four freezing days and nights, he and Pash read through them by candlelight, eating little, sleeping less. By the end, there was only one conclusion to draw: "Germany had no atom bomb and was not likely to have one in any reasonable time."

But at the start of 1945, with the war's end inevitable, Gerlach and Diebner continued to hope that their work would have some impact. They crisscrossed Germany, often at risk of strafing runs by Allied planes, distributing supplies and directing experiments in a last-ditch effort to obtain at least a working uranium machine. In a rock-hewn wine cellar in Haigerloch, a hillside village in southwest Germany, Heisenberg had established a laboratory and constructed a lattice of uranium cubes submerged in heavy water, similar to the one set up by Diebner before the evacuation from Berlin. Using 1.5 tons of uranium and heavy water, the machine produced the highest level of neutron multiplication yet achieved. By Heisenberg's calculations, he was sure to have a self-sustaining reactor if he could only obtain 50 percent more uranium and heavy water. He would get neither.

On June 15, 1944, Leif Tronstad was at home, looking out at a rain-soaked Hampstead Heath and drafting a plan for what he had coined Operation Sunshine. Since the Allied thrust into France just over a week before, it had become clear that there would be no invasion to free Norway. His countrymen would have to do it themselves. Tronstad, recently promoted to major, was determined to be on the ground. There was a danger that the Germans would implement a scorched-earth policy when they withdrew their 350,000 troops, as they had done when leaving Italy.

Wilson and Brun wanted him to stay in London, but he was not to be dissuaded. He had just finished a three-week course at Rheam's STS 17 sabotage school, receiving high marks ("first class in all respects"). He would no longer send other men to fight in his stead.

Just before midnight, he heard a tremendous roar passing over his house. Moments later, an explosion. The jet-powered V-1 rocket attacks had begun. Day after day, hundreds of German missiles landed on the city. All the while, Tronstad continued with his strategizing for Operation Sunshine. By the end of July, he had his approval from General Hansteen and assembled his team. Jens-Anton Poulsson would lead one division, with his best friend Claus Helberg as radio operator. Arne Kjelstrup would be in charge of another. Einar Skinnarland, who was in Norway, would be Tronstad's radio operator. A number of other Kompani Linge members, including Gunnar Syverstad, who had come to Britain for training after the ferry sabotage, would also join him. Tronstad would be in charge of protecting "the major industrial objectives" in the area, including its power stations, which supplied almost 60 percent of southern Norway's electricity.

On August 27, he finished his farewell letter to Bassa, to be given to her in the event of his death. Although he was returning to Norway at last, she could not know that he was there. Tronstad gave Gerd Vold Hurum, his faithful secretary, the small key to his office safe. "Please take care of my diaries," he said. Overwhelmed with emotion, she accepted the key. "When the war's over," he continued, "you must go and meet my family." With that, he left Kingston House, his office of almost four years.

At long last, on October 5, Tronstad returned to Norway, dropping by parachute into the Vidda. His "long exile" was over. When the other parachutists assembled, Tronstad called for a toast, and they drank whiskey from their metal flasks. Then they pitched their tents.

Over the next five months, Tronstad recruited a small army of Milorg resistance fighters that ultimately numbered twenty-two hundred men. His headquarters were a ten-foot-wide hut buried in the deep snow near Lake Møs. As the commander of Operation Sunshine, he skied back and forth across Telemark and the neighboring regions, from Kongsberg to the east, Notodden to the south, Rjukan to the north, and Rauland to the west, coordinating with London and Milorg and making sure that the two worked seamlessly together. He met secretly with management at Norsk Hydro and other Norwegian companies to ensure they were onboard when the time came to eject the Germans.

Poulsson, Kjelstrup, and others set up separate bases of operation; Haukelid and his band of fighters joined the operation as well. They coordinated drops of arms and supplies, trained resistance cells in firearms and explosives, and performed small-scale sabotages on arms dumps. They also infiltrated power stations, dams, and industrial plants, teaching the workers how to thwart any German attempts to destroy their buildings, including how to implode the roofs so that the precious machinery inside remained operable once the debris was cleared.

Tronstad ate, slept, hunted, and skied beside his men. Most of them had thought highly of him while he was their boss in London. In the wilds of the Telemark, their loyalty and respect deepened into something greater still.

By spring 1945, the time for action looked imminent. Nazi Germany was collapsing, and the march into Berlin would soon cut off the head of the snake. Throughout Norway, the sabotage of railway transports, ports, ships, and communication lines was hobbling the Wehrmacht and obstructing the removal of its troops to reinforce their defenses inside Germany itself. On the evening of March 11, Tronstad and two of his men, Syverstad and Jon Landsverk, were interrogating Torgeir Lognvik, a Nazi-appointed Norwegian sheriff. Tronstad wanted to know about his activities and those of other Nazi sympathizers, and they wanted to prevent him from informing the Gestapo about their underground activities in the district of Rauland. They debated whether or not to kill him, and Tronstad decided that Haukelid should hold him prisoner at Bamsebu.

Skinnarland had helped to capture Lognvik and erased the tracks to their location, a two-room cabin in the countryside not far from Lake

Møs. When this was done, he left for a rendezvous at a neighboring farm. Tronstad finished his questioning of Lognvik and was getting ready to leave to join Skinnarland when suddenly the door to the hut burst open. The sheriff's brother, Johans, entered with a gun and started shooting. Syverstad was struck in the head and, falling backward, knocked Landsverk out of the line of fire. Tronstad tried to rush Johans Lognvik, and in the bitter fight that followed two more shots rang out.

Tronstad dropped to the floor, killed either by the bullets fired by Johans or by the blows of a rifle Torgeir Lognvik had grabbed in the melee. The brothers escaped, locking the door from the outside. Finally Landsverk freed himself from the cabin and rushed to find Skinnarland. When the two returned to the hut, Syverstad was near death; there was no saving him. Looking from Tronstad's disfigured face to Syverstad's, Skinnarland was deeply shaken. All that blood. Then he collected himself. The sheriff would be back, with Germans.

Skinnarland and Landsverk moved quickly. They packed up all the papers and equipment, anything that might lead to other arrests. By the time they finished, Syverstad, who like Tronstad was married and the father of two young children, was dead. They carried the bodies down to the lake on sledges, cut a hole in the ice, and sank them before the Germans could take them. Then they left to alert the others, including Haukelid, warning that any attempt to avenge the killings might bring about a war with the local German garrison, a fight that Milorg was not ready to have.

Their quick actions prevented any roundups or intelligence leaks, but it did not take long for the Germans to find the bodies. Tronstad and Syverstad were dragged from the lake and brought to a nearby village to be inspected and photographed. The Germans then doused the bodies with petrol, set them on fire, and tossed the burned remains into a river. Skinnarland sent word to London of the tragic events, and Poulsson was put in charge of Operation Sunshine. He held the post until the end of the war, an end that Leif Tronstad did not see for himself.

The order to mobilize was given on May 8, 1945, the day Churchill declared victory over Germany from a balcony overlooking Whitehall to a throng of revelers. The forces throughout Norway, including deep in the heart of Telemark, went into action. After years of fighting as an underground army, they put on uniforms and simple armbands and took back Rjukan and the surrounding towns. They occupied Vemork and other power stations in the area, seized control of communication lines and key public buildings, and took responsibility for law and order, including the arrest of Norwegian traitors and SS officers. The soldiers in the German garrisons surrendered their weapons and went where they were told. Similar scenes played out across Norway. The invaders numbered almost four hundred thousand, Milorg roughly forty thousand. There could have been an ugly fight, but there was none. At last Norway was free, and parties broke out in the streets of Oslo and throughout the country. That night, at the royal estate of Skaugum, Reichskommissar Terboven ate a sandwich and read an English detective novel. Then, resigned to follow Hitler in death, at 11:00 p.m. he entered a bunker, where he drank half a bottle of brandy and lit a five-meter fuse that led to a box of explosives. The fuse was calculated to burn for eight minutes and twenty seconds. At 11:30 sharp, the explosion sounded across the estate.

Heinrich Fehlis attempted to flee. He was arrested in Porsgrunn, a southern harbor town, wearing the uniform of a Wehrmacht lieutenant. Before he could be interrogated and his identity revealed, he took poison and shot himself in the head. Other Gestapo officials, including Siegfried Fehmer, tried to slip away. They were caught and brought to trial for their crimes. Vidkun Quisling and General Nikolaus von Falkenhorst were also arrested and tried.

A month after the German surrender, his country secure, King Haakon VII stepped onto Norwegian territory: the pier in front of Oslo's Town Hall. In spite of a steady drizzle, fifty thousand Norwegians cheered and waved flags to celebrate his return. Among those present to honor him were Colonel John Wilson and over a hundred members of Kompani Linge, most of them wearing helmets or badges that named their operations. Grouse and Gunnerside were well represented. "Many times it may have looked dark," Haakon spoke soberly to the gathering. "But I never doubted Norway would get back her rights."

On Midsummer's Eve, June 23, 1945, a smaller but no less jubilant liberation celebration was underway at the Skinnarland hotel beside Lake Møs. The front of the hotel was decorated with silk parachutes, and there was an illustrated schedule for the three-day event. Over dinners of trout and reindeer steak and bottles of champagne, aquavit, and beer, those who fought in the resistance around Rjukan spoke of past battles and of their future, too.

Among the honored guests were the Hamarens, Hovdens, and Skindalens, Poulsson, Helberg, Haukelid, Kjelstrup, Sørlie, Lillian Syverstad, Ditlev Diseth, and Kjell Nielsen. Einar Skinnarland celebrated with his brothers Torstein and Olav and reunited with Gudveig. A few seats were left empty. Olav Skogen, who had survived his imprisonment at Dachau, had not yet returned to Rjukan. Leif Tronstad and Gunnar Syverstad never would.

A week after the celebration, on June 28, 187 members of Kompani Linge, with Poulsson and Rønneberg in the lead, paraded in uniform before King Haakon. Of their select unit, fifty-one had died during the war. The king paid tribute to the men and their clandestine work. The following day, Colonel Wilson disbanded the company, asking them to serve their country in peace as they had in war.

By early August, Norway was beginning to recover after the long occupation, and Bassa Tronstad was back at her rented house outside Oslo. Her husband's remains had been recovered from the river and buried in late May in a moving ceremony at an Oslo cemetery. Now, she was trying to make sense of the loss. The circumstances of his death she knew, but she had many questions about his time in London and about what had brought him to return to Norway. It was then that Gerd Vold Hurum came to her door. After offering her condolences, Gerd presented Bassa with eight diaries. There was also a letter. The diaries would take days to read, but the farewell letter was short and direct, and told Bassa all she needed to know.

"Dearest Bassa . . . I have the honor to lead an important expedition home, which will be of great importance to Norway's future. It is in line with the course I chose on April 9, 1940, to put all my effort and abilities toward our country's welfare . . . The war is singing its last verse, and it requires every effort from all who would call themselves men. You will understand that, won't you? We have had so many magical happy years, and my highest wish is to continue that happy life together. But should the Almighty have another course for me, know that my last thought was of you . . . Time is short, but if all will not go well, don't feel sorry for me. I am completely happy and thankful for what I have had in life even though I very much would like to live to help Norway back to its feet." He wished the best for Sidsel and Leif. He looked forward to seeing them all again. The letter was signed "Your beloved."

· · ·

In Farm Hall, a quiet country house outside Cambridge, ten Uranium Club scientists were waiting for a decision to be made about their fate. They had been held there since July 3, 1945, rounded up when the Nazi regime fell, along with their papers, laboratory equipment, and supplies of uranium and heavy water. Among them were Otto Hahn, Werner Heisenberg, Walther Gerlach, Paul Harteck, and Kurt Diebner. They spent their time reading in the library, tending the rose garden, playing bridge, and wondering if they would see their families again. Unbeknownst to them, every room in the house was bugged, and their every word was recorded on shellacked metal disks to be reviewed by British intelligence.

At 6:00 p.m. on August 6, 1945, a short BBC bulletin reported that an atomic bomb had been dropped on Japan by the American B-29 bomber *Enola Gay*. Major Terence Rittner, who was in charge of security at Farm Hall, went to Hahn's room to inform him. As Rittner reported to his superiors, the man who discovered fission was "shattered" by the news and "felt personally responsible for the deaths of hundreds of thousands of people." Rittner steadied him with a glass of gin.

The news staggered the other German scientists at Farm Hall as well. Their disbelief was followed by cynicism. Surely the Allies were bluffing the Japanese into surrender. Surely the Americans and the British were not able to build an atomic bomb. Their shock and doubts were soon answered by another BBC broadcast a few hours later: "Here is the news: It's dominated by the tremendous achievement of Allied scientists — the production of an atomic bomb."

Then followed a statement from Churchill: "The greatest destructive power devised by man went into action this morning . . . The bomb, dropped today on the Japanese war base of Hiroshima, was designed for a detonation equal to twenty-thousand tons of high-explosive . . . By God's mercy British and American science outpaced all German efforts . . . The possession of these powers by the Germans at any time might have altered the result of the war . . . Every effort was made by our Intelligence Service and by the Air Force to locate in Germany anything resembling the plants which were being created in the United States. In the winter of 1942–43 most gallant attacks were made in Norway on two occasions by small parties of volunteers from the British Commandos and Norwegian forces, at very heavy loss of life, upon stores of what is called 'heavy water,' an element in one of the possible processes. The second of these two attacks was completely successful." Churchill concluded, "This revelation of the secrets of nature, long mercifully withheld from man, should arouse the most solemn reflection in the mind and conscience of every human being capable of comprehension."

The scientists launched into heated conversation. "They can only have done that if they have uranium isotope separation," Hahn said.

Harteck countered, "That's not absolutely necessary. If they let a uranium engine run, they separate [plutonium]."

"An extremely complicated business for they must have an engine which will run for a long time," Hahn said. "If the Americans have a uranium bomb then you're all second-raters."

Heisenberg was still dumbfounded by the announcement. "Did they use the word 'uranium' in connection with this atomic bomb?" His fellow scientists responded: no.

Diebner interrupted: "We always thought we'd need two years for a bomb."

Deep into the night, the ten men continued their conversation in the bugged dining hall. Some expressed horror at the use of such a bomb by the Allies. Others lamented how far behind their own program had been. They argued the science and mechanics of how exactly the bomb was produced — and the tremendous investment that must have been made by the Americans.

For a time, they played the blame game. They had had too few staff. Not enough support or supplies, chiefly not enough heavy water. There was too much infighting among their scientists, not enough cooperation. They concentrated too much on uranium machines and moderators, too little on isotope separation. The meeting with Speer in June 1942 had ended any hope of an industrial program. The "official people," Diebner said, "were only interested in immediate results." Their institutes were smashed by bombing runs. They never had the chance.

They also debated the morality of using a weapon of such devastating power and whether or not they had ever intended to produce one. Some, like Heisenberg, were already putting together the framework of their own defense, conveniently justifying the failure of their efforts as a calculated strategy to keep Hitler from obtaining the bomb.

Epilogue

ON JANUARY 3, 1946, the ten Uranium Club scientists were released, free to return to Germany and to their scientific pursuits. In the years and decades afterward, they gave many interviews, wrote their memoirs, and contributed to biographies and other books about their war work. These were added to the thousands of secret reports, letters, and papers gathered up after the collapse of the Third Reich.

Many histories have been written.

At the start of 1942, the Germans and the Allies were roughly neck and neck in terms of their atomic theory and research. Then the Americans pushed ahead with the Manhattan Project, while German Army Ordnance, and then Speer, backed away from committing to such an expansive program.

R. V. Jones, a leading British intelligence officer whose war work focused on combating German technology, wrote, "A bad experiment on one side or the other was often the cause of divergence." If the Germans had not ruled out graphite as a moderator so early, would they have been the first to realize a self-sustaining reactor? Might this have convinced officials to allocate resources to an atomic bomb instead of to the V-1 and V-2 program? Should they have invested more time and effort into U-235 isotope separation instead of a heavy water reactor to produce plutonium for a bomb?

Some historians have concluded that the campaign against Vemork's supply, from cod-liver oil contamination by Brun and others, to Operation Gunnerside, to the American bombing raid, to the sinking of the D/F *Hydro*, was all for nothing. But if the Germans *had* fashioned a self-sustaining reactor with heavy water, what then? Diebner had believed he would have enough heavy water by the end of 1943 for a reactor. He would not have stopped there. "The obliteration of deuterium production in Norway," he wrote later in his memoir, "is one of the major reasons why Germany never obtained one."

Making history was never the aim of the Norwegian saboteurs, nor of the British sappers who were sent before them. After the war, the sacrifice of the British Royal Engineers and RAF crews of the ill-fated Operation Freshman was not forgotten. Thirty-seven bodies were recovered and buried at gravesites in Norway. Bill Bray's headstone reads, "To live in the hearts of those that loved me is not to die." The four sappers killed in Stavanger, whose bodies were dumped into the sea, were honored with a memorial close to where they died. Laurence Binyon's poem "For the Fallen" was read in English and in Norwegian translation at the ceremony: "They shall grow not old, as we that are left grow old: / Age shall not weary them, nor the years condemn. / At the going down of the sun and in the morning, / We will remember them." Memorials were also established for the Norwegian men, women, and children who perished in the American bombing raid at Vemork and in the sinking of the *Hydro*.

As time went on, those who had participated in the heavy water sabotages were pinned with medals from many grateful nations: Norway, Great Britain, Denmark, France, and the United States of America. But whenever they were asked about their most significant, defining action during the war, most of them mentioned operations other than Vemork.

For Joachim Rønneberg and Birger Strømsheim, they were most proud of Fieldfare, the operation they launched in March 1944 to prepare for the destruction of German supply lines in the Romsdal Valley, culminating in the blowing up of a key railway bridge. Knut Haugland thought his work setting up wireless radio links with London was his greatest achievement. Jens-Anton Poulsson considered his actions in Operation Sunshine the most important of his war against the Germans. It could be argued that Leif Tronstad would similarly have downplayed Vemork, had he lived, particularly since his diaries mention "the juice" only a couple of dozen times, among the many other operations he planned that fill their pages. Despite their feeling that their other operations during the German occupation deserved as much, if not more, attention, it was the commandos' actions against Vemork that attracted the most accolades. They were Norwegian heroes, international heroes.

Beyond the medals and memorials, the war marked the men of Grouse and Gunnerside in other ways, sometimes dark ones. They had seen friends die. Some had become killers. All of them had lived under the constant threat of discovery and death. At times in the years after the peace, they woke up in the middle of the night, imagining the enemy at the door, reaching for guns that were not there. Einar Skinnarland's children knew well not to approach their father suddenly. Some instincts never fade. A few of them turned to alcohol to dull memories they never asked for. Many simply sought solace in the place where they had once struggled to survive. The "smallness of being a human being in nature" settled Rønneberg. "You could sit down on a stone and let your thoughts fly away." Knut Haugland spent 101 days in 1947 as the radio operator on the *Kon-Tiki*, a simple raft that crossed the Pacific Ocean with only a six-man crew. Beyond offering great adventure, the journey exorcised his own demons. What nature and time could not mend, their friendship supported them through. Until the end of their lives, Kompani Linge members gathered often to share experiences that few others could understand.

Knut Haukelid dedicated his war memoir to his father, who had died in 1944 as a result of the deprivations he endured as a prisoner at Møllergata, and then at Grini. The dedication read, "He died without knowing why . . ." Although he was arrested for having illegal radio equipment in his warehouse, the truth was that Bjørgulf was targeted because of his son's actions, all of which Knut had had to keep hidden from him.

Skinnarland did not fight for accolades, nor for the pile of medals he received, which he kept in a drawer of junk in his basement. The deaths of Tronstad and Syverstad weighed heavily on him. Only in his last years did Skinnarland revisit his war diary and the long string of telegrams he had sent from the Vidda, and only then did he allow himself some pride over what he had endured for the sake of his country — and the world. He shared this with his family, among them his daughter, Kirvil. He and Haukelid, who ultimately reconciled with his wife, Bodil, both kept the promise they had made each other in the summer of 1943 by naming their daughters such.

Finally, Rønneberg, the leader of Gunnerside and the last surviving saboteur, who was ninety-six years old in 2016, often spoke eloquently about why he braved the North Sea to be trained in Britain and why he then returned, twice, by parachute, to Norway. "You have to fight for your freedom," he said. "And for peace. You have to fight for it every day, to keep it. It's like a glass boat; it's easy to break. It's easy to lose."

Acknowledgments

Now and again readers ask how long it takes me to research and write a book. I ballpark it at three years, depending on the availability of sources and narrative complexity. That amount of time sounds right to some folks, quick to others. The truth is, however, each of my books would have taken many more years if not for the community of people who contribute to the finished work. And the quality would have been far, far diminished without them.

First, I would like to applaud my research assistants who spearheaded my efforts in Norway (Windy Kester and Arne Holsen) and Germany (Almut Schoenfeld). They were tireless in helping me comb through archives and in tracking down individuals to interview.

Before I began my research, all but one of the saboteurs (Joachim Rønneberg) had passed away. I had the benefit of their recollections in numerous interviews, memoirs, and diaries, but even this bounty did not provide the kind of rich portrait I hoped to paint of each of them. Fortunately, their families offered to speak with me, sharing family lore as well as a number of documents that have never before been seen. A big thanks to the Haukelids (Kirvil, Bjørgulf, and Knut), Skinnarlands (Marielle, Kirvil, Ron, Inger-Berit Bakke), Hauglands (Trond, Torfinn, Torill), Poulssons (Unni, Mia), Tronstads (Leif Jr., Sidsel), and Finn Sørlie. In particular, I'd like to send my appreciation to Leif Tronstad Jr., who shared many, many hours of his time and family papers with me, as well as reading the final draft of the book. Also to Marielle Skinnarland for her generous attention to my exhaustive list of questions. And to the Hauglands, who put me up in their cabin outside Vemork and took me on a cross-country ski tour of the area (and even offered some of their father's gear for me to use). I also appreciate the insight of former Kompani Linge member Ragnar Ulstein and a handful of Norwegians who know this story well and provided much guidance to me, including Bjørn Iversen, Svein Vetle Trae, Berit Nøkleby, Asgeir Ueland, and Tor Nicolaysen. In particular, Svein Vetle offered me a wonderful weekend at his cabin, not to mention the maps for this book.

A work of history is often only as good as the primary sources available to the author. In this case, they were plentiful and rich. One of these was the Norges Hjemmefrontmuseum (NHM). It is a gem of an archive, and its staff are first-rate experts on the Norwegian homefront resistance. My appreciation to my guides there: Frode Faerøy, Ivar Kraglund, Benjamin Geissert, and Arnfinn Moland (who provided me with an unpublished interview with Gunnerside leader Rønneberg that ran over a hundred pages of pure gold). Rjukan's Norsk Industriarbeidermuseum (VM) also held great treasure, and I would have been lost without its director, Kjetil Djuve, and Ingelinn Kårvand. Also a big thanks to the staffs at the National Archives (UK) who fielded innumerable requests on my behalf, as well as those at the Imperial War Museum, Niels Bohr Library, and Rensselaer Institute, among others. Without my translators, Carl Stoll and Mark McNaught, much of this material would have been indecipherable.

Once the research is done — and the first draft complete, another community comes to my side. A shoutout to my early readers, Carl Bartoli, Henry Bartoli, John Tuohy, and Mike Faley, who clarified atomic physics as well as the intricacies of B-17 bombing runs. As always, my first-line editor, Liz O'Donnell of the Little Red Pen, shaped and shifted and refined almost every paragraph in the book. I can never sing her praises enough. Thanks to my literary agent — and consigliere in all things publishing — Eric Lupfer at William Morris Endeavor, and his colleagues Ashley Fox, Simon Trewin, and Raffaella De Angelis. Eric brought me home to my longtime publisher Houghton Mifflin Harcourt and my superb editor there, Susan Canavan. She's a great champion for my work — and a friend. Kudos also to the always wonderful Megan Wilson, the marketing wizard Carla Gray, Jenny Xu, and Melissa Dobson.

And finally, thanks to Diane and our precious girls Charlotte and Julia, who live through the ups, downs, and in-betweens of the life of an author. You make it all worth it.

Notes

ABBREVIATIONS

National Archives, Kew, UK (TNA)

National Archives, College Park, MD (NA)

Niels Bohr Library and Archives, College Park, MD (NB)

Norges Hjemmefrontmuseum, Oslo (NHM)

Norsk Industriarbeidermuseum, Rjukan (VM)

Leif Tronstad Archive, NTNU, Dorabiblioteket, Trondheim (DORA)

Bundesarchiv-Militärarchiv, Freiburg (Barch-MA)

Papers of Dan Kurzman, Howard Gotlieb Archival Research Center, Boston University (KA)

Imperial War Museum (IWM)

Papers of David Irving, German Atomic Bomb, British Online Archives (DIA)

Papers of Leif Tronstad, courtesy of Leif Tronstad Jr. (LTP)

Diary of Leif Tronstad, Papers of Leif Tronstad, courtesy of Leif Tronstad Jr. (LTD)

Papers of Einar Skinnarland, courtesy of Skinnarland Family (ESP)

Diary of Einar Skinnarland, courtesy of Skinnarland Family (ESD)

Interview with Joachim Rønneberg by Arnfinn Moland, NHM (Rønneberg Interview, Moland)

Poulsson, Jens, "General Report on Work of Advance Party by Swallow," NHM: Box 25 (Poulsson Report)

Haugland, Knut, "Wireless Service in the Grouse Group," NHM: SOE, Box 23 (Haugland Report)

Rønneberg, Joachim, "Operation Gunnerside Report," NHM: FOIV, Box D17 (Rønneberg Report)

Sørlie, Rolf, Unpublished Memoir, courtesy of Sørlie Family (Sørlie Memoir)

Brun, Jomar, Some Impressions from my work with Z, November 30, 1942, TNA: HS 8/955/ DISR (Brun Report)

PROLOGUE

page

xviii   *In a staggered line:* Draft BBC Talk by Lieutenant Rønneberg, TNA: HS 7/181;

Haukelid, 105–8; Rostøl and Amdal, 86; Rønneberg Report; Lunde, 99–101; Gallagher, 96–97.

xviii *Despite the distance:* Report: Vemork Power Station and Electrolysis Plant, October 30, 1942, TNA: DEFE 2/219; Adamson and Klem, 138; Draft BBC Talk by Lieutenant Rønneberg, TNA: HS 7/181.

Standing at the edge: Interview with Haukelid, DIA: DJ 31; Berg, 128; Interview with Poulsson, IWM: Oral History 27189.

"to blow up a good": Interview with Poulsson, IWM: 26625; Myklebust, 108.

xix *"went west":* Ragnar Ulstein, Author Interview.

Back in England: Speech at the Imperial War Museum, 1978, DORA: L-0001.

1. THE WATER

3 *On February 14, 1940:* Biographical Note, Papiers de Jacques Allier. Archives Nationales, Paris; Goldsmith, 84–88.

4 *Startled by the:* Letter from Rjukan, Vedr. Tungt Vann, January 11, 1940, VM: Box 4F-D17-98; Norsk Hydro Heavy Water Discussion with Bjarne Eriksen, May 24, 1945, TNA: CAB 126/171.

Short of the false: Top Secret Report by J.C.W., March 6, 1946, TNA: CAB 126/171.

"At any price": Goldsmith, 86.

5 *For thousands of years:* Norsk Hydro, Promotional Pamphlet, TNA: DEFE 2/221.

The river's flow: Vemork Power Station and Electrolysis Plant, NHM: FOIV, Box 78; Norsk Hydro Report, September 14, 1942, TNA: HS 2/184.

6 *The Amerian chemist:* Rhodes, 270; Brun, 9; Report by D. R. Augood, December 1954, VM: JBrun, Box 17.

"50 tons of": Per Dahl, *Heavy Water,* 41.

In 1933 Leif Tronstad: P. M. fra konferanse i Trondheim julen 1933, VM: Box 4F-D17-99; Brun, 10–13.

"Technology first, then": Tronstad Family, Author Interview.

An early working: Brun, 14–20; "Interrogation of G. Syverstad," TNA: HS 2/188; Per Dahl, *Heavy Water,* 41–48.

7 *Researchers found:* Njølstad, 60–61, 77–79; Brun, 9.

Vemork shipped: Olsen, 399; Advertisement for "Schweres Wasser," VM: JBrun, Box 2.

In June 1939: Brun, 15.

8 *"atoms and void":* Rhodes, 29.

"Could a proper detonator": Ibid., 44.

Then, in 1932: Interview with Dr. Alan Morton, IWM: 26662; Per Dahl, *Heavy Water,* 62; Bertrand Goldschmidt, "The Supplies of Norwegian Heavy Water to France and the Early Development of Atomic Energy," in Ole Grimnes, "The Allied Heavy Water Operations at Rjukan," (IFS Info, 1995).

9 *In December 1938:* Rhodes, 251–54.

Springboarding off an: Ibid., 256–60.

One physicist calculated: Karlsch, 32.

10 *"A little bomb":* Rhodes, 275.

By annexing Austria: Hargreaves, 11; Shirer, *The Rise and Fall of the Third Reich,* 599.

"This is not a question": Langworth, 270.

"It's about bombs": Karlsch, 34.

Thirty-four years of: "Notes on Captured German Reports on Nuclear Physics," TNA: AB 1/356; Bagge and Diebner, 157; Karlsch, 32.

11 *"malarkey":* Bagge and Diebner, 21.

When those among: Ibid., 23; Powers, 15.

"*possibility for the*": Letter from Harteck to Reich Ministry of War, April 24, 1939, Papers of Paul Harteck, Rensselaer Institute.

12   *Otto Hahn, on:* Interview with Heisenberg, DIA: DJ 31.
"*If there is*": Bagge and Diebner, 23; Powers, 16.
Ten days later: Powers, 14.

13   *Heisenberg made:* Heisenberg Report, "Die Möglichkeiten der technischen Energiegewinnung aus der Uranspaltung," NB: G-39 (German Reports on Atomic Energy); Cassidy, 422; Interview with Heisenberg, DIA: DJ 31; Per Dahl, *Heavy Water,* 52–54.
"*greater than the*": Heisenberg Report, NB: G-39.
On the subject: Ibid.; Cassidy, 422.
In recognition of: Interview with C. F. von Weizsäcker, Oral History, NB; Rosbaud Report, NB: Goudsmit Papers, III, B27, F42.

14   *By year's end:* Schaaf, 108; Letter from Heisenberg to Harteck, January 18, 1939, DIA: DJ 29; Letter to Rjukan Saltpeterfabriker, January 11, 1940, VM: Box 4F-D17-98; Letter from Harteck to Heisenberg, January 15, 1940, DIA: DJ 29; Walker, *German National Socialism,* 18–27.
By January 1940: Olsen, 399–400.
When Allier visited: Letter from Jomar Brun to Erik Lunde, October 28, 1968, VM: JBrun, Box 4; Brun, 16–18.

15   *After settling these:* Goldsmith, 87.
On March 9: Letter from Jomar Brun to Erik Lunde, October 28, 1968, VM: JBrun, Box 4; Per Dahl, *Heavy Water,* 108.
To smuggle out: Goldsmith, 86–89; Top Secret Report by J.C.W., March 6, 1946, TNA: CAB 126/145.

2. THE PROFESSOR

17   *To Trondheim they:* Haarr, 290–97; Fen, 34.
"*proceed toward Trondheim*": Haarr, 294.

18   *In an auditorium:* Tronstad Family, Author Interview; Njølstad, 15–17. The author is deeply indebted to Olav Njølstad, whose wonderful and informative biography informs much of what we know about Leif Tronstad.
One among them: Interview with Haukelid, DIA: DJ 31.
Tronstad informed: N. A. Sørensen, "Minnetale over Professor Leif Tronstad," LTP; Njølstad, 15–17.
"*sleeping government*": Njølstad, 15.
Only a few days: Haarr, 64.

19   "*What kind of*": Tronstad Family, Author Interview.
The day after: Petrow, 70–80.

20   *Pockets of resistance:* Fen, 50–51.
He was charged: N. A. Sørensen, "Minnetale over Professor Leif Tronstad," LTP; Njølstad, 18–19.
On May 1: Njølstad, 18–20.
"*We have cows there*": Tronstad Family, Author Interview.

21   *Three months before:* N. A. Sørensen, "Minnetale over Professor Leif Tronstad," LTP; Jomar Brun, "Leif Tronstad," *Det Kongelige Norske Videnskabers Selskab,* DORA: L-0001.
"*I work as a slave*": Letter from Tronstad to Josefine Larsen, November 8, 1924, LTP.
"*little angel*": Letter from Tronstad to Bassa, October 27, 1925, LTP.
"*to wait until*": Letter from Tronstad to Bassa, May 10, 1928, LTP.

22   *Emblazoned across the:* Tronstad Family, Author Interview.

"A great day today": July 23, 1932, LTD.

Talented not just: Njølstad, 55.

"If you like": Tronstad Family, Author Interview.

He was soon: Ibid.; Håkon Flood, "Falt for Sitt Land," DORA: L-0001; "Professor Leif Tronstad" (*Nature,* 1945), LTP.

On November 11: Report—"P. M. fra konferanse på Rjukan November 11, 1940 angående kapaciteten av tungtvannsanlegget," November 11, 1940, VM: Box 4F-D17-98; Letter from A. Enger to Aktieselskabet Rjukanfos, November 19, 1940, VM: Box 4F-D17-98.

23 *Shortly after Rjukan:* Letter to Aktieselskabet Rjukanfos, June 11, 1940, VM: Box 4F-D17-99; Brun Report.

over the course: Report—"P. M. fra konferanse på Rjukan November 11, 1940 angående kapaciteten av tungtvannsanlegget."

In private: Jan Reimers, "Leif Tronstad slik jeg kjente ham," NHM: Box 10B; Brun, 19–21; Njølstad, 30–32.

As for Brun: Brun Report.

24 *In March 1941:* Brun, 21–23; Letter to Aktieselskabet Rjukanfos, February 27, 1941, VM: Box 4F-D17-98.

"personally responsible": Brun Report.

Soon after, Alf: Letter from Bjorn Rørholt to Jomar Brun, March 3, 1985, VM: JBrun, Box 17; Jan Reimers, "Leif Tronstad slik jeg kjente ham," NHM: Box 10B.

Around this time: Brun, 22.

Realizing the importance: Njølstad, 36–40; Jan Reimers, "Leif Tronstad slik jeg kjente ham," NHM: Box 10B.

On the morning: Letter from Bjorn Rørholt to Jomar Brun, March 3, 1985, VM: JBrun, Box 17; Interview with Bjorn Rørholt. NHM: Box 16.

25 *"The Mailman must disappear":* Interview with Haakon Sørbye, NHM: Box 16.

"We must go": "Fortalt av Hans Kone, Edla Tronstad," February 1992, DORA: L-0001; Tronstad Family, Author Interview.

"Family, house and": September 22, 1941, LTD.

At 10:15 a.m.: September 23, 1941, LTD; Njølstad, 41.

"I'm not afraid": Tronstad Family, Author Interview.

In Oslo he: September 23–24, 1941, LTD.

26 *"preferential treatment":* Letter from Tronstad to Bassa, September 22, 1941, LTP.

"Good for a single journey": Leif Tronstad Passport, DORA: L-0001.

Aboard a bomber: October 19–21, 1941, LTD.

As arranged by SIS: Letter from Bjorn Rørholt to Jomar Brun, May 3, 1985, VM, JBrun, Box 17.

London, a city: October 21, 1941, LTD.

27 *On his first Sunday:* October 26, 1941, LTD.

The spy was: Kramish, 91–96; Powers, 282; Dorril, 134.

In fact, Tronstad soon: Powers, 53, 94.

Tronstad was open: Handwritten note, TNA: AB 1/355.

28 *One day he met:* October 21–December 1, 1941, LTD.

"acquainted with the particular": Unsigned Letter, October 23, 1941, TNA: AB 1/355.

He was introduced: Urey Report, December 1, 1941, NA: Bush-Conant Papers.

During these weeks: Meeting of the Technical Committee, Tube Alloys, December 11, 1941, TNA: CAB 126/46; Rjukan Report, December 20, 1941, TNA: HS 2/184; Memorandum on Operation Clairvoyant, January 1, 1942, TNA: HS 2/218; Letter from Tronstad to Perrin, November 30, 1943, LTP.

29 *"I want to be close":* November 6, 1941, LTD.

3. BONZO

30 *Before dawn:* TNA: HS 8/435, 14–17, 163–73; Jensen, 37–45; Lunde, 55–57.
    At first glance: Personnel File of Haukelid, TNA: HS 9/676/4.
    "siren of the fjords": Life, April 18, 1938.
31 *"Never give your":* Drummond, 56.
    From the time he: Haukelid Family, Author Interview.
    "This is your": Haukelid, 43.
    "Aim low. A bullet": Jensen, 41.
    "We killed so many": Bailey, 44.
    "Never smoke while": Jensen, 42.
    They blew up: History of the Training Section of SOE 1940–45, TNA: HS 8/435.
    day after day: Ibid.
32 *"One can go":* Myklebust, 54.
    "He is a cool and calculating": Personnel File of Haukelid, TNA: HS 9/676/4.
    That same year: Haukelid Family, Author Interview.
34 *One night, when:* Letter to Dan Kurzman, KA.
    At last, he returned: Haukelid Family, Author Interview.
    In early April: Haukelid, 16–20.
35 *"What are you doing?":* Haukelid Family, Author Interview.
    "Remain true": Johnson, 47.
    Then he got word: Haukelid, 21–25.
36 *Reichskommissar Terboven:* Fen, 63; Johnson, 129–34, 285–87; Petrow, 99–124; Ivar
    Kraglund Interview, IWM: 26660.
    Then, in September 1941: Nøkleby, Gestapo, 49–53, 165–69; Kjelstadli, 118–24.
    "to their knees": Nøkleby, Josef Terboven, 171–72.
37 *At the same time:* Kjelstadli, 124–26.
    Haukelid fled: Haukelid Family, Author Interview.
    The former lawyer: Fehmer Report/Interrogation, NHM: FII, Fehmer.
    Blond, six feet tall, intelligent: Resume of Interrogation of Tor Gulbrandsen, October
    10, 1942, TNA: HS 2/129.
    "He is in": Haukelid Family, Author Interview; Haukelid, 31.
    In fact, Haukelid: Haukelid, 42.
38 *There, in an attic-floor:* Ibid., 40–41; Ragnar Ulstein, Author Interview; Myklebust,
    46–48; Rostøl and Amdal, 50–53; Lunde, 54–56. In his memoir, Haukelid sparingly
    details his meeting with Linge, primarily recounting how impressed he was with the
    Army captain. Memoirs of Kompani Linge recruits have, almost to a man, a similar
    recollection of the initial meeting with the officer.
    "Are you married": Myklebust, 46–48.
    Marches through Stodham Park: History of the Training Section of SOE 1940–45,
    TNA: HS 8/435; TNA: HS 2/188; Jensen, 48–71; Haukelid, 43–44; Foot, 80.
39 *"This is war, not":* Rigden, 362.
40 *It was a merciless:* Haukelid, 44.
    "a really sound man": Personnel File of Haugland, TNA: HS 9/676/2. Some of the
    personnel file of Haukelid was mistakenly inserted into Haugland's file; thus the ref-
    erence.
    "We must organize": Dalton, 368.
    Culling staff and methods: Foot, 4–9; Stafford, 11–13.
    On January 14: Personnel File of Haugland, TNA: HS 9/676/2.
    Roughly 150 Norwegians: Jensen, 47; Myklebust, 66–67.
41 *While Haukelid:* Norwegian Section History, TNA: HS 7/174, p. 27; John Wilson,
    "Great Britain and the Norwegian Resistance," NHM: Box 50A; Petrow, 127–29.

The Lofoten debacle: Rønneberg Interview, Moland; "Minute to Minister," February 1942, TNA: HS 8/321.

A dozen members: Progress Report of SOE for week ending January 28, 1942, TNA: HS 8/220.

42   *Others, like Jens-Anton:* Personnel File of Poulsson, TNA: HS 9/1205/1; Poulsson, 59.

For two weeks: Haukelid, 44–45.

The company showed: January 31–February 3, 1942, LTD; Njølstad, 102–3; Kjelstadli, 176–81.

Wilson told them: Sæter, 41; "Special Confidential Report," TNA: HS 9/1605/3.

Wilson, who was: Wilson, 1–76.

43   *"We want a country":* February 1, 1942, LTD; Njølstad, 103.

4. THE DAM-KEEPER'S SON

44   *On Thursday, March:* March 1–17, 1942, ESD; Hauge, 82–83.

45   *For centuries:* ESP.

In the sixteenth: Sagafos, 24.

"shameless bodies of": Nøkleby, *Josef Terboven,* 38.

At the turn of: "Norsk Hydro," TNA: DEFE 2/220; Tranøy, 15.

Civilization may have: Skinnarland Notes, ESP; Skinnarland Family, Author Interview; Marielle Skinnarland, Author Interview.

46   *On April 9:* Skinnarland, *Hva Som Hendte,* ESP.

47   *He dated a young:* Bergens Tidende, February 1, 2015.

At the start of: Skinnarland Notes, ESP. In many histories of the Vemork action, Skinnarland is described as having in hand intelligence on the heavy water activities of the Germans at Vemork at the time he traveled to Britain. This is clearly not the case, as stated by Einar Skinnarland himself. He left for England with the intention of starting a wireless-transmission site to provide intelligence on German activity in the area, no doubt centering on Norsk Hydro.

The son of a ship's: "Cheese's Report," July 30, 1941, TNA: HS 2/150; John Wilson, "On Resistance in Norway," NHM: Box 50A.

48   *Two days after:* "Preliminary Report on Cheese's Return Journey," March 18, 1942, TNA: HS 2/151; Hauge, 90–93.

49   *"I'm afraid I've arrived":* Hauge, 94.

50   *"We have captured":* Teleprint, March 15, 1942, TNA: HS 2/151.

Through the night: Skinnarland Notes, ESP; Letter, May 29, 1942, NHM: SOE, Box 25; "Preliminary Report on Cheese's Return Journey," March 18, 1942, TNA: HS 2/151.

"Galtesund making for": Hauge, 108.

"killing and scorching": H. G. Wells, *The World Set Free* (New York: Dutton, 1914), 222.

51   *Since his first:* January 24, 1942, LTD. Tronstad recorded in his diary that he'd finally received an "assignment of great importance for country and people" from the defense minister.

After the failed: January 3, 1942; February 13, 1942; February 17, 1942; February 21, 1942; March 7, 1942, LTD.

"No sacrifice": January 1, 1942, LTD.

Then he learned: January 12, 1942; March 4, 1942, LTD; Tronstad Family, Author Interview.

52   *Tronstad prayed:* March 7, 1942, LTD.

"To begin with": Drummond, 19–20.

Now that he: "Clairvoyant," January 1, 1942, TNA: HS 2/218.

A team was: Rjukan, December 20, 1941, TNA: HS 2/184. There was also a draft of a plan for six men to blow up the pipelines and valves above the plant, but this was clearly nixed in favor of the bombing run, as confirmed by the recollections of Poulsson (Poulsson, 75).

53 *Before Skinnarland's arrival:* Operation Grouse, March 28, 1942, NHM: SOE, Box 22. *Skinnarland could prepare:* Ibid. In the original operational instructions, there was no mention of Skinnarland providing intelligence on heavy water. Nonetheless, from the moment Skinnarland landed, this was part of his activity. This is confirmed by everyone from Colonel Wilson ("Heavy Water Operations in Norway," NHM: Box 50A) to Skinnarland himself (Letter from Einar Skinnarland to Dan Kurzman, ESP). *"young heroes":* March 20, 1942, LTD.
On the clear: "Report on Operation Undertaken by 138 Squadron," March 29, 1942, TNA: HS 9/1370/8; Drummond, 21–26.
"flying barn door": Lunde, 77.
"presents for members": "Operation Grouse," March 28, 1942, NHM: SOE, Box 22.
54 *Over the course:* "Sergeant Einar Skinnarland," March 6, 1944, TNA: HS 9/1370/8.
"How you are": Jensen, 94–95.
On the first: Lunde, 73–74.
"dinghy in the": Jensen, 99.
"He showed great": Personnel File of Skinnarland, TNA: HS 9/1370/8.
"Feet together and": Drummond, 21.
55 *At 11:44 p.m.:* "Report on Operation Undertaken by 138 Squadron," March 29, 1942, TNA: HS 9/1370/8.
"We're going back": Interview with Einar Skinnarland, KA.
In the small: Marielle Skinnarland, Author Interview.

5. OPEN ROAD

56 *The Tube Alloys:* "Minutes of 4th Meeting of Technical Committee," April 23, 1942, TNA: CAB 126/46.
"employ a weapon": Kramish, 59.
more imminent, prompting: Clark, *Tizard,* 210–14.
"destroy life in a": Ibid., 214–17.
57 *"decipher the signals":* Churchill, *Their Finest Hour,* 338.
"even money": Lord Cherwell Minute to the Prime Minister," August 27, 1941, TNA: AB 1/170.
"Although personally I am": Text of Churchill's Statement, NA: Harrison-Bundy Papers.
"no time, labour, material": Letter from General Ismay to Lord President of the Council, September 4, 1941, TNA: CAB 126/330.
Two German pilots: Letter to Dr. Pye, September 11, 1941, TNA: AB 1/651.
One German émigré: Interview with Fritz Reichl, NB: Oral History.
"A tale has": Telex from R. Sutton Pratt, November 10, 1941, TNA: AB 1/651.
58 *"and we're working":* Powers, 124. The famous Heisenberg-Bohr meeting is one of enduring attraction for historians and dramatists alike. What was said, who said it, what they meant—these remain open questions. The author will leave it to such fine books as Thomas Powers's *Heisenberg's War* to parse the truth, but as Powers suggests, the British very likely learned of this conversation by spring 1942.
As early as: Gowing, 43.
After the occupation: Jablonski, 93–95.
But British and: Smyth, 38; Walker, *German National Socialism,* 22–23.
59 *"Since recent experiments":* Minutes of 4th Meeting of Technical Committee, April 23, 1942, TNA: CAB 126/46.

In the weeks: Letter from Wilson to Tronstad, May 1, 1942, NHM: Box 10/SIS A; Discussion with Professor Tronstad, May 1, 1942, VM: JBrun, Box 4; May 11–14, 1942, LTD.

"our juice": Letter from Tronstad to Brun, May 15, 1942, LTP.

60 *"We must know":* Letter from Tronstad to Wergeland, May 15, 1942, LTP.

When these letters: May 14, 1942, LTD.

Almost a month: Nøkleby, *Josef Terboven,* 202; Reports on Televåg, TNA: HS 2/136; Herrington, 336.

"If they can": Nøkleby, *Josef Terboven,* 202.

61 *If Lillian Syverstad:* Lillian Gabrielson, Author Interview, ESP.

On the night: Einar Skinnarland, Rapport avgitt I Oslo, September 6, 1942, NHM: SOE, Box 23B; Skogen, 45; ESP; *Bergens Tidende,* February 1, 2015.

62 *On October 3:* Letter to Paul Harteck from Erhard Schöpke, October 30, 1951, Papers of Paul Harteck, Rensselaer Institute; Concerning the Journey to Norsk Hydro in Oslo and Rjukan, a Report by Paul Harteck, NB: G-341.

"Long live Hydro": Finn Sørlie, Author Interview.

After Brun gave: Letter to Paul Harteck from Erhard Schöpke, October 30, 1951, Papers of Paul Harteck, Rensselaer Institute; Concerning the Journey to Norsk Hydro in Oslo and Rjukan, a Report by Paul Harteck, NB: G-341; Brun Report; Schöpke Report, August 3, 1943, NB: G-341.

63 *In January 1942:* Brun Report; Brun, 24–28.

On returning home: Minutes of a Meeting with Norsk Hydro, May 27, 1942, NB: G-341; Letter to Paul Harteck from Erhard Schöpke, October 30, 1951, Papers of Paul Harteck, Rensselaer Institute.

64 *"depend too much":* Drummond, 25.

Another from Brun: Report sent to London, Summer 1942, VM: JBrun, Box 6a; Brun, 28–31.

On June 4: Speer, 269–71; Bagge and Diebner, 29–31; Cassidy, 455–57; Macrakis, 173–75; Powers, 142–50; Roane, 48–49, 78.

Thanks to the Nazis': Irving, 72; Walker, *German National Socialism,* 26–27. There has been much written about the erroneous calculations of Walther Bothe on graphite as a moderator, namely that Bothe's findings contributed to the Germans' concentrating solely on heavy water. Making a well-sourced, thorough explanation of why the German program would have pursued heavy water regardless of Bothe's work, Mark Walker writes that Heisenberg figured a reactor using graphite required "much more uranium and much more moderator than a heavy water device" and that Army Ordnance determined that "boron and cadmium-free carbon of sufficient purity could be produced, but only at prohibitive costs."

65 *Over seventy scientists:* Werner Heisenberg, "Research in Germany on the Technical Application of Atomic Energy," *Nature* (August 16, 1947), NB: Goudsmit Papers, III, B10, F94; C. F. von Weizsäcker, "A Possibility to Produce Energy from U-238," 1940, NB: Goudsmit Papers, III, B10, F95; Walther Bothe, "Die Diffusionslänge für thermische Neutronen in Kohle," 1940–1941, Deutsches Museum Archiv.

"an open road": Interview with Werner Heisenberg, DIA: DJ 31.

But two months: Bagge and Diebner, 28–29; Nagel, 77.

"In the present": Bagge and Diebner, 29–32.

66 *The very same:* Vortragsfolge, February 26, 1942, NB: Goudsmit Papers, III, B25, F13.

"Research in the": Goebbels, 140.

"power ships, possibly": Karlsch, 87–89. For many years this speech by Heisenberg was lost to historians. In 2005 Rainer Karlsch found a copy in a Russian archive, revealing indeed that Heisenberg had promoted a plutonium bomb, but did not see it as imminently achievable.

67   *Basic research on:* Walker, *German National Socialism,* 32.
     Prime Minister Winston Churchill: Sandys, 149–51; Meacham, 180–84; Moran, 50–57.
68   *Earlier that week:* Memorandum Report on Proposed Experiments with Uranium, NA: Bush-Conant Papers.
     A "Uranium Committee": Smyth, 72–84.
     "Nobody can tell": Letter to Vannevar Bush from Leó Szilárd, May 26, 1942, NA: Bush-Conant Papers; Letter to Compton from Leó Szilárd, June 1, 1942, NA: Bush-Conant Papers.
     "Ok. V.B.": Letter to President Roosevelt from Vannevar Bush, June 17, 1942, NA: Bush-Conant Papers.
69   *On June 20:* Meacham, 183–84.
     "What if the enemy": Churchill, *Hinge of Fate,* 380.
     A few days: Note on Mr. Norman Brooke, Deputy Secretary of War Cabinet Office, July 3, 1942, TNA: HS 2/184; Akers Discussion with Norman Brooke, June 30, 1942, NHM: Box 16.

6. COMMANDO ORDER

73   *"the greatest opportunity":* Poulsson, 76.
     Over the next: SOE Group B Training Syllabus, TNA: HS 7/52–54.
74   *"Remember: the best":* "Opening Address," TNA: HS 7/52–54a.
     "Much more intelligent": Poulsson Personnel File, TNA: HS 9/1205/1.
     Poulsson had been: Poulsson Family, Author Interview; Interview with Poulsson, IWM: 27189.
75   *"The saddest day":* Poulsson, 31; Interview with Poulsson, IWM: 27189.
     "mud and stone": Poulsson, 49–59. Quotes in this passage are from Poulsson's diary, excerpted in his memoir on the Vemork raid.
     After concluding his: Interview with Poulsson, IWM: 27189; Poulsson, 80–85.
76   *Reassured by their:* Letter from Malcolm Munthe to Poulsson, Haukelid, Helberg, and Kjelstrup, June 13, 1942, TNA: HS 2/172.
     "fit for duty": Knut Haukelid Personnel File, TNA: HS 9/676/4. It is unclear from the records exactly when this accident occurred — and when Haukelid was removed from the roster. One doctor's report is dated mid-August, another mid-July. Poulsson states that Helberg replaced Haukelid. Another member of the party, Gjestland, was also removed from the list.
     The inventory list: Stores Ready to Be Packed for Grouse I, NHM: SOE, Box 22; Gallagher, 19.
77   *"small independent groups":* Operation Instructions for Grouse, August 31, 1942, NHM: Box 25.
     That same day: August 31, 1942, LTD.
     And indeed, since: Letter from Keyes to Prime Minister, October 14, 1941, DEFE 2/698.
     Since Churchill's return: Mann, 104, 146, 165–68.
78   *The four men:* Lurgan Report, September 3, 1942, TNA: HS 2/184.
     Tronstad was desperate: September 3, 1942, LTD.
     But at times: October 24, 1942, LTD.
79   *After celebrating his:* March 27–August 30, 1942, LTD.
     Earlier in the: Nøkleby, *Josef Terboven,* 197–99; Warbey, 140–44.
     "War makes the": August 7, 1942, LTD.
     "Somewhere in England": Letters from Poulsson to Haukelid, September 10, 1942, TNA: HS 2/172.
     Once they connected: Letter from Munthe to Gjestland, August 8, 1942, TNA: HS 2/172.

80 *"Of course we"*: Letter from Poulsson/Helberg to Haukelid, September 29, 1942, TNA: HS 2/172.

General Nikolaus von Falkenhorst: Freshman Report, November 14, 1942, NHM: FOIV, Box D17.

With a face: Falkenhorst Personnel File, PERS 6–24, Barch-MA; Bericht des Genralobersten v. Falkenhorst, ZA1–1749, Barch-MA; Petrow, 31–34.

81 *"equipped with automatic"*: Freshman Report, November 14, 1942, NHM: Box 10B.

"I have sincerely": Nøkleby, *Josef Terboven*, 212.

"to atone for several": Ibid., 213.

The Swedish border: Kjelstadli, 166–68; Nøkleby, *Gestapo*, 175–77.

Terboven intensified: Kjelstadli, 154–56; Nøkleby, *Josef Terboven*, 215–16.

82 *"Henceforth all enemy"*: German Order to Kill Captured Allied Commandos and Parachutists, Report FF-2127, TNA: WO 331/7. Although dated October 18, 1942, this report states "the order was distributed to regimental commanders and staff officers of corresponding rank on October 10." There was also a communiqué to the Wehrmacht on October 7 that basically stated the same.

7. MAKE A GOOD JOB

83 *When Colonel Wilson*: Interview with Haugland, IWM: 26624.

"Finally": Sæter, 45.

At Chiltern Court: Ibid., 56.

Rather than fomenting: Freshman — Appendix A, October 17, 1942, NHM: Box 25.

84 *Then Wilson led*: Sæter, 56–57; Interview with Haugland, IWM, 26624.

"This is Piccadilly": Freshman — Appendix A, October 17, 1942, NHM: Box 25.

85 *"This mission is"*: Nota tang Freshman, June 30/03, NHM: Box 25; Interview with Poulsson, IWM: 27189; Myklebust, 88–89.

"Make a good": Sæter, 57.

At STS 26's: Rønneberg Interview, Moland.

"valuable work": Ibid.

"vocational school for": Ibid.

Rønneberg was from: Ibid.

"If you were": Myklebust, 21.

86 *At twenty*: Rønneberg Interview, Moland; Interview with Rønneberg, IWM: 27187.

"If you only": Myklebust, 12.

"We're off to": Rønneberg Interview, Moland.

87 *"Number one, go!"*: Air Transport Operation Report, October 18, 1942, TNA: HS 2/185; Gallagher, 20–21.

The Vidda spread: Topography of Hardangervidda Report, October 13, 1942, TNA: HS 2/184; Mears, 47–49; Adamson and Klem, 141–42.

88 *As he scanned*: Poulsson Report; Gallagher, 22; Berg, 104.

They made camp: Grouse Equipment, NHM: SOE, Box 22; Myklebust, 86; Lauritzen, 32.

"There's a new": Halvorsen, *Den Norske Turistforening årbok 1947.*

"You don't jump": Sæter, 59.

89 *The men spent*: Poulsson, "General Report," VM: JBrun, Box 4; Haugland Report; Interview with Haugland, IWM: 26624; Gallagher, 24–25.

8. KEEN AS MUSTARD

90 *On October 20*: Report from Skinnarland, November 1, 1942, NHM: SOE, Box 23; Ueland, 60–61; Message from Stockholm, June 15, 1942, TNA: HS 2/172.

"*This is the latest*": Operation Grouse Instruction, NHM: SOE, Box 22.

91 "*I'll see you*": Recollections of Tomy Brun, LTP; Letter from Brun to Commander Thorsen, September 8, 1984, VM: JBrun, Box 17; Per Dahl, *Heavy Water*, 167–69.

Winston Churchill himself: November 3–7, 1942, LTD.

Vemork was no: Personal for Captain Tronstad, October 28, 1942, NHM: Box 10/SIS A; Freshman Report, November 14, 1942, NHM: FOIV, Box D17.

92 *At 9:00 a.m. sharp*: Drew, 84–85; Henniker, *Memoirs of a Junior Officer*, 1, 22–188. With respect to Operation Freshman, the book by Drew et al. served as an invaluable resource and should be consulted by anybody looking for more information into this part of the story.

"*keen as mustard*": Note written by Mark Henniker, given to Peter Yeates, 1983, KA.

Henniker had reluctantly: Minutes of Meeting in COHQ, October 26, 1942, TNA: DEFE 2/224; Note written by Mark Henniker, given to Peter Yeates, 1983, KA.

93 *Every single sapper*: Notes, KA; *Yorkshire Evening Post*, August 15, 1984; Drew, 113–25.

Henniker instructed them: Mark Henniker, Report on Operation Freshman, November 23, 1942, TNA: DEFE 2/224.

In their first: Report by Group Captain Tom Cooper, 1942, TNA: AIR 20/11930; Drew, 87–89; Interview Notes, KA.

"*You aren't to know*": Drew, 87–89.

Although the Germans: Lynch, 8–21.

Group Captain Tom Cooper: Ibid., 196–97.

94 *It was the duration*: Report by Group Captain Tom Cooper, 1942, TNA: AIR 20/11930; Wilson, 84.

Combined Operations had: Freshman Plan, October 14, 1942, TNA: DEFE 2/224; Notes on Practicability of Operation, October 30, 1942, TNA: DEFE 2/224; Freshman Outline Plan, October 13, 1942, TNA: DEFE 2/224; Notes on Operation Freshman, October 17, 1942, TNA: DEFE 2/224; Mark Henniker, Report on Operation Freshman, November 23, 1942, TNA: DEFE 2/224. The records of reports and minutes of these meetings reveal the fascinating iterative process the planners of Combined Operations went through in preparing for Operation Freshman. Hindsight vision is perfect, but one cannot fault these men for lack of consideration or preparation.

"*In all probability*": Minutes of Meeting on Operation Freshman, October 14, 1942, TNA: DEFE 2/224.

The SOE suggested: Operation Freshman, Outline Plan, October 14, 1943, NHM: Box 10C; Report on 38 Wing Operation Order #5 — "Operation Freshman," December 8, 1942, TNA: DEFE 2/219.

95 "*I've just been*": Special Report on Escape Routes from Vemork to Swedish Frontier, October 12, 1942, NHM: Box 10/SIS C.

Suggestions were made: Freshman Report from Barstow, November 3, 1942, TNA: DEFE 2/224; Letter from A.P.1 for C.A.P., October 31, 1942, TNA: DEFE 2/219; Letter from Colonel Gubbins to Major General Haydon, October 30, 1942, TNA: DEFE 2/219.

"*It is of*": Letter from Mountbatten to A.O.C.-in-C., October 29, 1942, TNA: DEFE 2/219.

"*of the highest priority*": Notes on Operation Freshman, October 17, 1942, TNA: DEFE 2/219.

At noon on: Freshman Training, October 27, 1942, TNA: DEFE 2/219; Drew, 92.

96 *If the British*: Plant Installation and Proposed Demolition, November 16, 1942, TNA: DEFE 2/224; Vemork Power Station and Electrolysis Plant Report, October 30, 1942, TNA: DEFE 2/219; Operation Lurgan, Preliminary Technical Report, TNA: HS 2/185.

Knut Haugland was: Poulsson Report; Haugland Report; Interview with Haugland, IWM: 27212; Interview with Poulsson, IWM: 26625; Interview with Haugland, IWM: 26624; Poulsson, 91–99; Gallagher, 24–27; Sæter, 57–62.

98   *"We are fairly":* Poulsson Report.

He had always: Haugland Family, Author Interview; Sæter, 9–40; Interview with Haugland, IWM: 26624.

99   *On the call to:* Sæter, 26.

In Oslo, he: Personnel File of Haugland, TNA: HS 9/676/2.

"Quiet, keen, hardworking": Ibid.

Now, thirteen days: Haugland Report; Interview with Haugland, IWM: 27212.

100   *Eventually Haugland's teammates:* Haukelid, 97–99.

"Helberg proved the": Poulsson Report.

At the dam: Claus Helberg, "Report About Einar Skinnarland," July 30, 1943, NHM: SOE, Box 23.

With new snowfall: Njølstad, 99; Berg, 114.

101   *The next day, Haugland:* Report; Sæter, 45, 62; Interview with Haugland, IWM: 27212; Interview with Haugland, IWM: 26624.

"Happy landing in spite": Message from Grouse Primus, November 9, 1942, TNA: HS 2/172.

9. AN UNCERTAIN FATE

102   *In his fifth-floor:* Hurum, 114–16.

"We were very": Message to Grouse, November 2, 1942, NHM: SOE, Box 22. In the text, the author references Tronstad returning this message. In the archives, there are few references indicating who penciled the responses for Grendon Hall Home Station operators to transmit, but those that exist cite either Tronstad or Wilson.

Fearing the worst: Private cypher from Stockholm, November 8, 1942, TNA: HS 2/172.

"Battery run down": Message from Grouse, November 9, 1942, TNA: DEFE 2/220.

103   *"Stick to it":* From Colonel Wilson to Grouse, November 9, 1942, TNA: HS 2/184.

"nice, flat ground": Message from Grouse, November 10, 1942, TNA: DEFE 2/220.

On November 12: Letter from Jomar Brun to Arnold Kramish, August 6, 1986, VM: JBrun, Box 6a; Letter from Jomar Brun to Bjørn Rørholt, May 25, 1985, VM: JBrun, Box 17; November 12, 1942, LTD.

"all necessary measures": Niederschrift — Besuch von Regierungs-Baurat Dr. Diebner, September 2, 1942, VM: Box 4F/D17/98.

Confident that Suess: Hans Suess, "Virus House: Comments and Reminiscences," *Bulletin of Atomic Scientists,* June 1968; Brun, 29–30.

Vemork had: Freshman Report, November 17, 1942, TNA: DEFE 2/219.

104   *Despite Tronstad's:* November 12, 1942, LTD.

"super bombs": Clark, *Tizard,* 215.

While this mission: Progress Report for SN Section, November 3, 1942, NHM: SOE, Box 3A.

"The little boy is": Letter from Bassa to Leif, October 11, 1942, LTP.

"I am well": Letter from Leif to Bassa, October 4, 1942, LTP.

On November 15: Minutes of Meeting held at 154 Chiltern Court, November 15, 1942, TNA: DEFE 2/224; November 15, 1942 and November 20, 1942, LTD; Freshman — Translations of Messages, November 15, 1942, TNA: HS 2/184.

105   *"handsome and undoubtedly":* November 15, 1942, LTD.

"Good policy for": Freshman Report, November 17, 1942, TNA: DEFE 2/219.

In a dark: Interview with Haugland, IWM: 26624; Myklebust, 100–101.

106   *In the three:* Poulsson Report; Claus Helberg, "Report about Einar Skinnarland," July
       30, 1943, NHM: SOE, Box 23.
       Haugland was the most: Interview with Haugland, IWM: 26624; Sæter, 65–66.
       "Larger lakes like": Message from Grouse, November 17, 1942, TNA: DEFE
       2/219.
107   *That same day in Scotland:* Personnel File of Haukelid, TNA: HS 9/676/4.
       "Which is it": Note written by Mark Henniker, given to Peter Yeates, 1983, KA; Hen-
       niker, *Image of War,* 95–98.
108   *"dark and light":* Henniker, *Image of War,* 95–98.
       "it is too": Note from C.C.O., Operation Freshman, November 18, 1942, TNA: DEFE
       2/224.
       Churchill had already: Memorandum to Prime Minister, November 17, 1942, TNA:
       DEFE 2/224.
109   *Cooper had his own:* Petterssen, Forecasting for the Freshman Operation, November
       22, 1942, NHM, FOIV, Box D17. It is worth noting that Petterssen was the meteorolo-
       gist who General Eisenhower called on to set the date of the D-day landing. His call
       to postpone the invasion by a day because of weather likely saved innumerable lives.
       Wallis Jackson, Bill: Drew, 93–103; Interview with Michael Douglas, IWM: 31404; Re-
       port by Group Captain Tom Cooper, 1942, TNA: AIR 20/11930.
       "I need 250": Drew, 97.
110   *"Mamie, if you":* Letter from Wallis Jackson, November 18, 1942, KA.
       "A few lines": Drew, 118–19.
       "Whatever happens": Report by Group Captain Tom Cooper, 1942, TNA: AIR
       20/11930.
       They wore steel: Freshman Report — Appendix A — Standard Gear, TNA: DEFE
       2/219.
       Most of them: Drew, 103.
       The floor beneath: Ibid., 105.
111   *After a slight:* Freshman Message List, November 18–20, 1942, TNA, DEFE 2/219;
       Report on 38 Wing Operation Order No. 5, December 8, 1942, TNA, DEFE 2/219.
       Including aircrews: Note written by Mark Henniker, given to Peter Yeates, 1983, KA.
       There is disparity in the archives with respect to the time standards used by the
       Freshman crews and the Grouse team. To prevent confusion, the author used Nor-
       wegian standard time, even when referring to the takeoff of the planes from Skitten.
       "Two small birds": November 19, 1942, LTD.
       After Haugland acknowledged: Freshman — Appendix A, October 17, 1942, NHM:
       Box 25; Interview with Haugland, IWM: 27212; Interview with Poulsson, IWM:
       27189; Poulsson Report.

10. THE LOST

113   *"I hear the Rebecca":* Interview with Knut Haugland, IWM: 26624; Interview with
       Poulsson, IWM: 26625; TNA: HS 2/190; Sæter, 66–67; Poulsson Report.
114   *Over the next:* Poulsson Report.
       Flying with the moon: Report on 38 Wing Operation Order No. 5, December 8, 1942,
       TNA: DEFE 2/219; Letter from Colonel Wilson to Colonel Head, January 21, 1943,
       TNA: DEFE 2/224; Drew, 127–31.
116   *Flight Lieutenant Parkinson:* Report on 38 Wing Operation Order No. 5, December
       8, 1942, TNA: DEFE 2/219; Eyewitness Report — The Planes that Were Wrecked in
       the Egersund District, April–June 1943, NHM: SOE, Box 23; Report from Johannes
       Munkejord, KA; Berglyd, 59–61. Although the exact course of Halifax B is unknown,
       it is clear the plane reached the site, since Haugland heard the Eureka/Rebecca tone,

and Halifax A's was broken. Further, there were reports from Norwegians around Egersund who reported that the plane came around several times before crashing, obviously searching for its lost glider.

Four miles away: Report from Anne Lima, March 13, 1944, NHM: SOE, Box 23; Report to Chief of Police Rogaland from Lensmann in Helland, November 21, 1942, TNA: WO 331/18; Statement of Lensmann Trond Hovland, 1945, TNA: WO 331/18; Statement by Tellef Tellefsen, June 1945, TNA: WO 331/18; Case No. UK-G/B. 476, United Nations War Crime Commission Against Von Behrens and Probst, TNA: WO 331/387.

117   *Through the night:* Freshman Message List, November 18–20, 1942, TNA: DEFE 2/219; Drew, 169.

118   *That same afternoon:* November 20, 1942, LTD.

In silence: Drummond, 51–52.

Wilson made it: John Wilson, "On Resistance in Norway," NHM: Box 50A.

119   *"Thank God for":* Ibid.

"We consider that": Letter from Gubbins to Haydon, November 20, 1942, TNA: DEFE 2/219.

"with the same": Letter from Hansteen to Mountbatten, November 21, 1942, TNA: DEFE 2/219.

"Your work has": Message to Grouse, November 20, 1942, NHM: SOE, Box 22.

"During the night": BBC Monitoring Service — Freshman, November 21, 1942, TNA: DEFE 2/224.

Tronstad was certain: November 21, 1942, LTD.

"Alas": Minute by the Prime Minister, November 22, 1942, TNA: DEFE 2/219.

After Lieutenant Allen: Statement of Lensmann Trond Hovland, 1945, TNA: WO 331/18.

120   *Walther Schrottberger:* Letter from Major Rawlings to PS&W Branch, July 2, 1945, TNA: WO 331/18.

Given that their: Wigner, 447–48.

"no quarter should": German Order to Kill Captured Allied Commandos and Parachutists, Report FF-2127, TNA: WO 331/7.

While orders on: Shooting by the Germans of Allied Personnel Captured in Norway, January 14, 1944, TNA: HS 2/184; Case No. UK-G/B. 476, United Nations War Crime Commission Against Von Behrens and Probst, TNA: WO 331/387; Statement by Werner Siemsen, July 6, 1947, TNA: WO 331/387; Report on the Interrogation of Colonel Oberst, September 12, 1945, TNA: WO 331/387; Statement by Cid Gunner, June 29, 1945, TNA: WO 331/18; Statement by Michael Spahn, June 29, 1945, TNA: WO 331/18; Statement by Rolf Greve, June 14, 1945, TNA: WO 331/18.

Then, in the late: Berglyd, 50–52; Statement of Kurt Hagedorn, August 31, 1945, TNA: WO 331/387; Statement by Cid Gunner, June 29, 1945, TNA: WO 331/18; Letter from Major Rawlings to PS&W Branch, July 2, 1945, TNA: WO 331/18; Statement by Tellef Tellefsen, June 1945, TNA: WO 331/18.

121   *"Towing aircraft's":* Tagesmeldung, November 20, 1942, RW 39/39, Barch-MA; BDS in Olso Berichtet, November 21, 1942, DIA: DJ 31; Irving, 139–42.

The thirty-six-year-old: Brauteset, 32.

Born in the industrial city: Fehlis Personnel Files, VBS 286/6400009794, Bundesarchiv, Berlin.

122   *He ordered them:* Interrogation of Wilhelm Esser, July 10, 1945, TNA: WO 331/386.

James Cairncross: Drew, 111–14.

After the crash: Berglyd, 63–79; Statement of Ravn Tollefsen, date unknown, TNA: WO 331/386; Statement of Martin Fylgjesdal, August 3, 1945, TNA: WO 331/386; Statement of Sigurd Stangeland, July 25, 1945.

A German patrol: Statement of Fritz Seeling, November 6, 1945; Statement of Fritz Feuerlein, September 28, 1945, TNA: WO 331/386; Statement of Kurt Seulen, date unknown, TNA: WO 331/386; Statement of Erich Hoffmann, December 12, 1945. Four of the participants in these atrocious acts were captured after the war. As one might imagine, their accounts are contradictory, likely owing to an attempt to paint themselves in the best light.

11. THE INSTRUCTOR

127 *"Sabotage troops were":* Report from Wilson (SN), November 21, 1942, TNA: HS 2/184.
    Reports from London: Poulsson Report; Message from Grouse, November 23, 1942, NHM: SOE, Box 22.
    For now, the four: Poulsson Report.
128 *After dividing up:* Poulsson, 109–15; Sæter, 74.
129 *"You're to be":* Joachim Rønneberg, "Operation Gunnerside" (IFS Info, 1995); Rønneberg Interview, Moland; Interview with Rønneberg, IWM: 27187; Gallagher, 40.
    When Martin Linge: Ragnar Ulstein, Author Interview.
    "He had a quality": Myklebust, 84–85.
130 *"Where are we":* December 1, 1942, LTD; Minutes of Meeting Held at Norway House, November 26, 1942, TNA: HS 2/185; Interview with Joachim Rønneberg, KA.
    Birger Strømsheim was: Personnel File of Strømsheim, TNA: HS 9/1424/2; Rønneberg Interview, Moland.
    "reliable as a rock": Ibid.
    Next was Fredrik: Personnel File of Kayser, TNA: HS 9/824/2; Lunde, 1–68.
131 *Third was Kasper:* Personnel File of Idland, TNA: HS 9/774/4; Rostøl and Amdal, 1–62.
    His fourth choice: Personnel File of Storhaug, TNA: HS 9/1420/7; Lunde, 88.
    The fifth member: Ragnar Ulstein, Author Interview; Rønneberg Interview, Moland; Myklebust, 108–11.
132 *"Now, I do":* Rostøl and Amdal, 75.
    He then outlined: Myklebust, 107–9.
    In a clearing: Nøkleby, *Gestapo,* 67–68; Wright, 110.
    Gestapo officer Wilhelm Esser: Interrogation of Wilhelm Esser, July 10, 1945, TNA: WO 331/386; Statement of Erik Dahle, August 15, 1945, TNA: WO 331/383; Statement of Hans Behncke, August 14, 1945, TNA: WO 331/383.
133 *"special liquids":* "Bericht über Sabotageunternehmen Lysefjord-Egersund, December 27, 1942," RW 39/40, Barch-MA.
    at the Kummersdorf: Nagel, 45–47; Abschrift: Uran-Bomben, May 8, 1943, DIA: DJ 29.
    In late fall: "Bericht über einer Würfelversuch mit Uranoxyd und Paraffin," G-125, Deutsches Museum Archiv; Interview with Georg Hartwig, NB: Oral History; Walker, *German National Socialism,* 95–97.
    "dreadful drudgery": Nagel, 73.
134 *"If only all":* Ibid.
    Since the meeting: Interview with Erich Bagge, DIA: DJ 29.
    In early summer: Irving, 117–18; Walker, *German National Socialism,* 84.
    Then, on June 23: "Bericht über zwei Unfalle beim Umgang mit Uranmetall," G-135, Deutsches Museum Archiv; Per Dahl, *Heavy Water,* 188–90.
    The disaster did not: "Bericht über einer Würfelversuch mit Uranoxyd und Paraffin," G-125, Deutsches Museum Archiv; Interview with Georg Hartwig, NB: Oral History; Bagge and Diebner, 25; Karlsch, 73, 98–100; Walker, *German National Socialism,* 97.

135   *On December 2:* Rhodes, 401, 438–40.
      "Nothing very spectacular": Wigner, 447.

12. THOSE LOUTS WON'T CATCH US

136   *Minutes before dawn:* "Gestapo Lager Razzia og Unntagstilstand i Rjukan," NHM:
      FOIV, Box D17; Sørlie Memoir.
      Rolf Sørlie: Sørlie Memoir.
137   *Hans and Elen:* Skinnarland, *Hva Som Hendte,* ESP.
      Knut Haukelid arrived: December 8, 1942, LTD.
      "strayed into protective": Njølstad, 173.
      Although Haukelid did: Ragnar Ulstein, Author Interview; Myklebust, 108–9.
      "Heavy water is very": Haukelid, 74; Interview with Haukelid, DIA: DJ 31.
138   *He had come: Instruks for Bonzo,* December 18, 1942, NHM: FOIV, Box D17; Letter
      from Malcolm Munthe to Gjestland, August 8, 1942, TNA: HS 2/172.
      "They will do all": Haukelid, 75.
      "Our working conditions": Message from Swallow, December 9, 1942, NHM: FOIV,
      Box D17.
139   *"highest possible priority":* Method of Clearing Traffic, October 30, 1942, TNA: HS
      2/172.
      "Four Gestapo are": Skogen, 12–14; Friend Report—Øystein Jahren, NHM: SOE,
      Box 23B.
      At that same: Skinnarland Notes, ESP; Marielle Skinnarland, Author Interview; De-
      cember 10, 1942, ESD; Report by Gunleik Skogen, December 1, 1943, TNA: HS 2/174;
      Ueland, 117–20.
140   *In the late:* Rønneberg Interview, Moland.
      "What the hell's": Ibid.
      In the ten: Ibid.; Myklebust, 110–17; Gunnerside—Operating Instructions, Decem-
      ber 15, 1942, NHM: FOIV, Box D17; Rønneberg Report.
142   *"Well, you just":* Rønneberg Interview, Moland.
      "Seven people properly": O'Connor, 47–48.
      Rheam wanted his: Ibid., 45.
      The Gunnerside team: Orientering vedr. Gunnerside. December 11, 1942, NHM:
      FOIV, Box D17; Gunnerside—Operating Instructions, December 15, 1942, NHM:
      FOIV, Box D17; Myklebust, 119–21; Rostøl and Amdal, 74–78; Lunde, 88–90; Rønne-
      berg Interview, Moland.
143   *When they were:* Rønneberg Interview, Moland; Lunde, 89; Interview with Haukelid,
      DIA: DJ 31.
      "That's doomed from": Rønneberg Interview, Moland.
144   *"Who's there?":* Sørlie Memoir; Skogen, 29–31; Sæter, 74.
145   *Each of the:* Poulsson Report; Sæter, 74–75.
      "It's full of vitamins": Gallagher, 80; Sæter, 75.
      "active part": Message to Swallow, December 13, 1942, NHM: SOE, Box 22.
      Misfortune hounded them: Poulsson Report; Sæter, 74–75; Interview with Helberg,
      IWM: 26623; Mears, 101.
146   *Skinnarland provided them:* December 10–18, 1942, ESD.
      On December 17: Message to Swallow, December 17, 1942, NHM: SOE, Box 22; His-
      tory of Grouse/Swallow Eureka, December 1943, NHM: SOE, Box 23.
      "If the conditions": Letter from George Rheam, December 18, 1942, TNA: HS
      2/185.
      "For the sake of": Handwritten briefing notes, December 14, 1942, NHM: FOIV, Box
      D17; Rostøl and Amdal, 76. There are quite a few versions of Tronstad's parting re-

marks to the Gunnerside team. This quote combines material in his briefing notes and Rostøl and Amdal's account.
"You won't get rid": Myklebust, 127.

13. RULES OF THE HUNTER

147   *The four bearded*: Poulsson, 19–21; Poulsson Report. In many of the histories, this cabin is referred to as Svensbu, a later name. In cipher messages at the time with Home Station, the cabin is referred to as Fetterhyatta (Cousin's Cabin). Further, there remains some confusion on whether Poulsson traveled to the cabin by himself a few days before the others, or at the same time.
"En route home": Poulsson, 90.

148   *The canoe was now*: Interview with Poulsson, IWM: 27189; Svein Vetle Trae, Author Interview.
"Just wait until": Gallagher, 50.
Days passed: Interview with Poulsson, IWM: 27189; Gallagher, 50–51.

149   *"Let's go home"*: Haukelid, 77.
The men ruminated: Myklebust, 126–28.
One way or: Instruks for Bonzo, December 18, 1942, NHM: FOIV, Box D17.
"Crisp and clear": Gallagher, 51–52.
"Your rifle is a weapon": "Regler og forskrifter," courtesy of Mia Poulsson.

150   *His team had*: Interview with Poulsson, IWM: 27189.
"They are like": Ingstad, 156.
After zigzagging to: Poulsson, 20–22; Gallagher, 49–65; Svein Vetle Trae, Author Interview. For this narrative of the hunt, the author drew from *Assault in Norway*. Gallagher wrote a masterful account of this first successful kill that saved Grouse, and Poulsson clearly helped inform him on the details.

153   *The next night*: Sæter, 75; Poulsson, 22–23.

154   *"She'll Be Coming"*: Poulsson, 24.
In his office: Letter to Fehlis, December 14, 1942, VM: A-1108/Ak, Box 1. This is one of a series of letters between Norsk Hydro and the Germans to obtain the release of Skinnarland, Jahren, and Skogen. They were all denied.
He had recently: Nøkleby, *Josef Terboven*, 242.
Every week, 99.5 percent pure: Memorandum from N. Stephansen, June 1943, NHM: Box 10/SISA.
Once the crates: Schöpke Report, August 6, 1943, NB: G-341.

155   *"Our security teams"*: Bemerkungen zum Schutz der We-Wi-Betriebe, December 20, 1942, RW 39/40, Barch-MA.
On the morning of: Memo of "Arresterte funksjonarer og arbeidere ved våre bedrifter," April 1, 1942, VM: A-1108/AK, Box 1.
A thirty-minute, steep: Marielle Skinnarland, Author Interview; Skinnarland Notes, ESP; Kjell Nielsen Remembrance, NHM: Box 10B.
"great steak and pancakes": December 24, 1942, ESD.

156   *On December 27*: December 27, 1942, ESD; Marielle Skinnarland, Author Interview.
Inside, he found: Poulsson, 116.

14. THE LONELY, DARK WAR

157   *Marstrander's ship hit*: Hauge, 122; Progress Report for SN Section for Period January 2–9, 1943, NHM: SOE, Box 3A.
"We grow harder": January 6, 1943, LTD.
Over the holidays: December 23, 1942–January 1, 1943, LTD.

Bassa had written: Letter from Bassa to Tronstad, December 20, 1942, LTP.

"Do you think": Letter from Bassa to Tronstad, December 29, 1942, LTP.

"We've been mostly": Letter from Sidsel to Tronstad, January 3, 1943, LTP.

158   *"big and pretty":* Letter from Tronstad to Sidsel, LTP. From references in the letter, this note was written soon after Christmas 1942.

To cripple the: December 16, 1942, LTD.

In a small: Skogen, 1–74. All quotes and descriptions of the horrendous conditions faced by Skogen come from his fine memoir. His recollections of torture at the hands of the Gestapo match those of many survivors from Møllergata 19 and Grini.

160   *"enhanced interrogation":* Nøkleby, *Gestapo,* 59–65. As Nøkleby relates in her study of the Gestapo, this is the term the Germans used.

On January 19: Statement of Oscar Hans, August 11, 1945, TNA: WO 331/383; Affidavit in Respect of the Case of Able Seaman R. P. Evans, TNA: WO 331/383; Statement of Alfred Zeidler, TNA: WO 331/383; Affidavit of Erik Dahle, TNA: WO 331/18; War Crimes — Operation Freshman (Trandum), November 28, 1945, TNA: WO 331/17; Interrogation of Wilhelm Esser, July 10, 1945, TNA: WO 331/386.

161   *Huddled in his:* Poulsson, 124–25. This scene of a typical morning was recounted by Poulsson in a diary entry from which this passage was adapted. He did not state the specific date, but it was clearly within the January moon phase in which a drop could occur.

162   *"Murky weather":* Poulsson, 125.

Sometimes they spent: Poulsson Report.

Through sources in: Interrogation of Lt. Skinnarland, July 27, 1945, TNA: HS 9/1370/8; Helge Dahl, *Rjukan,* 284.

"Weather still bad": Message to Swallow, January 16, 1943, NHM: FOIV, Box D17.

There were petty: Report from Claus Helberg, July 10, 1943, NHM: SOE, Box 23; Interview with Poulsson, IWM: 26625; Interview with Haugland, IWM: 26624; Lauritzen, 63; Berg, 120–22.

163   *"As number three":* Lauritzen, 63.

When they ran: Interview with Helberg, IWM: 26623.

Through the window: Haukelid, 15–16.

164   *Rønneberg had used:* Rønneberg Interview, Moland; Letter from Tronstad to Rønneberg, January 1943, TNA: HS 2/185; Rostøl and Amdal, 78.

Now, sitting in: Handwritten letter from Rønneberg to Tronstad, January 26, 1943, NHM: FOIV, Box D17; Air Transport Operation Report, January 23, 1943, TNA: HS 2/131; Summary of Meeting Gunnerside Abortive Sortie, January 26, 1943, TNA: HS 2/185; Letter from Flight Lieutenant Ventry to Captain Adamson, January 25, 1943, TNA: HS 2/185; Lunde, 92–93; Haukelid, 38; Rønneberg Interview, Moland.

"sniff our way": Myklebust, 130.

165   *That night, on:* Claus Helberg, "Report about Einar Skinnarland," July 30, 1943, NHM: SOE, Box 23; Poulsson Report; Poulsson, 120–22.

"Deeply regret weather": Message to Swallow, January 28, 1943, NHM: FOIV, Box D17.

"Jan. 29 — Skinnarland": Poulsson Report; Poulsson, 129.

Skinnarland mostly jotted: January 28–February 13, 1943, ESD.

At the approach: Interview with Haugland, IWM: 26624.

166   *"What was the":* Poulsson, 127.

15. THE STORM

167   *After the failed:* Report on "Crispie," TNA: HS 2/185; Letter from Rønneberg to Tronstad, December 29, 1942, NHM: FOIV, Box D17; Rønneberg Interview, Moland.

"German troops have": Messages from Grouse, February 8–10, 1943, NHM: FOIV, Box D17.

Since the Freshman: Report from Rjukan, December 1942, TNA: HS 2/186; Poullsson Report.

168   *Paul Rosbaud*: Njølstad, 251–52; Kramish, 129.

Harald Wergeland: Rosbaud Report, NB: Goudsmit Papers, III, B27, F42; Kramish, 188–89; Per Dahl, *Heavy Water*, 164; Hinsley, 123–27.

"not to be": "On Memorandum of February 6th, 1943 submitted by N. Stephansen on the production of D20 of Norsk Hydro," NHM: Box 10/SIS/A.

"intended to make": Obituary of Njål Hole, written by Jomar Brun, VM: IA4FB, Box 13; Letter from Njål Hole to Tronstad, January 19, 1943, NHM: Box 10/SIS/A.

169   *Rain battered*: Most Secret Report of Operations Undertaken by 138 Squadron on Night February 16–17, 1943, TNA: HS 2/131; Haukelid, 81.

"Whatever you do": Bailey, 140.

After four days: Interview with Haukelid, DIA: DJ 31; Rønneberg Interview, Moland.

"We'll find our": Myklebust, 132–33.

"Ten minutes": Haukelid, 81; Rostøl and Amdal, 45.

170   *"We may be"*: Haukelid, 83.

Then the team: Rønneberg Interview, Moland; Rønneberg Report.

171   *"We have to"*: Rønneberg Interview, Moland.

The huge storm: Poullsson Report; Interview with Helberg, IWM: 26623.

172   *Rønneberg took the*: Rønneberg Interview, Moland.

With no wireless set: Gallagher, 70–71.

173   *"Same weather. Storm"*: Rønneberg Report.

In the middle: Gallagher, 71–72; Rønneberg Interview, Moland.

"The storm raged": Rønneberg Report.

174   *As quickly as*: Ibid; Gallagher, 73.

At 1:00 p.m. the: Rønneberg Interview, Moland; Myklebust, 137–39; Haukelid, 84–85; Rostøl and Amdal, 80–81. All quotes from this section were derived from an assembly of these sources. They largely matched one another.

176   *Kristiansen immediately proved*: Rønneberg Interview, Moland.

At the entrance: Haukelid, 86–87.

177   *Haukelid crept through*: Ibid., 88.

"Dr. Livingstone, I presume": Drummond, 69.

Crowded into Fetter: Interview with Poullsson, IWM: 27189.

"tobacco directly imported": Haukelid, 88.

"Stay on the": Rønneberg Interview, Moland; Rønneberg Report.

16. BEST-LAID PLANS

179   *After a restless*: Orientering vedr. Gunnerside, December 11, 1942, NHM: FOIV, Box D17; Rønneberg Interview, Moland; Sørlie Memoir; Haukelid, 97–104; Poullsson, 135–41; Lunde, 96–98; Berg, 125–26; Halvorsen, *Den Norske Turistforening årbok 1947*; Sæter, 84–86; Interview with Poullsson, IWM: 27189; Interview with Rønneberg, IWM: 27187; Rønneberg Report; Interview with Poullsson, IWM: 26625; Interview with Helberg, IWM: 26623; February 19, 1943, ESD; Letters from/to Poullsson and Rønneberg, NHM: Box 25; Letters from/to Poullsson and Helberg, NHM: Box 25; Notes from Poullsson, NHM: Box 25; Interview with Haukelid, DIA: DJ 31. The many original sources given here reflect an enduring controversy among members of the sabotage operation as to who suggested what in terms of the overall planning. However, one thing is clear: Tronstad recommended the approach across the gorge as offering the team the best chance of success.

181   *"Is that edible?"*: Haukelid, 95–96.
     On Thursday, February 25: Sørlie Memoir; Finn Sørlie, Author Interview.
     "I'm glad it's": Sørlie Memoir.

182   *Throughout Friday*: Rønneberg Report; Interview with Helberg, IWM: 26623.
     Helberg began: Gallagher, 85–86; Poulsson Notes on Colonel Wilson's Book, November 2003, NHM: Box 25.

183   *"If trees are"*: Interview with Haukelid, DIA: DJ 31; Interview with Poulsson, IWM: 27189; Poulsson Notes on *Blood and Water* manuscript, NHM: Box 25; Orientering vedr. Gunnerside, December 11, 1942, NHM: FOIV, Box D17.
     The central concept: Bericht über einen Versuch mit Würfeln aus Uran-Metall und Schwerem Eis, G-212, NB: Goudsmit Papers, III/B25/F16; Per Dahl, *Heavy Water*, 210; Nagel, 81–82.
     "If you make": Ermenc, 109–11; Interview with Professor Paul Harteck, DIA: DJ 29.
     Paul Harteck had: Letter from Harteck to Rust, June 26, 1942, Papers of Paul Harteck, Rensselaer Institute.
     Diebner had also: Sagafos, 123; Letter from Rjukan Saltpeterfabriker, March 2, 1942, VM: Box 4F/D17/98; Notes on Irving Manuscript Draft, NHM: Box 10B; Harteck Report: "Besichtigung des Elektrolysewerkes Sinigo bei Meran," December 1, 1942, DIA: DJ 29; Walker, *German National Socialism*, 119.

184   *For his next*: Schöpke Report, August 3, 1943, NB: G-341; Nagel, 80–81.
     Interest in the: Karlsch, 45–53, 126; Nagel, 42–44; Irving, 77, 125–26, 153–55; Walker, *German National Socialism*, 88. There has been much written about the June 4, 1942 meeting with Speer sounding the death knell to the Nazi atomic program. If the meeting had gone the other way, it might have prompted an immediate Manhattan Project–like intensity to the program, vastly improving the odds of a German bomb. That said, after June 1942, the project still held a high DE rating (*Dringlichkeitsentwicklung*), the highest priority rating for material/manpower, and there were a plethora of powerful patrons at the ready if progress could be shown.
     An hour after: Halvorsen, *Den Norske Turistforening årbok 1947*; Interview with Helberg, IWM: 26623.

185   *"It's possible"*: Interview with Helberg, IWM: 26623.
     Now they could: Again, the author refers to the plethora of contradicting original sources from an earlier endnote (*"After a restless"*; see previous page) to divine who suggested what and supported whom on the Gunnerside operational plan.

186   *"Nonsense"*: Myklebust, 150–51; Rostøl and Amdal, 84–85.
     That same afternoon: February 27, 1943, LTD.
     "Everything in order": Message from Swallow, February 25, 1943, NHM: FOIV, Box D17.
     the Carhampton operation: Carhampton Report, January 25, 1943, TNA: HS 2/130; Herrington, 157–58; Hauge, 133–58.

187   *When Tronstad arrived*: February 27, 1943, LTD.
     "Do you want": Nøkleby, *Josef Terboven*, 240.
     Sitting outside the: Haukelid, 104–5.
     "We were in": Interview with Lillean Tangstad, KA; Rostøl and Amdal, 85.

188   *"God save the King"*: Interview with Lillean Tangstad, KA.

17. THE CLIMB

189   *At 8:00 p.m.*: Rønneberg Report; Rønneberg Interview, Moland; Interview with Poulsson, IWM: 27189; Draft of Rønneberg BBC Speech, TNA: HS 7/181; Interview with Haukelid, DIA: DJ 31; Interview with Helberg, IWM: 26623; Interview with Poulsson, IWM: 26625; Haukelid, 102–8; Poulsson, 143–46; Gallagher, 96–110; Rostøl

and Amdal, 86–88; Lunde, 99–102; Berg, 127–30; Myklebust, 150–57. The events of the night of the Gunnerside mission, February 27–28, have been chronicled many times, both in interviews, memoirs, and reports (not to mention books). For this chapter, these are the sources from which the author drew his own account, most of them primary — or drawn from participant recollections. Unless there is a direct quote or distinct piece of information needing sourcing, the author will not note any further specific references in this chapter.

"*act on their*": Rønneberg Report.

They had practiced: Rønneberg Report; Rigden, 252–61, 316–22.

191   "*All right, let's*": Gallagher, 100.

192   *Each man took:* Ibid., 103–5.

193   *In their cabin:* February 27, 1943, ESD; Sæter, 86.

       Well-trained radio: Sæter, 44–45.

194   *Skinnarland could not:* Marielle Skinnarland, Author Interview.

       Careful with each: Haukelid, 108; Interview with Haukelid, DIA: DJ 31.

195   *The saboteurs returned:* Rønneberg Interview, Moland.

       "*In a few*": Draft of Rønneberg BBC Speech, TNA: HS 7/181.

       These were his: Interview with Haukelid, DIA: DJ 31.

196   "*Good luck*": Haukelid, 108.

       "*Good spot*": Gallagher, 109.

       The four saboteurs: Drawing of Gunnerside Approach/Retreat by Jomar Brun, VM: JBrun, Box 6a.

197   "*Locked*": Lunde, 102.

18. SABOTAGE

198   *Rønneberg rechecked:* Rønneberg Report; Rønneberg Interview, Moland; Interview with Poulsson, IWM: 27189; Draft of Rønneberg BBC Speech, TNA: HS 7/181; Interview with Haukelid, DIA: DJ 31; Interview with Helberg, IWM: 26623; Interview with Poulsson, IWM: 26625; Haukelid, 102–8; Poulsson, 143–46; Gallagher, 96–110; Rostøl and Amdal, 86–88; Lunde, 99–102; Berg, 127–30; Myklebust, 150–57. As with the previous chapter, the sourcing for the Gunnerside sabotage was drawn from these references unless specific attribution is necessary.

       At Brickendonbury: Brun, 71–72. Jomar Brun provided the information on the cable tunnel. In fact, he once used the tunnel himself to repair a cable.

       "*Here it is*": Rønneberg Interview, Moland; Gallagher, 110.

199   *NO ADMITTANCE EXCEPT:* Directions Report, November 15, 1942, NHM: FOIV, Box D17.

       "*Put your hands*": Rostøl and Amdal, 102–3; Gallagher, 112.

200   "*Watch out. Otherwise*": Rostøl and Amdal, 103.

201   "*We can light*": Myklebust, 157.

       "*Where are my*": Rønneberg Interview, Moland; Rønneberg, "Operation Gunnerside" (IFS Info, 1995).

202   *Close to finishing:* Extract from Report by Director Bjarne Nilssen, VM: JBrun, Box 6a.

       "*Up the stairs*": Rønneberg, "Operation Gunnerside" (IFS Info, 1995).

203   "*Is that what*": Interview with Poulsson, IWM: 26625.

204   "*No*": Poulsson, 147; Interview with Haukelid, DIA: DJ 31.

       "*Piccadilly*": Haukelid, 113.

205   *As he moved:* Myklebust, 162–63.

       Back at Vemork: Alf Larsen, Rapport over Hendelsen i Høykoncentreringsanlegget på Vemork 28. Febr. 1943, VM: JBrun, Box 17; Bjarne Nilssen, Vedr. Sabotage i tungt-vannsanlegget på Vemork, March 1, 1943, NHM: Box 25.

206   *"normal like we"*: Ibid.
      As the sirens: Bjarne Nilssen, P. M. Sabotasje Vemork, VM: JBrun, Box 6a; Rapport
      vedrørende anlegg for fremstilling av Tungt vann ved Vemork Vannstoff-fabrikk,
      Rjukan, September 14, 1943, NHM: FOIV, Box D17.

19. THE MOST SPLENDID COUP

209   *The nine saboteurs*: Rønneberg Interview, Moland.
      Both local boys: Drummond, 87.
      The sirens continued: Haukelid, 114–15; Rønneberg Interview, Moland; Rønneberg
      Report; Halvorsen, *Den Norske Turistforening årbok 1970*.
210   *The men sat*: Rønneberg Interview, Moland.
      Helberg prepared to: Interview with Helberg, IWM: 26623.
      If there was: Gallagher, 127.
      The other men: Poulsson, 149–50.
211   *Lying in his*: Interview with Poulsson, IWM: 26625.
      As they made: Poulsson, 150.
      When his car: Rapport til her politimesteren I Rjukan, June 23, 1945. Papers of Bjørn
      Iversen.
      As soon as: Bjarne Nilssen, p.m. Sabotasje Vemork, VM: JBrun, Box 6a; Bericht über
      Konsul Ing. E. Schöpkes Reise und Besprechung, March 13, 1943, NB: G-341.
212   *He kept his*: Interview with Larsen, DIA: DJ 31.
      "three strongly built": Bjarne Nilssen, P.M. Sabotasje Vemork, VM: JBrun, Box 6a;
      Bjarne Nilssen, Vedr. Sabotage i tungtvannsanlegget på Vemork, March 1, 1943,
      NHM: Box 25.
      "sharp coercive measures": Til Rjukans befolkningt, February 28, 1943, VM: JBrun,
      Box 6a.
213   *"an installation of"*: Feindnachtrichtenblatt Nr. 28 — 21.2. bis 9.3.1943, RW 39/44,
      Barch-MA; Irving, 166.
      While Muggenthaler waited: Bjarne Nilssen, P.M. Sabotasje Vemork, VM: JBrun, Box
      6a; Report, Gunnerside, April 14, 1943, TNA: HS 2/186. It should be noted that in
      most historical accounts General Rediess and Terboven were stated to have come
      to Rjukan on February 28, 1943, just hours after the sabotage. The report by Nilssen
      never states their presence, and his account is exhaustive.
      Early the next: Falkenhorst Note, February 28, 1943, RW 39/43, Barch-MA.
      "most splendid coup": Message from Swallow, March 10, 1943, NHM: FOIV, Box D17.
      "When you have": Interview with Larsen, DIA: DJ 31; Interview with Haukelid, DIA:
      DJ 31; Haukelid, 125–26; Gallagher, 131–33; Bjarne Nilssen, P.M. Sabotasje Vemork,
      VM: JBrun, Box 6a. The description of this visit by Falkenhorst was assembled from
      these sources, each with their own slightly different version of his conversation/in-
      teraction with Glaase, but all with the same thrust.
214   *A manhunt for*: Heinrich Himmler's Telephone Log, March 1, 1943, RG242, Roll 25,
      NARA. As referenced in Manuscript Notes, Irving, NHM: Box 10B.
      As for the: Letter from Eberling to OKHWa Forsch, March 2, 1943, NB: G-341.
      ". . . perpetrated against Norsk": Swedish Home Service, March 1, 1943, TNA: HS
      2/185.
      The Nazi atomic: March 1, 1943, LTD.
      A wireless message: Hauge, 156.
215   *"High-concentration plant"*: Rønneberg Report.
      "Give our best": Haukelid, 119.
      "Well, Arne": Gallagher, 137.

216   *After thirty miles:* Interview with Poulsson, IWM: 27189; Poulsson, 156–57; Gallagher, 138–40. All quotes and descriptions in this scene are from these sources.

217   *The wireless radio:* March 2–5, 1943, ESD. In his diary, Skinnarland mentions Gunnerside on March 5, but in a cryptic way (that is also likely a typo): "Operasjonen Gunnerside iorden." Potentially, "Operation Gunnerside in order." As his daughter Marielle Skinnarland corresponded with the author, this entry intimates that Skinnarland found out about the sabotage from local farmers in Lie. However, against the clear account from Haukelid in his memoir that the news he shared with Skinnarland about the sabotage came indeed as news, the author followed the Haukelid version.

      Approaching Skårbu: Haukelid, 120–21.

      "Don't worry, Knut": Ibid., 121.

218   *Haugland tried to:* March 6–11, 1943, ESD; Interview with Haugland, IWM: 26624. It is clear from Skinnarland's diary that he received the news on this date; however, the first message to London was not sent until the tenth. On this date and the next, Skinnarland makes short mention of issues with connection and the radio's oscillator.

      "You can bet": Haukelid, 122.

      At noon on: Rønneberg Report.

      Rønneberg had always: Rønneberg Interview, Moland; Myklebust, 166–69; Mears, 180–81.

219   *The next day:* Rønneberg Report.

20. THE HUNT

220   *In Møllergata 19:* Skogen, 96–105; Report by Gunleik Skogen, December 1, 1943, TNA: HS 2/174.

221   *"Until now you":* Skogen, 107.

      "Operation carried out": Message from Swallow, March 10, 1943, NHM: FOIV, Box D17.

      Tronstad was moved: Hurum, 123.

      "Heartiest congratulations": Message to Swallow, March 10, 1943, NHM: FOIV, Box D17.

      Two days later: SOE and Heavy Water, March 1943, TNA: HS 2/185; Minutes of ANCC Meeting, March 12, 1943, TNA: HS 2/138; Myklebust, 201–2; SOE Progress Report, March 15, 1943, TNA: HS 8/223.

222   *"It's justified":* Tronstad, Note on Heavy Water, March 18, 1943, LTP.

      Sir John Anderson: March 15, 1943, LTD.

      "England's once-proud": Hauge, 157.

      "sacrificed enough for": March 3–6, 1943, LTD.

      He arranged: SOE Progress Report, March 15, 1943, TNA: HS 8/223; Njølstad, 222–24.

223   *All the while:* Precis of a Meeting between Professor Goldschmidt, Professor Tronstad, and Lt. Commander Welsh, March 15, 1943, LTP.

      The Danish physicist: Letter from Eric Welsh to Tronstad, January 16, 1943, LTP.

      "heartbeats": March 18, 1943, LTD.

      On March 13: Rønneberg Report; Rønneberg Interview, Moland; Myklebust, 173–88. The march to Sweden was one of the most remarkable aspects of the Gunnerside operation. Rønneberg recounts it in detail in his report, but a better, longer chronicle comes from Gunnar Myklebust in his biography of the saboteur. These are the three primary sources for the description of this retreat given here.

224   *"You must all":* Rostøl and Amdal, 97–99.

      Under the light: Rønneberg Report; Mears, 182–85. On the survival skills of the saboteurs, Mears offers fine insight.

225   *"All right, let's"*: Rønneberg Interview, Moland.
       "Guys": Myklebust, 182–83.
       As night fell: Interview with Rønneberg, IWM: 27187; Rønneberg Interview, Moland;
       Rønneberg Report.
226   *When Knut Haukelid*: Haukelid, 129; Interrogation of Knut Haukelid, July 25, 1945,
       TNA: HS 9/676/4.
       "When this war": Haukelid, 129–30.
       Over two weeks had: Ibid., 124–31.
227   *"It's not safe"*: Berg, 137.
       Kjelstrup left: Report by Arne Kjelstrup, October 30, 1943, NHM: SOE, Box 23.
       After reports pointed: Report "Angår Aksjonen på Hardangervidda," July 17, 1946,
       NHM: Box 10B; Report on the Interrogation of Major Ernst Lutter, July 5, 1945,
       NHM: Box 16; Tätigkeitsbericht AOK/Ic, April 1943, RW 39/44, Barch-MA; Ueland,
       191–93; Helge Dahl, *Rjukan*, 291. Figures on those involved in the March–April 1943
       razzia range everywhere from two thousand to twelve thousand men. The key con-
       fusion likely rests around the fact that there were three actions conducted at roughly
       the same time, as outlined by Major Lutter, who participated in them. Exact num-
       bers have yet to be uncovered, but Lutter states three thousand men in the Hardan-
       gervidda, another with an "even more formidable army" for regions south and west
       of the Vidda, and the third with two thousand men in northern Norway near Trond-
       heim.
228   *A Norwegian hunter*: Haugland Report; Lunde, 109.
       "Seven men were": Report, "Vemork kraftstasjon," March 24, 1943, NHM: FOIV, Box
       D17.
       If enemy commandos: Report, "Angår aksjonen mot Hardangervidda I tiden 23/3/ til
       8/4/43," NHM: Box 10B; Report, "Unternehmen Adler," March 30, 1943, NHM: Box
       10B; Technique of the Agent in the European Field, TNA: HS 2/229; Report by Arne
       Kjelstrup, October 30, 1943, NHM: SOE, Box 23; Ueland, 161–63; Kjelstadli, 264.
       Although the Vidda: Report on the Interrogation of Major Ernst Lutter, July 5, 1945,
       NHM: Box 16; Report from Swedish Telegraph Agency, March 29, 1943, NHM:
       FOIV, Box D17; Report from Fenrik Haugland, September 23, 1943, NHM: SOE, Box
       23.

21. PHANTOMS OF THE VIDDA

230   *As soon as*: Berg, 137–38.
       "The whole district": Drummond, 104.
       Haukelid knew: Haukelid, 130–31.
       A good tracker: Ibid., 134.
231   *In the late*: Interview with Helberg, IWM: 26623.
       After separating from: Poulsson Report; Sørlie Memoir; Report, Claus Urbye Hel-
       berg, April 19, 1943, TNA: HS 2/186.
232   *Peering out through*: Interview with Helberg, IWM: 26623; Report by Claus Helberg,
       June 28, 1943, NHM: SOE, Box 23; Interrogation of Sergeant Helberg, July 23, 1943,
       TNA: HS 9/689/6. The narrative of Helberg's dramatic escape between March 25–30
       derives primarily from these four sources. The author has also consulted Gallagher,
       149–63; Interview with Claus Helberg, KA; Ueland, 194–201, 212–16. Any quotes or
       other select material will be separated out in the endnotes.
233   *"Halt! Arms up"*: Interview with Helberg, IWM: 26623.
234   *"You have to"*: Ibid.
       The boat for: There remains some discrepancy in the sources about what hotel Hel-
       berg stayed in during his time in Dalen. Some refer to the Bandak Tourist Hotel, oth-

ers to the Dalen Hotel. Given the Dalen Hotel was the nicest hotel in the town as well as the fact that the Germans set up headquarters there, the author sided with Dalen.

235 *The conversation was:* Letter from Wehrmachtsbefehlshaber in Norwegen, May 15, 1943, RW 4/639, Barch-MA.

"No," Hassel: Berit Nøkleby, "Uforskammet opptreden mot Terboven," *Aftenposten,* February 25, 1983.

236 *"As you can see":* Gallagher, 159.

"You sit there": Berit Nøkleby, "Uforskammet opptreden mot Terboven," *Aftenposten,* February 25, 1983.

238 *"Every day, you":* Report, "Aksjonen på Hardangervidda," March 23–April 8, 1943, NHM: Box 10B.

Some stores of explosives: Wochenbericht für die Woche vom 29.3–4.4.43, RW 39/45, Barch-MA.

High in the: March 20–April 20, 1943, ESD.

Skinnarland and Haugland: Haugland Report.

239 *Though exposed on:* Skinnarland Notes, ESP; April 1–19, 1943, ESD.

Best of all: Skinnarland, *Hva Som Hendte,* ESP.

The following morning: Haukelid, 137–39; April 16–19, ESD; Haugland Report.

Skinnarland was: Haugland Report; Sæter, 97–98.

22. A NATIONAL SPORT

241 *In mid-April 1943:* Bericht über Konsul Ing. E. Schöpkes Reise und Besprechungen, March 13, 1943, NB: G-341; Rapport vedrörende anlegg for fremstilling av Tungt vann ved Vemork Vannstoff-fabrikk, Rjukan, September 14, 1943, NHM: FOIV, Box D17; Per Dahl, *Heavy Water,* 211–12; Bericht von Konsul Schöpke über die Besprechungen am 17. und 18.6.1943, NB: G-341.

"national Norwegian sport": Letter from Ebeling tò OKH Wa Forsch, March 2, 1943, NB: G-341.

Others, including Bjarne: Bjarne Nilssen, P. M. Sabotasje Vemork, VM: JBrun, Box 6a.

"swift decision": Letter from Ebeling to OKH Wa Forsch, March 2, 1943, NB: G-341; Olsen, 417.

provided any materials: Andersen, 400–404; Report, Gunnerside, April 15, 1943, TNA: HS 2/186.

By the time: Bericht über Konsul Ing. E. Schöpkes Reise und Besprechungen, March 13, 1943, NB: G-341; Rapport vedrörende anlegg for fremstilling av Tungt vann ved Vemork Vannstoff-fabrikk, Rjukan, September 14, 1943, NHM: FOIV, Box D17; Per Dahl, *Heavy Water,* 211–12.

242 *While this was:* Olsen, 417; Bericht Schutz von We-Wi-Betrieben, April 19, 1943, RW 39/45, Barch-MA.

Three weeks later: Niederschrift über die Besprechung am 7.5.1943 i.d. PTR, NB: G-341.

"rumors abound in": Karlsch, 162–63.

"uranium bombs": Abschrift, Allgemein verständliche Grundlagen zur Kernphysik, May 8, 1943, DIA: DJ 29.

243 *Only the previous day:* Schriften der Deutschen Akademie DNR Luftfahrtforschung, NB: Goudsmit Papers, III/B27/F29.

He wanted to: Niederschrift über die Besprechung am 7.5.1943 i.d. PTR, NB: G-341.

His team's most: Bericht über einen Versuch mit Würfeln aus Uran-Metall und Schwerem Eis. G-212, NB: Goudsmit Papers, IV/B25/F16; Irving, 174–75.

"too small to": Niederschrift über die Besprechung am 7.5.1943 i.d. PTR, NB: G-341.

Diebner had his: Interview with Paul Harteck, NB: Oral History; Interview with Georg Hartwig, NB: Oral History.

244   *"doer, a driver"*: Ingstad, 159.
      "biggest sonuvabitch": Nichols, 108.
      In the hills: Rhodes, 451, 486, 497.
      All these efforts: Groves, 186–91.
      In April he: Private Cipher Message for Field Marshal Dill from C.A.S., April 7, 1943,
      NHM: FOII, Box 61.
245   *"You might be"*: Kurzman, 186.
      "colossal amounts": Preliminary Statement concerning the possibility of the use of ra-
      dioactive material in warfare, July 1, 1943, TNA: CAB 98/47.
      On the morning of: Bush, Memorandum of Conference with the President, June 24,
      1943, NA: Bush-Conant Papers; Powers, 210–11.
      "strain every nerve": Letter from James Conant to General Groves, December 9, 1942,
      NA: Bush-Conant Papers.
      "going very aggressively": Bush, Memorandum of Conference with the President,
      June 24, 1943, NA: Bush-Conant Papers.
      Earlier in the: Haukelid, 149; Haukelid Family, Author Interview; Interrogation of
      Knut Haukelid, July 25, 1945, TNA: HS 9/676/4; Report by Arne Kjelstrup, October
      30, 1943, NHM: SOE, Box 23; Report from Bonzo, NHM: SOE, Box 23.
246   *Haukelid and Kjelstrup*: Haukelid, 152; Berg, 142–43.
247   *Haukelid took a*: Interrogation of Lieutenant Einar Skinnarland, July 27, 1945, TNA:
      HS 9/1370/8; Skinnarland, *Hva Som Hendte*, ESP.
      "Bonzo was waiting": June 18–20, 1943, ESD.
      On one of: Letter from Einar Skinnarland to Kirvil Skinnarland, December 1998,
      ESP.
      Skinnarland remained for: June 28–July 8, 1943, ESD; Skinnarland Notes, ESP.
248   *"Vemork reckons on"*: Message from Swallow, July 8, 1943, NHM: FOIV, Box D17.
      "What rewards are": Prime Minister's Personal Minute, April 14, 1943, TNA: HS
      2/190.
      any renewed deliveries: Report, "Lurgan," July 4, 1943, LTP; Letter from Brun to
      Thomas Powers, October 11, 1988, VM: JBrun, Box 17.
      "tackle the juice issue": July 13, 1943, LTD.
249   *In his report*: Tronstad, "Notat vedr. X," July 19, 1943, NHM: Box 10/SIS B; Brun, 73–77.
      On July 21: Letter to Tronstad from Wilson, July 16, 1943, DORA: Correspondence
      1937–45.
      Later, Selborne: July 21, 1943, LTD; Myklebust, 214–15; Poulsson, 160–63; Rønneberg
      Interview, Moland.
      Not seventy-two: Kjelstadli, 201–4; Olsen, 410–12; Sagafos, 105–110.

23. TARGET LIST

251   *On August 4*: Report, "Meeting Held at Rjukan on the 4th August 1943," NHM: Box
      10/SIS B; Olsen, 406, 418; Per Dahl, *Heavy Water*, 212.
252   *In June 199 kilograms*: Alf Larsen, Vemork production figures, DIA: DJ 31; Letter
      from Harteck to Diebner, February 16, 1944, NB: G-341.
      "personal conviction": Report, "Meeting Held at Rjukan on the 4th August 1943,"
      NHM: Box 10/SIS B.
      Within a few: Kjelstadli, 260; Andersen, 422.
      "usual output": Message from Swallow, August 4, 1943, NHM: SOE, Box 22.
253   *"With care"*: Message from Swallow, August 9, 1943, NHM: SOE, Box 22.
      Over the next: Messages from Swallow, August 7–22, 1943, TNA: HS 2/187.
      To sustain himself: July–September 1943, ESD; Letter from Skinnarland to Dan Kurz-
      man, May 12, 1997, ESP; Marielle Skinnarland, Author Interview.

"*shot a stone*": September 16, 1943, ESD.

During this period: Bergens Tidende, February 1, 2015.

He was still: July 11–17, 1943, ESD; Marielle Skinnarland, Author Interview.

254 "*everything required*": Letter from A. R. Boyle to Wilson, August 9, 1943, TNA: HS 2/187.

"*local action*": Message to Swallow, August 10, 1943, including handwritten comments from Tronstad and Wilson, TNA: HS 2/187.

Now more than: Tronstad, Report with reference to attacks at Rjukan and Vemork, TNA: HS 8/955/DISR.

Late in 1942: Kjelstadli, 200–205.

255 *Most pointedly, Tronstad:* Tronstad, "Notat vedr. X," July 19, 1943, NHM: Box 10/SIS B.

Njål Hole, Tronstad's young: Letter from Hole to Tronstad, September 1, 1943, NHM: Box 10/SIS B.

"*of vital importance*": Stephansen Report, October 21, 1943, TNA: HS 2/187.

"*We have to do*": Letter from Tronstad to Sidsel, August 20, 1943, LTP.

256 *Two days after:* August 20, 1943, LTD.

"*I would propose*": Perrin Report, "Norway and Production of Heavy Water," August 20, 1943, TNA: AIR 8/1767.

"*My name is Knut*": Haukelid, 169; Haukelid Family, Author Interview.

"*Remember: Keep your*": Interrogation of Knut Haukelid, July 25, 1945, TNA: HS 9/676/4; Report by Bonzo, via Arne Kristoffersen (Kjelstrup), October 1943, NHM: SOE, Box 23.

Once, by a: Berg, 143–44.

257 "*Six housewives*": Request for Packing of Stores in Containers, Swallow Two, August 24, 1943, TNA: HS 2/131.

After midnight on: Operational Report, Swallow Two, September 21/22, 1943, TNA: HS 2/131.

"*How'd it go?*": Haukelid, 157.

Haukelid and Kjelstrup searched: Ibid.; Report, Swallow Two drop, November 18, 1943, TNA: HS 2/131.

258 *But the full meals:* Berg, 141, 144–47.

A couple of: Haukelid, 158; October 13, 1943, ESD.

Skinnarland welcomed him: Skinnarland Note, ESP.

"*The powers that*": Letter from Lt. Colonel Sporborg to Brigadier Mockler-Ferryman, October 5, 1943, TNA: HS 2/218; Tube Alloys Technical Committee Meeting, September 19, 1943, TNA: CAB 126/46; Letter from L. C. Hollis to CAS (Chief of Air Staff, Sir Charles Portal), October 18, 1943, TNA: AIR 8/1767.

259 *Using intelligence:* Report, Heavy Water Production at Vemork, October 16, 1943, TNA: HS 2/218.

Armed with the: Letter from L. C. Hollis to Sir Charles Portal, October 18, 1943, TNA: AIR 8/1767; Letter from Sir Charles Portal to L. C. Hollis, October 20, 1943, TNA: AIR 8/1767.

From his base: Parton, 155.

"*We'll bomb them*": Ibid., 130.

Groves continued to: Groves, 189.

"*When the weather*": Kurzman, 188; Letter from Sir Charles Portal to Brigadier Hollis, October 20, 1943, TNA: AIR 8/1767.

Throughout all these: Report with Reference to Attacks at Rjukan and Vemork, TNA: HS 8/955/DISR.

"*in a special*": Njølstad, 264–65.

260 "*should be changed*": Minutes of the 24th ANCC, November 11, 1943, TNA: HS 2/138.

24. COWBOY RUN

261 *At 3:00 a.m. on:* Interview with Owen Roane, KA.
     At the same: Freeman, 7–15.
262 *Given the distance:* Roane, 96.
     "special explosive": Interview with Owen Roane, KA.
     "milk run": Bennett, 15.
     Although Roane: Interview with Owen Roane, KA; Roane, 1–14.
     The average lifespan in: Harry Crosby, Jan Riddling, and Michael P. Faley, "History of
     the 100th Bomb Group," United States Air Force Military Heritage Database, http://
     www.8thairforce.com/legacy_100thbomb.htm.
263 *On one mission:* Roane, 29–80.
     "I'm coming in": Michael Faley, "Owen Roane: The Last Cowboy," *Splasher Six* 29
     (Fall 1998).
     At 5:00 a.m. Roane: Freeman, 16–18, 244–45.
     The ground-crew chief: Roane, 95–101. In Roane's book, he included transcripts of
     the reports from Command Pilot Bennett, Lead Pilot Roane, as well as the naviga-
     tor, bombardier, and the Third Division air commander.
     "Ready to go": Interview with Owen Roane, KA.
     After checking the: Roane, 95–101.
264 *Three hundred and:* Bomber Command Narrative of Operations, 131st Operation,
     November 16, 1943, TNA: AIR 40/481.
     After some time: Bennett, 16–21.
     They crossed the: Ibid., 17.
265 *"Make a large":* Kurzman, 197. Of the histories on the Eighth Air Force attack on
     Vemork, Dan Kurzman succeeded best in interviewing some of the pilots and crews,
     giving a thorough account of the events of November 16.
     A B-17 in: Ibid., 19–20.
     In total, 176: Bomber Command Narrative of Operations, 131st Operation, Novem-
     ber 16, 1943, TNA: AIR 40/481.
     On a farm: Haukelid, 177; November 16, 1943, ESD.
266 *At 11:33 a.m.:* Nielsen, Kjell, "Notat angående omtalen av fergeaksjonen på Rjukan i
     Februar 1944," NHM: Box 10B.
     "There are even": Rapport fra luftvernlederen ingeniør Fredriksen over flyangrepet på
     Vemork Kraftstasjon og Vemork Fabrikkompleks, November 16, 1943, VM: A-1108/
     AK, Box 1.
     "Run to your": Report by Unnamed Witness, KA.
     The Ninety-Fifth: Mears, 95–101; Bomber Command Narrative of Operations 131st
     Operation, November 16, 1943, TNA: AIR 40/481; Quotes on Tuesday's 8th AAF
     Heavy Bomber Operations, November 16, 1943, TNA: AIR 2/8002.
267 *In total, 711:* Bomber Command Narrative of Operations, 131st Operation, Novem-
     ber 16, 1943, TNA: AIR 40/481; Report, "Norway: Result of USAAF raid on Rjukan,
     Vemork," December 28, 1943, TNA: AIR 40/481.
     Just as the: Attack on Fertilizer Works at Rjukan by USAAF, December 9, 1943, TNA:
     AIR 2/8002; 8th Air Force Command Provisional Report, November 18, 1943, TNA:
     AIR 40/481.
268 *"My God, what's":* Kurzman, 202.
     A former member: Ibid., 202–4; Nielsen, Kjell, "Notat angående omtalen av fergeak-
     sjonen på Rjukan i Februar 1944," NHM: Box 10B.
     In Rjukan, four: Ømkomme under bombingen, November 16, 1943, VM: A-1108/AK,
     Box 1; Olsen, 419.
269 *SS officer Muggenthaler:* Letter from Muggenthaler to Befehlshaber der SS und des
     SD, November 17, 1943, R70/32, Bundesarchiv, Berlin.

"SH-200 high-concentration": Fernschreiben an den Chef der Sicherheitspolizei und des SD, Kaltenbrunner, November 18, 1943, R70/32, Bundesarchiv, Berlin.

25. NOTHING WITHOUT SACRIFICE

273   *On the day:* Njølstad, 274–75.

"Hope that the": November 16, 1943, LTD.

The Allies had: Notat vedrörende angrepene på Rjukan og Vemork, November 16, 1943, NHM: Box 10/SIS B.

274   *"On leaving Norway":* Letter from Hagen (Brun) to Perrin, November 1943, VM: JBrun, Box 2.

When Tronstad returned: Aide-Memoire, "The bombing of industrial targets in Norway," TNA: AIR 2/8002; Notat vedrörende angrepene på Rjukan og Vemork, November 16, 1943, NHM: Box 10/SIS B; Letter from Trygve Lie, January 29, 1943, TNA: AIR 2/8002; Letter from A. W. Street, December 22, 1943, TNA: AIR 2/8002. There are a series of illuminating letters in this folder (Air Attacks on Targets in Norway, TNA: AIR 2/8002), a worthy starting point for anybody interested in the back-and-forth imbroglio betweeen British, American, and Norwegian officials.

From messages received: Letter from Hole to Tronstad, December 16, 1943, LTP.

"filled with explosive": November 5, 1943, LTD.

"decided to abandon": Cable from S. D. Felkin, December 22, 1943, TNA: HS 2/187.

This unconfirmed intelligence: Notat vedrörende angrepene på Rjukan og Vemork, November 16, 1943, NHM: Box 10/SIS B.

275   *"It was a":* Njølstad, 270–71.

"It is my": Letter from Tronstad to Bassa, August 23, 1943, LTP.

"We get nothing": Letter from Tronstad to Bassa, December 8, 1943, LTP.

"It is hard": April 30, 1943, LTD. This reflection matches perfectly the one expressed by Malcolm Munthe, his British colleague, who on May 3, 1943, sent Tronstad a letter that stated, "As you perhaps have known — I have for some time past felt — very keenly, when sending out into the field some of my Norwegian friends — the need to once again try to do my bit in the active side of the war myself." Letter from Munthe to Tronstad, May 3, 1943, DORA: Correspondence 1937–45.

Throughout November: Walker, *German National Socialism,* 100–102.

For their G-III: Bericht über die Neutronenvermehrung einer Anordnung von Uranwürfeln und Schwerem Wasser (GIII), Deutsches Museum Archiv; Nagel, 90–92; Irving, 190–92.

276   *"Given the relatively":* Bericht über die Neutronenvermehrung einer Anordnung von Uranwürfeln und Schwerem Wasser (GIII), Deutsches Museum Archiv.

Straightaway his team: Bagge and Diebner, 35.

His success came: Letter from Göring to Esau, December 2, 1943, NB: Goudsmit Papers, III/B27/F30; Per Dahl, *Heavy Water,* 219.

Tall, with a: Irving, 200; Per Dahl, *Heavy Water,* 220–21.

"the emperor of": Karlsch, 104–5.

"In my opinion": Nagel, 94.

Prior to his: Karlsch, 106; Walker, *German National Socialism,* 130–31.

277   *On December 11:* Protokoll über die in Norsk Hydro Buro, Oslo, December 11, 1943, NB: G-341.

"expose the company's": Protokoll über die in Norsk Hydro Buro, Oslo, December 11, 1943, NB: G-341.

At the same: Mark Walker and Rainer Karlsch, "New Light on Hitler's Bomb" (*Physics World,* June 1, 2005); Nagel, 92–93; Irving, 213–17. There is little doubt Diebner was working on this effort. The bigger controversy revolves around its success or failure. Irving skirts away from the answer, but Karlsch argues in his book that Diebner and crew successfully tested such weapons. Whether they did — or did not — the author agrees with the Karlsch/Walker article that says: "What is important is the revelation that a small group of scientists working in the last desperate months of the war were *trying* to do this."

278 *"We are sending"*: Messages from/to Swallow, December 19, 1943–January 1, 1943, NHM: SOE, Box 22.

Bunkered *down*: Haukelid, 171–72; December 25, 1943–January 1, 1944, ESD.

Haukelid was somber: Ording, 255; Haukelid, 159.

The two men: Marielle Skinnarland, Author Interview.

279 *"I don't understand"*: Haukelid, 166–67.

Despite these squabbles: Ibid.; Marielle Skinnarland, Author Interview.

"It is reported": Njølstad, 288–89; Message from London, January 29, 1943, TNA: HS 2/188.

Although the American: Report on Rjukan, January 1, 1944, TNA: HS 2/188.

"secret weapon forge": Translation of Extract from Swedish Newspaper, "Brilliant Coup Against Hitler's Secret Weapon," November 23, 1943, TNA: HS 2/188.

Like Sørlie, most: Sørlie Memoir.

280 *Viten told*: Ibid.; January 30, 1943, ESD.

The wind blew: Sørlie Memoir.

26. FIVE KILOS OF FISH

281 *On February 1*: Rolf Sørlie, Report on Milorg at Rjukan, May 12, 1944, NHM: SOE, Box 23; Sørlie Memoir; Løken, 102; Drummond, 152; Report by Sheriff Foss, January 1944, NHM: SOE, Box 23.

Some workmen had: Interrogation of Gunnar Syverstad, April 5, 1944, TNA: HS 2/188.

282 *The next day*: Sørlie Memoir.

Sørlie delivered his: Message from Swallow, February 2, 1943, NHM: Box 10/SIS B.

"Do you always": Sørlie Memoir.

At last, an: Message from Swallow, February 3, 1943, TNA: HS 2/174.

283 *"Did you meet"*: Haukelid, 178.

Sørlie reported: Rolf Sørlie, Report on Milorg at Rjukan, May 12, 1944, NHM: SOE, Box 23; Message from Swallow, February 5, 1944, TNA: HS 2/174.

"We will probably": Message from Swallow, February 6, 1943, NHM: SOE, Box 23.

Sørlie returned: Haukelid, 181.

After a week: February 6, 1944, LTD.

284 *Over the past*: Njølstad, 298–99; Messages to/from Swallow, February 1–7, 1943, NHM: SOE, Box 23.

In the end: Letter from Tronstad to Wilson, February 7, 1944, TNA: HS 2/188; February 7, 1942, LTD; Letter from Welsh to Tronstad, February 8, 1944, LTP.

"We are interested": Message to Swallow, February 8, 1944, TNA: HS 2/188.

"We will do": February 7, 1943, LTD.

285 *"Einar, you awake?"*: Drummond, 156.

Sørlie arrived at: Haukelid, 182–83; Account given by Engineer Larsen of the transaction during the attack on Tinnsjø ferry, February 20, 1944, NHM: SOE, Box 23.

286 *First they*: Haukelid, Report on the Sinking of the Ferry *Hydro*, February 20, 1944, NHM: SOE, Box 23; Account given by Engineer Larsen of the transaction during the attack on Tinnsjø Ferry, February 20, 1944, NHM: SOE, Box 23. These two accounts are the best distillation of the thinking behind the various options. The same break-

down can be found in a number of other sources, including Haukelid's memoir as well.

Back at Nilsbu: Message from Swallow, February 9, 1943, NHM: SOE, Box 23; Marielle Skinnarland, Author Interview; Skinnarland Notes, ESP.

287 *"Agree to sinking":* Message to Swallow, February 10, 1943, NHM: SOE, Box 23.

"You have to": Sørlie Memoir.

The letter was a blow: Haukelid Family, Author Interview; Haukelid, 182; Sørlie Memoir; Message from Swallow, February 12, 1943, NHM: SOE, Box 23; Letter to SNA, February 17, 1944, NHM: SOE, Box 23.

288 *On February 13:* February 13, 1944, ESD; Sørlie Memoir.

Olav Skindalen met: Haukelid, Report on the Sinking of the Ferry *Hydro,* February 20, 1944, NHM: SOE, Box 23; Sørlie Memoir.

The next day: Kjell Nielsen, "Notat angående omtalen av fergeaksjonen på Rjukan i Februar 1944," NHM: Box 10B; Interrogation of Gunnar Syverstad, April 5, 1944, TNA: HS 2/188; Haukelid, Report on the Sinking of the Ferry *Hydro,* February 20, 1944, NHM: SOE, Box 23.

289 *"effect of the":* Message from Swallow, February 16, 1944, NHM: SOE, Box 23.

They returned to: Haukelid, Report on the Sinking of the Ferry *Hydro,* February 20, 1944, NHM: SOE, Box 23; Sørlie Memoir.

After dinner that: Sørlie Memoir.

290 *A heavy snow:* February 16, 1943, ESD.

"The matter has": Message to Swallow, February 16, 1943, NHM: SOE, Box 23.

Skinnarland did not: Marielle Skinnarland, Author Interview; Skinnarland Notes, ESP; Haukelid, 185–86.

27. THE MAN WITH THE VIOLIN

291 *The following day:* Interview with Alf Larsen, DIA: DJ 31; Interview with Knut Haukelid, DIA: DJ 31.

"I know it's": Drummond, 160.

It was one: Haukelid, 187.

Haukelid continued: Haukelid, Report on the Sinking of the Ferry *Hydro,* February 20, 1944, NHM: SOE, Box 23.

292 *"place a time":* Ibid.

Next they needed: Kjell Nielsen, "Notat angående omtalen av fergeaksjonen på Rjukan i Februar 1944," NHM: Box 10B; Interrogation of Gunnar Syverstad, April 5, 1944; Account given by Engineer Larsen of the transaction during the attack on Tinnsjø Ferry, February 20, 1944, NHM: SOE, Box 23; Sørlie Memoir.

The pensioner had: Diseth, Friends Report, NHM: SOE, Box 23B; Haukelid, Report on the Sinking of the Ferry *Hydro,* February 20, 1944, NHM: SOE, Box 23; Gallagher, 175–76.

293 *The next day:* Interview with Haukelid, DIA: DJ 31.

The ferry: Payton and Lepperød; Interrogation of Gunnar Syverstad, April 5, 1944, TNA: HS 2/188; Irving, 203.

When Haukelid boarded: Haukelid, Report on the Sinking of the Ferry *Hydro,* February 20, 1944, NHM: SOE, Box 23; Haukelid, 187–88; Interview with Haukelid, DIA: DJ 31; Interview with Haukelid, IWM: Oral History.

294 *It turned out:* Interview with Knut Lier-Hansen, KA; Report by Gunleik Skogen, December 1, 1943, TNA: HS 2/174; Knut Lier-Hansen, Friends Report, NHM: SOE, Box 23B.

295 *Haukelid liked Lier-Hansen:* Haukelid, 189. There is some murkiness in regard to

when Lier-Hansen joined the team. In his own interview, Lier-Hansen puts himself fairly center stage as early as February 10, but this runs against most other accounts, which do not have him coming aboard until the final days (and relate the worry of still being a man down). Sørlie and Haukelid both recount this version. Given the assembly of evidence, February 18 looks to be the most likely scenario. Haukelid states explicitly they met this day in his after-action report and Lier-Hansen is not mentioned in any of the meetings that Larsen, Syverstad, or Nielsen participated in prior to that date.

"smashing": Larsen, 1242–49.

"Anything that can": Ibid., 1249–50.

296   *Fehlis had sent*: Irving, 205–6.

On February 18: February 18, 1944, LTD.

According to a: Letter to Michael Perrin, February 15, 1944, TNA: HS 8/955/DISR.

When Einar Skinnarland: February 10, 1944, LTD.

Brun pleaded: Brun, 85–86.

Wilson sent: John Wilson, "On Resistance in Norway," NHM: Box 50A; Interview with Michael Perrin, DIA: DJ 31.

297   *After his hike*: February 18, 1944, LTD.

Two pops ripped: Haukelid, 188.

"At least they": Drummond, 162.

The explosive time: Haukelid, Report on the Sinking of the Ferry *Hydro*, February 20, 1944, NHM: SOE, Box 23; Gallagher, 176; Interview with Haukelid, DIA: DJ 31.

298   *Over in Vemork*: Letter to Welsh, March 20, 1944, TNA: HS 2/188.

"unnamed destination": Account given by Engineer Larsen of the transaction during the attack on Tinnsjø Ferry, February 20, 1944, NHM: SOE, Box 23.

28. A 10:45 ALARM

299   *Shortly before midnight*: Haukelid, 191; Message from Swallow, March 30, 1944, NHM: Box 10.

The two saboteurs: Interview with Haukelid, DIA: DJ 31; Interview with Larsen, DIA: DJ 31.

"You dumb brute": Drummond, 165.

300   *"I'm sorry"*: Kurzman, 224.

They had already: Gallagher, 179.

The lone car moved: Haukelid, Report on the Sinking of the Ferry *Hydro*, February 20, 1944, NHM: SOE, Box 23; Interview with Haukelid, DIA: DJ 31; Interview with Knut Lier-Hansen, KA; Haukelid, 191–93; Drummond, 167–70; Gallagher, 179–82; Account given by Engineer Larsen of the transaction during the attack on Tinnsjø Ferry, February 20, 1944, NHM: SOE, Box 23; Sørlie Memoir. Other than specifically noted quotes, the narrative of the placing of the explosives aboard the *Hydro* is drawn from these sources collectively.

301   *"Is that you, Knut"*: Gallagher, 181; Haukelid, Report on the Sinking of the Ferry *Hydro*, February 20, 1944, NHM: SOE, Box 23; Sikkerhetspoliti Rapport av John Berg, February 21, 1944, NHM: Box 10B.

303   *"I'll be back"*: Drummond, 171.

On Sunday morning: Vedr. D/F *Hydro* forlis February 20, 1944, VM: IA4FB, Box 13; Irving, 209.

In his childhood: Lillian Gabrielson, Author Interview.

"That's the chief": Haukelid, 195.

304   *Up on the*: Interview with Lier-Hansen, KA.

Captain Erling: Rapport Sørensen, February 21, 1944, VM: IA4FB, Box 13; Gallagher, 184.

Just before 10:45: Rapport Sørensen, February 21, 1944, VM: IA4FB, Box 13; *Aftenposten,* February 23, 1944, TNA: HS 2/188; *Fritt Folk,* February 23, 1944, TNA: HS 2/188; *Rjukan Dagblad,* February 22, 1944, TNA: HS 2/188; Interview with Eva Gulbrandsen, KA; Omkomne D/F Hydro, February 20, 1944, VM: IA4FB, Box 13.

"Steer toward land": Rapport Sørensen, February 21, 1944, VM: IA4FB, Box 13.

"A bomb": Interview with Eva Gulbrandsen, KA.

305   *"I can't help":* Ibid.

306   *Of the fifty-three:* Haukelid, 197; Omkomne D/F *Hydro,* February 20, 1944, VM: IA4FB, Box 13. The exact numbers on the ferry that day are somewhat obscured by the fact that the ticket clerk — and his records — were lost in the sinking. The final accounting comes from an intelligence report found by Knut Haukelid after the war in the records of the military governor of Norway.

Rolf Sørlie spent: Sørlie Memoir.

"What have I": Ibid.

At the Hamarens': February 21, 1944, ESD.

As soon as: Message from Swallow, February 22, 1944, NHM: Box 10/SIS B.

307   *"If you disappear":* Interrogation of Gunnar Syverstad, March 25, 1944, NHM: Box 10.

When the Gestapo: Sikkerhetspoliti Rapport av John Berg, February 21, 1944, NHM: Box 10B; Gudbrandsen, Rapport til Lederen av Statspolitiet, February 23, 1944, NHM: Box 10B.

Knut Haukelid was: Haukelid, 195–202; Haukelid Family, Author Interview.

308   *On February 26:* February 26–March 7, 1944, LTD.

309   *As for Vemork:* April 13, 1944, LTD.

29. VICTORY

310   *By the end:* Irving, 217–19.

"precarious position": Bericht uber die Arbeiten auf Kernphysikalischen Gebiet, NB: Goudsmit, IV/B25/F13.

311   *"Very soon I shall":* Karlsch, 166–67.

In July, a bomb: May 12, 1944, LTD.

In August, Colonel: Irving, 246–47, 258–60; Alsos Mission Report, DIA: DJ 31.

"Germany had no": Goudsmit, 71

They crisscrossed: Interview with Georg Hartwig, NB: Oral History; Nagel, 129–30; Cassidy, 496.

312   *On June 15, 1944:* June 10, 1944, LTD.

Since the Allied: Njølstad, 331–35.

"first class in all": Finishing Report, STS 17, DORA: Correspondence 1937–45.

Jens-Anton Poulsson: Report of Operation Sunshine, TNA: HS 2/171.

"the major industrial objectives": Appendix A, Sunshine Action Plans, NHM: Box 10C.

On August 27, Tronstad: Letter from Tronstad to Bassa, August 27, 1944, LTP.

"Please take care": Speech by Gerd Hurum Truls, October 10, 1987, LTP.

"long exile": August 27, 1944, LTD.

When the other: Njølstad, 358.

313   *Over the next:* Report of Operation Sunshine, TNA: HS 2/171.

On the night of: Skinnarland Report on the Deaths of Major Tronstad and Sergeant Syverstad, March 16, 1945, TNA: HS 2/171; Njølstad, 410–24.

314 *The order to:* Report of Operation Sunshine, TNA: HS 2/171; Herrington, 283–85; Military Homefront Survey, December 1, 1944, NHM: SOE, Box 4.
315 *That night, at:* Nøkleby, *Josef Terboven,* 291–93.
      Heinrich Fehlis: Klykken, Frits, "Saken Fehlis," Porsgunn Folkebibliotek.
      A month after: Colonel Wilson, Diary of a Scandinavian Tour, TNA: HS 9/1605/3.
      "Many times it may": Chicago Tribune, June 8, 1945.
      On Midsummer's Eve, June 23: Skinnarland, *Hva Som Hendte,* ESP; Freds Og Midt-sommerskal, June 23, 1945, ESP.
316 *A week after:* Colonel Wilson, Diary of a Scandinavian Tour, TNA: HS 9/1605/3.
      By early August: Njølstad, 426–29.
      "Dearest Bassa . . . I have": Letter from Tronstad to Bassa, August 27, 1944, LTP.
317 *In Farm Hall, a quiet:* Powers, 434–35; Bagge and Diebner, 51–55.
      "shattered": Frank, 70.
      "Here is the news": Bernstein and Cassidy, Appendix C.
318 *"They can only":* Ibid., 115–18.

EPILOGUE

319 *"A bad experiment":* R. V. Jones, "Thicker Than Heavy Water," *Chemistry and Industry,* August 26, 1967.
320 *"The obliteration of ":* Bagge and Diebner, 35.
      "To live in the": Drew, 205.
      "They shall grow not": Ibid., 222.
      But whenever they: This conclusion is drawn by the author from the memoirs, interviews, and diaries of these individuals. Some, like Poulsson and Rønneberg, state this unequivocally.
321 *Einar Skinnarland's:* Marielle Skinnarland, Author Interview.
      "smallness of being": Interview with Rønneberg, IWM: 27187.
      Knut Haugland spent: Haugland Family, Author Interview.
      "He died without": Haukelid Family, Author Interview.
      Skinnarland did not: Marielle Skinnarland, Author Interview.
322 *"You have to fight":* "If Hitler Had the Bomb," transcript from documentary at the Norsk Industriarbeidermuseum, Vemork.

Bibliography

On first blush, a bibliography often looks like a dry recounting of sources, some primary, others secondary. The list below does not do justice to the excitement that came with researching *The Winter Fortress*, from finding the diaries of Leif Tronstad and Einar Skinnarland, to the top-secret SOE files, to interviews with the families of the saboteurs, to the unpublished manuscripts of Colonel John Wilson, Rolf Sørlie, and the Skinnarland family during the war. Over the course of my research, I read through hundreds of books, some in English, some in Norwegian and German, many listed here. But the lion's share of this story was constructed from reminiscences and interviews and memoirs of the key individuals as well as correspondence, action reports, diaries, and other archival papers written at the time these events unfolded.

ARCHIVES

United States

Howard Gotlieb Archival Research Center, Boston University (Boston, MA)
National Archives (College Park, MD)
Niels Bohr Library and Archives (College Park, MD)
Rensselaer Polytechnic Institute, Archives and Special Collections (Troy, NY)

Norway

Norges Hjemmefrontmuseum (Oslo)
Norges Teknisk-Naturvitenskapelige Universitet Arkiv (Trondheim)
Norsk Industriarbeidermuseum (Vemork)

Germany

Bundesarchiv (Berlin)
Bundesarchiv (Freiburg)
Deutsches Museum Archiv (Munich)

United Kingdom

British Online Archives (Wakefield)
Imperial War Museum (London)
National Archives (Kew)

France

Archives Nationale (Paris)

PERSONAL PAPERS

Leif Tronstad (courtesy of Leif Tronstad Jr.)
Knut Haugland (courtesy of Trond, Torfinn, and Torill Haugland)
Einar Skinnarland (courtesy of Marielle and Kirvil Skinnarland)
Jens-Anton Poulsson (courtesy of Mia and Unni Poulsson)
Knut Haukelid (courtesy of Bjørgulf, Kirvil, and Knut Haukelid)
David Irving (courtesy of British Online Archives)
Rolf Sørlie (courtesy of Finn Sørlie)
Bjørn Iversen

INTERVIEWS

Tronstad Family (Oslo)
Haugland Family (Oslo, Rjukan)
Haukelid Family (Oslo)
Poulsson Family (Oslo)
Skinnarland Family (Oslo, Rjukan, United States)
Finn Sørlie (Oslo)
Ragnar Ulstein (United States)
Svein Vetle Trae (Telemark)
Lillian Gabrielson (Oslo)

NORWEGIAN- AND GERMAN-LANGUAGE BOOKS AND ARTICLES

Andersen, Ketil. *Hydros historie 1905–1945*. Bind 1, *Flaggskip i fremmed eie: Hydro 1905–1945*. Oslo: Pax Forlag, 2005.

Bagge, Erich, Kurt Diebner, and Kenneth Jay. *Von der Uranspaltung bis Calder Hall*. Hamburg: Rowohlt, 1957.

Berg, John. *Soldaten som ikke ville gi seg: Lingekaren Arne Kjelstrup, 1940–45*. Oslo: Metope, 1986.

Bergfald, Odd. *Hellmuth Reinhard: Soldat eller morder?* Oslo: Schibsted, 1967.

Bøhn, Per. *IMI: Norsk innsats i kampen om atomkraften*. Trondheim: F. Bruns Bokhandels Forlag, 1946.

Brauteset, Steinar. *Gestapo-offiseren Fehmer: Milorgs farligste fiende*. Oslo: Cappelen, 1986.

Brun, Jomar. *Brennpunkt Vemork, 1940–1945*. Oslo: Universitetsforlaget, 1985.

Dahl, Helge. *Rjukan.* Bind 2, *Fra 1920 til 1980.* Rjukan: Tinn kommune, 1983.

Drew, Ion et al. *Tause helter: Operasjon Freshman og andre falne.* Stavanger, Rogaland, Norway: Hertervig Akademisk, 2011.

Fjeldbu, Sigmund. *Et lite sted på verdenskartet — Rjukan 1940–1950.* Oslo: Tiden, 1980.

Gestapo i Norge: Mennene, midlene og metodene. Oslo: Norsk kunstforlag, 1946.

Halvorsen, Odd. *Den Norske Turistforening årbok 1947.* Oslo: Den Norske Turistforening, 1947.

———. *Den Norske Turistforening årbok 1970.* Oslo: Den Norske Turistforening, 1970.

Hurum, Gerd Vold. *En Kvinne ved navn "Truls": Fra motstandskamp til Kon-Tiki.* Oslo: Wings, 2006.

Jensen, Erling. *Kompani Linge.* Bind 1. Oslo: Gyldendal, 1949.

Karlsch, Rainer. *Hitlers Bombe: Die geheime Geschichte der deutschen Kernwaffenversuche.* Munich: Deutsche Verlags-Anstalt, 2005.

Kjelstadli, Sverre. *Hjemmestyrkene.* Oslo: Bokstav & Bilde, 1959.

Larsen, Stein, Beatrice Sandberg, and Volker Dahm, eds. *Meldungen aus Norwegen, 1940–1945: Die geheimen Lageberichte des Befehlshabers der Sicherheitspolizei und des SD in Norwegen.* Munich: R. Oldenbourg Verlag, 2008.

Lauritzen, Per. *Claus Helberg: Veiviser i krig og fred.* Oslo: Den Norske Turistforening, 1999.

Løken, Roar. "Militær motstand i Milorgs D. 16, 1940–1945." PhD diss., Universitet i Oslo, 1976.

Lunde, Kjell Harald. *Sabotøren: Et portrett av mennesket og krigshelten Fredrik Kayser.* Bergen, Norway: Alma Mater, 1997.

Myklebust, Gunnar. *Tungtvannssabotøren: Joachim H. Rønneberg, Linge-kar og fjellmann.* Oslo: Aschehoug, 2011.

Nagel, Günter. *Atomversuche in Deutschland: Geheime Uranarbeiten in Gottow, Oranienburg, und Stadtilm.* Zella-Mehlis, Germany: Heinrich-Jung-Verlagsgesellschaft, 2002.

Njølstad, Olav. *Professor Tronstads krig.* Oslo: Aschehoug, 2012.

Nøkleby, Berit. *Gestapo: Tysk politi i Norge, 1940–45.* Oslo: Aschehoug, 2003.

———. *Josef Terboven: Hitlers mann i Norge.* Oslo: Gyldendal, 1992.

———. "Uforskammet opptreden mot Terboven." *Aftenposten,* February 25, 1983.

Olsen, Kristofer Anker. *Norsk Hydro gjennom 50 år.* Oslo: Norsk Hydro-Elektrisk Kvælstofaktieselskab, 1955.

Ording, Arne, Johnson Gudrun, and Johan Garder. *Våre falne, 1939–1945, Annen Bok.* Oslo: Grøndahl, 1950.

Payton, Gary, and Trond Lepperød. *Rjukanbanen: På sporet av et industrieventyr.* Rjukan: Maana Forlag, 1995.

Piekalkiewicz, Janusz. *Spione, Agenten, Soldaten: Geheime Kommandos im Zweiten Weltkrieg.* Munich: Schweizer Volks-Buchgemeinde, 1969.

Rostøl, Jack, and Nils Helge Amdal. *Tungtvannssabotør: Kasper Idland, fra krig til kamp.* Sandnes, Rogaland, Norway: Commentum, 2011.

Sæter, Svein. *Operatøren: Knut Haugland's egen beretning, Tungtvann, Gestapo, Kon-Tiki.* Oslo: Cappelen Damm, 2008.

Schaaf, Michael. "Der Physikochemiker Paul Harteck." PhD diss., Historisches Institut der Universität Stuttgart, 1999.

Schramm, Percy. *Hitler als militärischer Führer: Erkenntnisse und Erfahrungen aus dem Kriegstagebuch des Oberkommandoes der Wehrmacht.* Frankfurt am Main: Athenäum Verlag, 1965.

Skogen, Olav. *Ensom krig mot Gestapo.* Oslo: Aschehoug Pocket, 2009.

Tranøy, Joar. *Oppvekst i samhold og konflikt.* Oslo: Maana Forlag, 2007.

Ueland, Asgeir. *Tungtvannsaksjonen: Historien om den største sabotasjeoperasjonen på norsk jord.* Oslo: Gyldendal, 2013.

Veum, Erik. *Nådeløse nordmenn: Statspolitiet, 1941–1945.* Oslo: Kagge Forlag, 2012.

ENGLISH-LANGUAGE BOOKS, ARTICLES,
AND DOCUMENTARY TRANSCRIPTS

Adamson, Hans Christian, and Per Klem. *Blood on the Midnight Sun.* New York: Norton, 1964.

Baden-Powell, Dorothy. *Operation Jupiter: SOE's Secret War in Norway.* London: Hale, 1982.

Bailey, Roderick. *Forgotten Voices of the Secret War: An Inside History of Special Operations During the Second World War.* London: Ebury, 2008.

Bennett, John. *Letters from England.* San Antonio, TX: Hertzog, 1945.

Berglyd, Jostein. *Operation Freshman: The Hunt for Hitler's Heavy Water.* Stockholm: Leandoer & Ekholm, 2006.

Bernstein, Jeremy, and David Cassidy. *Hitler's Uranium Club: The Secret Recordings at Farm Hall.* New York: Springer Science & Business Media, 2013.

Beyerchen, Alan. *Scientists Under Hitler: Politics and the Physics Community in the Third Reich.* New Haven, CT: Yale University Press, 1977.

Casimir, Hendrik. *Haphazard Reality: Half a Century of Science.* New York: Harper & Row, 1984.

Cassidy, David C. *Uncertainty: The Life and Science of Werner Heisenberg.* New York: W. H. Freeman, 1992.

Churchill, Winston. *The Churchill War Papers: The Ever-Widening War, 1941.* Ed. Martin Gilbert. New York: Norton, 2001.

——. *The Hinge of Fate.* Vol. 4 of *The Second World War.* Boston: Houghton Mifflin, 1950.

——. *Their Finest Hour.* Vol. 2 of *The Second World War.* New York: Mariner Books, 1978.

Clark, Ronald. *The Birth of the Bomb.* London: Phoenix House, 1961.

——. *Tizard.* Cambridge, MA: MIT Press, 1965.

Compton, Arthur Holly. *Atomic Quest: A Personal Narrative.* Oxford, UK: Oxford University Press, 1956.

Cookridge, E. H. *Set Europe Ablaze: The Inside Story of Special Operations Executive.* New York: Thomas Crowell, 1967.

Cooper, D. F. "Operation Freshman." *The Royal Engineers Journal,* March 1946.

Cruickshank, Charles. *SOE in Scandinavia.* Oxford, UK: Oxford University Press, 1986.

Dahl, Per F. *Heavy Water and the Wartime Race for Nuclear Energy.* Bristol, UK: Institute of Physics Publishing, 1999.

Dalton, Hugh. *The Fateful Years: Memoirs 1931–45.* London: Frederick Muller, 1957.

Dank, Milton. *The Glider Gang: An Eyewitness History of World War II Glider Combat.* Philadelphia: J. B. Lippincott, 1977.

Dipple, John. *Two Against Hitler: Stealing the Nazis' Best-Kept Secrets.* New York: Praeger, 1992.

Dorril, Stephen. *MI6: Inside the Covert World of Her Majesty's Secret Intelligence Service.* New York: Free Press, 2002.

Drummond, John D. *But for These Men.* New York: Award Books, 1962.

Ermenc, Joseph, ed. *Atomic Bomb Scientists: Memoirs, 1939–45.* Meckler, 1989.

Fen, Åke. *Nazis in Norway.* London: Penguin Books, 1942.

Fine, Lenore, and Jesse A. Remington. *The Corps of Engineers: Construction in the United States.* Washington, DC: Office of the Chief of Military History, 1972.

Foot, M.R.D. *SOE: An Outline History of the Special Operations Executive, 1940–46.* London: Pimlico, 1999.

Frank, Charles. *Operation Epsilon: The Farm Hall Transcripts.* Oakland: University of California Press, 1993.

Freeman, Roger A. *The Mighty Eighth War Manual.* London: Jane's, 1985.

Gallagher, Thomas. *Assault in Norway: Sabotaging the Nazi Nuclear Program.* Guildford, CT: Lyons Press, 1975.

Gjelsvik, Tore. *Norwegian Resistance, 1940–1945.* Trans. by Thomas Kingston Derry. London: C. Hurst, 1979.

Goebbels, Josef. *The Goebbels Diaries.* London: H. Hamilton, 1948.

Goldsmith, Maurice. *Frédéric Joliot-Curie: A Biography.* London: Lawrence and Wisehart, 1976.

Goudsmit, Samuel A. *Alsos.* Woodbury, NY: AIP Press, 1996.

Gowing, Margaret. *Britain and Atomic Energy, 1939–1945.* London: Macmillan, 1982.

Groves, Leslie. *Now It Can Be Told: The Story of the Manhattan Project.* New York: Da Capo Press, 1983.

Haarr, Geirr H. *The German Invasion of Norway: April 1940.* Annapolis, MD: Naval Institute Press, 2009.

Hargreaves, Richard. *Blitzkrieg Unleashed: The German Invasion of Poland.* Mechanicsburg, PA: Stackpole Books, 2008.

Hauge, E. O. *Salt-Water Thief.* London: Duckworth, 1958.

Haukelid, Knut. *Skis Against the Atom.* Minot, ND: North American Heritage Press, 1989.

Heisenberg, Werner. *Physics and Beyond: Encounters and Conversations.* Translated by Arnold J. Pomerans. New York: Harper & Row, 1971.

Henniker, Mark. *An Image of War.* London: L. Cooper, 1987.

———. *Memoirs of a Junior Officer.* Edinburgh: Blackwood & Sons, 1951.

Hentschel, Klaus, ed. *Physics and National Socialism: An Anthology of Primary Sources.* Basel: Birkhäuser Verlag, 1996.

Herrington, Ian. "The Special Operations Executive in Norway 1940–45," PhD diss, De Montfort University, 2004.

Hewins, Ralph. *Quisling: Prophet Without Honor.* London: W. H. Allen, 1965.

Hinsley, F. H. *British Intelligence in the Second World War.* Vol. 2. Cambridge, UK: Cambridge University Press, 1981.

Hitler's Sunken Secret. PBS Documentary Transcript. November 8, 2005.

Humble, Richard. *Hitler's Generals.* London: Barker, 1973.

Ingstad, Helge. *The Land of Feast and Famine.* Translated by Eugene Gay-Tifft. Montreal: McGill-Queen's University Press, 1992.

International Military Tribunal. *Trial of Major War Criminals.* Vol. 27. Nuremburg: IMT, 1947.

Irving, David. *The German Atomic Bomb: The History of Nuclear Research in Nazi Germany.* New York: Perseus Books, 1983.

Jablonski, Edward. *Double Strike: The Epic Air Raids on Regensburg-Schweinfurt, August 17, 1943.* Garden City, NY: Doubleday, 1974.

Jeffery, Keith. *The Secret History of MI6.* London: Penguin Press, 2010.

Johnson, Amanda. *Norway: Her Invasion and Occupation.* Decatur, GA: Bowen Press, 1948.

Jones, R. V. "Thicker Than Heavy Water," *Chemistry and Industry,* August 26, 1967.

Jones, Reginald. *The Wizard War: British Scientific Intelligence, 1939–1945.* New York: Coward, McCann & Geoghegan, 1978.

Jungk, Robert. *Brighter Than a Thousand Suns: A Personal History of the Atomic Scientists.* Translated by James Cleugh. New York: Harcourt Brace, 1958.

Knudsen, H. Franklin. *I Was Quisling's Secretary.* London: Britons, 1967.

Kramish, Arnold. *The Griffin.* London: Macmillan, 1986.

Kurzman, Dan. *Blood and Water: Sabotaging Hitler's Bomb.* New York: Henry Holt, 1997.

Langworth, Richard M., ed. *Churchill by Himself: The Definitive Collection of Quotations.* New York: Public Affairs, 2008.

Lee, Sabine, ed. *Sir Rudolf Peierls: Selected Private and Scientific Correspondence.* Vol. 1. Singapore: World Scientific, 2009.

Lynch, Tim. *Silent Skies: Gliders at War 1939–1945.* Barnsley, UK: Pen & Sword, 2008.

Macrakis, Kristie. *Surviving the Swastika: Scientific Research in Nazi Germany.* New York: Oxford University Press, 1993.

Mann, Matthew. "British Policy and Strategy Towards Norway, 1941–44." PhD diss., University of London, 1998.

Marks, Leo. *Between Silk and Cyanide: A Codemaker's War.* New York: Free Press, 1998.

Meacham, Jon. *Franklin and Winston: An Intimate Portrait of an Epic Friendship.* New York: Random House, 2004.

Mears, Ray. *The Real Heroes of Telemark: The True Story of the Secret Mission to Stop Hitler's Atomic Bomb.* London: Coronet, 2003.

Moore, Ruth. *Niels Bohr: The Man, His Science, and the World They Changed.* New York: Knopf, 1966.

Moran, Charles McMoran Wilson. *Winston Churchill: The Struggle for Survival, 1940–1965; Taken from the Diaries of Lord Moran.* London: Constable, 1966.

Nichols, K. D. *The Road to Trinity.* New York: William Morrow, 1987.

Njølstad, Olav, et al. *The Race for Norwegian Heavy Water, 1940–1945.* Oslo: Institutt for Forsvarsstudier, 1995.

O'Connor, Bernard. *Churchill's School for Saboteurs: Station 17.* Gloucestershire, UK: Amberley, 2013.

Palmstrøm, Finn, and Rolf Torgersen. *Preliminary Report on Germany's Crimes Against Norway.* Oslo: Grøndal & Son, 1945.

Parton, James. *"Air Force Spoken Here": General Ira Eaker and the Command of the Air.* Bethesda, MD: Adler & Adler, 1986.

Peierls, Rudolf. *Bird of Passage: Recollections of a Physicist.* Princeton, NJ: Princeton University Press, 2014.

Perquin, Jean-Louise. *The Clandestine Radio Operators: SOE, BCRA, OSS.* Paris: Histoire and Collections, 2011.

Persico, Joseph E. *Roosevelt's Secret War: FDR and World War II Espionage.* New York: Random House, 2002.

Petrow, Richard. *The Bitter Years: The Invasion and Occupation of Denmark and Norway, April 1940–May 1945.* New York: William Morrow, 1974.

Poulsson, Jens-Anton. *The Heavy Water Raid: The Race for the Atom Bomb 1942–44.* Oslo: Orion Forlag AS, 2009.

Powers, Thomas. *Heisenberg's War: The Secret History of the German Bomb.* New York: Da Capo Press, 1993.

Rhodes, Richard. *The Making of the Atomic Bomb.* New York: Simon & Schuster, 1996.

Rigden, Denis. *How to Be a Spy: World War II SOE Training Manual.* Toronto: Dundurn, 2004.

Riste, Olav, and Berit Nøkleby. *Norway 1940–45: The Resistance Movement.* Oslo: Johan Grundt Tanum Forlag, 1970.

Roane, Owen. *A Year in the Life of a Cowboy with the Bloody 100th.* The Woodlands, TX: Mackenzie Curtis Publishing, 1995.

Rose, Paul Lawrence. *Heisenberg and the Nazi Atomic Bomb Project, 1939–1945.* Berkeley: University of California Press, 2002.

Sagafos, Ole Johan. *Progress of a Different Nature.* Oslo: Pax Forlag, 2006.

Sandys, Celia. *Chasing Churchill: Travels with Winston Churchill.* London: HarperCollins, 2004.

Seaman, Mark, ed. *Special Operations Executive: A New Instrument of War.* London: Routledge, 2006.

Shirer, William. *Berlin Diary: The Journal of a Foreign Correspondent, 1934–1941.* Baltimore, MD: Johns Hopkins University Press, 2002.

——. *The Rise and Fall of the Third Reich.* New York: Simon & Schuster, 1990.

Smyth, H. D. *Atomic Energy for Military Purposes.* York, PA: Maple Press, 1945.

Speer, Albert. *Inside the Third Reich.* New York: Macmillan, 1981.

Stafford, David. *Secret Agent: The True Story of the Covert War Against Hitler.* New York: Overlook Press, 2001.

Suess, Hans. "Virus House: Comments and Reminiscences." *Bulletin of Atomic Scientists* 24 (June 1968): 36–39.

Walker, Mark, and Rainer Karlsch. "New Light on Hitler's Bomb." *Physics World* 18 (June 2005): 15–18.

Walker, Mark. *German National Socialism and the Quest for Nuclear Power.* Cambridge, UK: Cambridge University Press, 1989.

——. *Nazi Science: Myth, Truth, and the German Atomic Bomb.* New York: Plenum Press, 1995.

Warbey, William. *Look to Norway.* London: Secker & Warburg, 1945.

Werner-Hagen, Knut. "Mission 'Moonlight' — Norway 1944." Website of the Austrian Armed Forces, http://www.bundesheer.at/truppendienst/ausgaben/artikel.php?id=886, March 2009.

Wiggan, Richard. *Operation Freshman: The Rjukan Heavy Water Raid, 1942.* London: William Kimber, 1986.

Wigner, Eugene Paul. *The Collected Works.* New York: Springer Science and Business Media, 2001.

Williams, Robert Chadwell. *Klaus Fuchs, Atom Spy.* Cambridge, MA: Harvard University Press, 1987.

Wilson, John Skinner. *Memoirs of a Varied Life.* London: Imperial War Museum, John Skinner Wilson Collection, unpublished.

Worm-Müller, Jacob. *Norway Revolts Against the Nazis.* London: Lindsay Drummond, 1941.

Wright, Myrtle. *Norwegian Diary, 1940–1945.* London: Friends Peace and International Relations Committee, 1974.

Index